FRUITS OF TROPICAL AND SUBTROPICAL ORIGIN

Composition, Properties and Uses

FRUITS OF TROPICAL AND SUBTROPICAL ORIGIN

Composition, Properties and Uses

Edited by

STEVEN NAGY
Scientific Research Department
Florida Department of Citrus
Lake Alfred, Florida

PHILIP E. SHAW
U.S. Citrus and Subtropical Products
Laboratory
Winter Haven, Florida

WILFRED F. WARDOWSKI
University of Florida, IFAS
Citrus Research and Education Center
Lake Alfred, Florida

FSS

FLORIDA SCIENCE SOURCE, INC.
LAKE ALFRED, FLORIDA

FLORIDA SCIENCE SOURCE, INC.
P. O. Box 927
Lake Alfred, Florida 33850-0927

Library of Congress Catalog Card Number: 89-92429
ISBN 0-944961-00-2

FRUITS OF TROPICAL AND SUBTROPICAL ORIGIN/edited by Steven Nagy, Philip E. Shaw and Wilfred F. Wardowski.

Current printing (last digit): 10 9 8 7 6 5 4 3 2 1
Printed in the United States of America

PREFACE

There is a considerable variety of fruit grown throughout the tropical and subtropical regions of the world. Nearly everyone is familiar with citrus, bananas, pineapples, dates, figs, mangos, papayas, and avocados, but few people recognize the potential value of a multitude of lesser known fruit. Many fruits are grown within limited geographical confines and, generally, are known only by inhabitants of that area. Perhaps one major reason why so many tropical and subtropical fruits were not distributed and domesticated worldwide was simply due to being overlooked by Portuguese, Spanish and Dutch traders during their seafaring explorations between the 15th and 18th centuries.

Lack of acreage expansion and accessibility to foreign markets for many of these lesser known tropical and subtropical fruits has been caused by several factors, namely, (1) restrictive climate and growing conditions, (2) ineffective or poor agricultural techniques, (3) limited postharvest knowledge for harvesting, handling and transporting, (4) lack of agrobusiness initiative to stimulate production, marketing and research, and (5) lack of knowledge of food value and functionality of processed products. Slowly but surely many tropical and subtropical fruits are becoming known and finding their way into commercial marketing channels. Fruits that were only distributed by ethnic markets ten years ago are now distributed by major market chains. These emerging fruits include pomegranate, carambola, calamondin, cherimoya, sugar apple and the spondias. Fruits that have considerable potential but are still relegated to limited ethnic and regional markets include marula, breadfruit, jackfruit, granadilla, pejibaye and mountain papaya. Finally, there are fruits that are relatively obscure but possess potential for commercialization; these include biriba, mangaba, cupuassu, bacuri, araza and jaboticaba.

Warm climates generally produce fruits with higher contents of vitamins and minerals than fruits of cool or temperate climates. In addition, many tropical and subtropical fruits possess delightfully strong attractive flavors, interesting functional properties, and are used in a variety of food preparations. As examples, breadfruit and jackfruit are two underutilized tropical fruits with considerable potential. They are rich in starch, proteins, minerals and many vitamins. Products derived from breadfruit are considerable and include fruit slices, pickles, spiced curries, flour for making papad, and as a fermented paste for breads. Ripe jackfruit are canned in syrup, made into chutney, preserves, candy syrup, jelly, pectin, and a multitude of other food uses. Jackfruit is considered the poorman's food in India owing to its many culinary uses. In the Philippines, demand for ripe jackfruit has increased 100-fold and it has become one of the most expensive fruits. Jackfruit imparts a strong, sweet aromatic odor, and in shredded form is used as an ingredient in many type desserts.

Another interesting, but virtually unknown fruit is the cupuassu of Brazil. The cupuassu has no common English name as it has never been transplanted from its origin in eastern Amazonia to other parts of the humid tropics. Apparently, the Amerindians of Brazil did not do much to domesticate this pre-Columbian crop. The cupuassu is one of the most popular fruits in eastern Amazonia. The pulp of cupuassu is used in nectars, ice creams, jellies, puree, canned pulp and even in pizza, while cupuassu seed is used to make chocolate (also known as cupulate).

The purpose of this book is to enhance awareness of the value for lesser known fruits. Although each fruit was considered equally important, the extent of coverage varied. There is a paucity of information on these lesser known tropical and subtropical fruits, and much of the information that is published is in sources not readily available to many research workers. Thus, this book should be a valuable reference source in an area of increasing worldwide interest—the search for new and exotic foods. Basically, the areas covered were fruit history and distribution, botanical descriptions and varieties, maturity stages, harvesting, handling and postharvest treatments, chemical composition, nutritional significance, flavor attributes, functional properties and food uses.

Nineteen scientists collaborated with us in covering 26 tropical and subtropical fruits. *Fruits of Tropical and Subtropical Origin* should be of considerable interest to food technologists, nutritionists, biochemists, consumers, students and agrobusiness entrepreneurs.

Steven Nagy
Philip E. Shaw
Wilfred F. Wardowski

Contributors

AHRENS, M. JOSEPH. Department of Vegetable Crops, University of California, Davis, California.

CAMPBELL, CARL W. Tropical Research and Education Center, University of Florida, Homestead, Florida.

CLEMENT, CHARLES R. Departamento Ciencias Agronomicas, Instituto Nacional de Pesquisas de Amazonia, Manaus, AM, Brazil.

CRANE, JONATHAN, H. Tropical Research and Education Center, University of Florida, Homestead, Florida.

DONADIO, LUIZ CARLOS. FCAV/UNESP, Campus de Jaboticabal, Jaboticabal, Sao Paulo, Brazil.

DURIGAN, JOSE FERNANDO. FCAV/UNESP, Campus de Jaboticabal, Jaboticabal, Sao Paulo, Brazil.

HOLTZHAUSEN, LUCAS C. Division of Food Science and Technology, CSIR, Pretoria, South Africa.

KUMAR, G. N. MOHAN. Department of Plant Science, University of Alberta, Edmonton, Canada.

LEAL, FREDDY. Facultad de Agronomia, Universidad Central de Venezuela, Maracay, Venezuela.

LEON, JORGE. Apartado 480, San Pedro de Montes de Oca, San Jose, Costa Rica.

LIZANA, L. ANTONIO. Facultad de Ciencias y Forestales, Universidad de Chile, Casilla, Santiago, Chile.

MABESA, LINDA B. Institute of Food Science and Technology, College of Agriculture, University of the Philippines at Los Banos, College, Laguno, Philippines.

NAGY, STEVEN. Scientific Research Department, Florida Department of Citrus, CREC, Lake Alfred, Florida.

NARAIN, NARENDRA. Departamento de Tecnologia, Quimica e de Alimentos, Universidade Federal da Paraiba, Joao Pessoa, Brazil.

NARASHIMHAN, PATTISAPU. Central Food Technology Research Institute, Mysore, India.

REGINATO, GABINO. Facultad de Ciencias y Forestales, Universidad de Chile, Casilla, Santiago, Chile.

SHAW, PHILIP E. United States Citrus and Subtropical Products Laboratory, United States Department of Agriculture, Winter Haven, Florida.

VAN WYK, PIETER J. Division of Food Science and Technology, CSIR, Pretoria, South Africa.

VENTURIERI, GIORGINI A. Departamento Ciencias Agronomicas, Instituto Nacional de Pesquisas da Amazonia, Manaus, AM, Brazil.

WARDOWSKI, WILFRED F. Citrus Research and Education Center, University of Florida, Lake Alfred, Florida.

WEINERT, INGRID. Division of Food Science and Technology, CSIR, Pretoria, South Africa.

WILSON, III, CHARLES W. United States Citrus and Subtropical Products Laboratory, United States Department of Agriculture, Winter Haven, Florida.

Contents

1

Origin and Distribution of Tropical and Subtropical Fruits

Jonathan H. Crane
Carl W. Campbell

This chapter will review the origin and distribution, botanical description, adaptation and cultural conditions, current importance and future prospects of 26 tropical and/or subtropical fruit crops. The authors hope the format used will expedite the readers familiarity and identification of these fruits. In a previous book 21 different tropical fruit crops were reviewed by Knight (1980) in a similar manner.

Some of these fruit crops (e.g., cashew and pomegranate) are familiar to people in the temperate zone, while others like carambola and sugar apple are now becoming familiar. Crops like breadfruit, lucuma, calamansi, and pejibaye, while well known in some areas of the world, are virtually unknown to North Americans and Europeans.

ANACARDIACEAE

The Anacardiaceae includes about 59 genera and 400 species of trees or shrubs possessing resinous bark and alternate, simple, or compound leaves (Bailey 1950). Species are found in the temperate and tropical zones throughout the world.

Some plants in the Anacardiaceae are of considerable economic importance. These include: mango (*Mangifera indica* L.), cashew (*Anacardium occidentale* L.), and pistachio (*Pistacia vera* L.). Others, like ambarella (*Spondias cytherea* Sonn.), and amra (*S.pinnata* Kurz) are less known but of regional importance.

CASHEW (*Anacardium occidentale* L.)

Origin and Distribution

The cashew is native to tropical South America from Mexico to Peru, Brazil and the West Indies (Martin *et al.* 1987; Purseglove 1969). However, cashews are grown throughout the American, Asian, and African tropics (Martin *et al.* 1987; Morton 1987; Purseglove 1969).

Botanical Description

Cashew trees are medium-sized, low branched and spreading evergreen trees to 12 m in height with resinous sap. The leaves are alternate, simple, leathery, oval or obovate shaped, rounded or notched at the apex and tapering at the base; lamina 6-20 cm long and 4-15 cm wide, veins prominent, 10-20 lateral veins spreading; petiole 1-2 cm long, swollen at the base and flattened on the upper surface (Bailey 1950; Purseglove 1969). The inflorescence is usually a many-flowered terminal panicle to 20 cm long. Flowers are fragrant and pink or red; bracts pubescent; polygamous with male and hermaphrodite flowers in the same inflorescence; 5 sepals, narrow, green, 0.5 cm long; petals green changing to red, 1 cm long, and reflexed in open flowers with 10 stamens (Bailey 1950; Purseglove 1969). Male and hermaphrodite flowers have 9 short (0.4 cm) and 1 long (1.2 cm) stamen with red anthers. The ovary is 1-locular with single ovule; style simple to 1.2 cm long. The fruit is a kidney-shaped achene (nut) about 2.5-3.0 cm long and 2.5 cm wide, grayish-brown with a hard pericarp borne on an enlarged pedicel, receptacle, and disc called the cashew apple (false fruit) (Bailey 1950; Martin *et al.* 1987; Purseglove 1969). The cashew apple (Fig. 1.1) is pyriform to rhomboid shaped (Mukherjee 1956), 5-20 cm in length, and 4-8 cm wide, with a shiny red, yellow, or red and yellow, thin, waxy skin; flesh soft, juicy, pale yellow, astringent, and acid to subacid (Bailey 1950; Martin *et al.* 1987; Morton 1987; Purseglove 1969). Seed is kidney-shaped with reddish-brown testa, 2 large white cotyledons, and small embryo (Purseglove 1969).

Adaptations and Cultural Considerations

Cashew trees grow best in hot tropical climates below 1,000 m elevation. Trees are tolerant of various soil types, and periodic drought (Martin *et al.* 1987; Purseglove 1969; Sturz 1985), but are intolerant of cold (Ruehle 1967) and waterlogged soils (Sturz 1985). Trees are propagated by seed, grafting, air layering, and cuttings (Martin *et al.* 1987; Nambiar 1977; Purseglove 1969). Seedling trees produce fruit that vary in shape, size, and weight of nuts, and shape, size, and color of the false fruit

FIG. 1.1, CASHEW APPLE, KIDNEY-SHAPED NUT AND LEAVES

(apple) (Nambiar 1977). Fruit production begins in 4-5 years from seed and 2-3 years from vegetative propagation.

Hermaphroditic flowers are self and/or cross pollinated by flies, ants, and other insects (Purseglove 1969) and the period from fruit set to fruit maturity ranges from about 60-90 days depending upon genetic, climatic, and cultural factors (Nambiar 1977; Purseglove 1969). Seedling trees vary in bearing habit, percent fruit set (1 to 24%), and yields (0.5 to 50 kg per tree) (Sturz 1985).

Cashew flowers are attacked by several insects including thrips (*Rhynthothrips revenis* G.) and mealy bugs (*Ferrisia virgata* (Ckll.)), and fungal pathogens (*Gloeosporium mangiferae* and *Phomopsis anacardii*) (Joseph 1979). In Australia, anthracnose (*Colletotrichum* spp.), bananaspotting bug (*Amblypelta* spp.), red banded thrips (*Selenothrips*), black beetle (*Rhyparida* spp.) and stem borers (*Bombatelia* spp.) attack cashew trees (Sturz 1985).

Current Importance and Future Prospects

The cashew apple is an extremely fragile fruit and is generally considered a by-product of the cashew nut industry (Morton 1987). However, the cashew apple is eaten by local inhabitants wherever it is grown (e.g., India, Latin America) and juices, wine, syrups, and preserves are made by local people (Morton 1987; W. Popenoe 1974; Purseglove 1969; Sturz 1985). Whole cashew apples are sometimes preserved in syrup and sold (Morton 1987).

Data on cashew apple production are not available nor is there much data on its export and import. The United States imported a total of 16.2 MT of frozen cashew apple from the Dominican Republic and Guatemala in 1985 (Pearrow 1988).

Although the cashew apple has moderate appeal as a fresh fruit (Martin *et al.* 1987) it is doubtful if production or consumption of the cashew apple itself will increase substantially. This is because the nut is much more valuable and the cashew apple has a very limited postharvest storage life (see Chapter 2).

MARULA (*Sclerocarya* birrea subsp. caffra)

Origin and Distribution

Native to tropical south and eastern Africa, marula trees are found throughout Africa (Bailey 1950) and have been distributed to Madagascar and Australia (Martin *et al.* 1987). Traditionally harvested from wild stands, marula orchards have recently been established in southern Africa (see Chapter 3).

Botanical Description

Marula trees are medium to large trees (9-18 m) with a single trunk (see Chapter 3). Marula leaves are compound, glabrous alternate, and 15-31 cm long; leaflets petiolate, 3.8-5 cm long with short oblique tapering points and acute bases with the upper surface darker green than lower surface (Bailey 1950). Trees are dioecious with the male inflorescence spicate; flowers, short pedicelled with 12-15 stamens and dark red to pink or white (Bailey 1950). Female inflorescence is a raceme: flowers long pedicelled; 4 sepals, imbricate; 4 petals, oblong, obtuse, spreading, reflexed, imbricate, disk depressed; ovary subglobose and 2-3 celled. Fruit is a suborbicular drupe, shell leathery, 3-4 cm in diameter (Bailey 1950) (see Chapter 3).

Adaptation and Cultural Considerations

Marula trees require tropical or subtropical dry savanna conditions (Martin *et al.* 1987). Little is known of their cultural requirements although trees are considered drought tolerant. Marula trees are propagated by seed and cuttings (Martin *et al.* 1987).

Current Importance and Future Prospects

The commercial South African marula fruit industry is in its infancy. Currently fruit and nut production consists of harvesting from wild stands of trees. Weinert *et al.* (Chapter 3) estimate 600 MT of marula juice was processed in South Africa in 1986 and that the potential of the industry is much greater. This is because the fruit and nut can be eaten fresh, the fruit pulp made into alcoholic beverages and jams, and the crushed nut into flour.

IMBU (*Spondias tuberosa* Arruda)

Origin and Distribution

Native to Brazil, the imbu has been distributed throughout South America (Bailey 1950; Kennard and Winters 1960; Martin *et al.* 1987; W. Popenoe 1974). Introduced into the United States in 1914, it does not appear to be well adapted to Florida conditions (W. Popenoe 1974; Ruehle 1967).

Botanical Description

Trees are spreading, up to 6 m in height, with swollen roots (Bailey 1950; W. Popenoe 1974; Martin *et al.* 1987). Leaves are 10.2-15.2 cm long, compound, with 5-9 oblong-ovate, 2.5-4.5 cm long and 1.6-2.5 cm wide leaflets (Bailey 1950). Inflorescence is a panicle, 10.2-15.2 cm long; flowers white, with 4-5 sepals and valvate petals, 8-10 stamens, and 3-5 styles (W. Popenoe 1974). The fruit is oval, 2.5-3.8 m long, in clusters of 3-4 with light yellow to greenish-yellow skin and whitish flesh. Seeds are oblong-ovate, 1.3-1.9 cm long and smooth (Bailey 1950; Martin *et al.* 1987; W. Popenoe 1974).

Adaptation and Cultural Considerations

Imbu is native to the dry plains of northeastern Brazil (W. Popenoe 1974), and is adapted to a hot tropical climate and very drought tolerant (Martin *et al.* 1987). Trees are generally frost susceptible, although they have been reported to withstand –2.0°C (W. Popenoe 1974). Flowering occurs in April-May in south Florida (Martin *et al.* 1987).

Imbu is propagated by seed and cuttings (Martin *et al.* 1987; W Popenoe 1974). Improved selections should be vegetatively propagated

(Sturz 1985). Imbu has been veneer grafted on *Spondias purpurea L.* and W. Popenoe (1974) reports imbu can be inarched successfully onto ambarella (*Spondias cytherea* Sonnerat.).

Current Importance and Future Prospects

Several authors consider the imbu the best of the *Spondias* spp. and that it is underutilized (W. Popenoe 1974; Martin *et al.* 1987). Currently imbu is harvested from wild stands and sold locally in Brazilian markets (Leon and Shaw, Chapter 4). No production or export and import figures are available.

RED OR PURPLE MOMBIN (*Spondias purpurea* L.)

Origin and Distribution

The red mombin is native to tropical America and is found growing in Mexico, Central and South America, and other tropical areas of the world like the Philippines, Nigeria, and the Caribbean (Kennard and Winters 1960; Martin *et al.* 1987; W. Popenoe 1974). Red mombin (Fig. 1.2) has been introduced to California, Hawaii and Florida but is grown

FIG. 1.2. RED MOMBIN

there only in home gardens (Bailey 1950; W. Popenoe 1974; Sturrock 1940).

Botanical Description

Trees are small (to 7.6 m) and usually spreading, deciduous, with a thick trunk and stiff branches (Bailey 1950; Morton 1987; W. Popenoe 1974). Leaves are compound, alternate, and 12.7-30.5 cm long. Leaves have 5-23 oblong-elliptic leaflets; each leaflet is 2.5-3.8 cm long with an entire or subserrate margin (Bailey 1950; Morton 1987; Ruehle 1967). Inflorescences are produced from small unbranched racemes about 1.3 cm long on wood of the previous season (Kennard and Winters 1960). Flowers are polygamous and yellowish or purplish-red with 4-5 petals (Bailey 1950; Campbell and Sauls 1980; Morton 1987). Fruit shape is variable, oblong-obovoid to pyriform in shape (2.5-5.0 cm long); borne singly or in clusters of 2-3 and from yellow to deep red (Morton 1987; Popenoe 1979; Ruehle 1967). Fruit flesh is yellow and aromatic with a juicy, subacid, sweet to sour taste and surrounds a rough, oblong-elliptical, 1.2-1.9 cm long stone containing seeds (Bailey 1950; Campbell and Sauls 1980; Martin *et al.* 1987; Morton 1987; W. Popenoe 1974; Ruehle 1967).

Adaptation and Cultural Considerations

Red mombin trees are adapted to hot tropical lowlands (sea level to 1,828 m) (Morton 1987; W. Popenoe 1974) with wet to relatively dry climates (Martin *et al.* 1987). Trees are also cold sensitive (Ruehle 1967; Sturrock 1940). Red mombin trees appear to be tolerant of a wide range of soil types (Martin *et al.* 1987; W. Popenoe 1974).

Trees are propagated by hardwood (Kennard and Winters 1960; Campbell and Sauls 1980; Martin *et al.* 1987; Ruehle 1967) and softwood cuttings (Marler 1989). Fruit production from seed takes 4-5 years and 2-3 years from cuttings (Martin *et al.* 1987). In south Florida, the red mombin flowers in March-April before the new leaves appear (Popenoe 1979) and fruit take 120 or more days to mature (Martin *et al.* 1987; Ruehle 1967). A number of cultivars have been selected in tropical America, but generally are known only in local areas.

Current Importance and Future Prospects

World-wide production, export and import figures for red mombin are not available. Red mombin is a commercial crop of local importance in Central America, Mexico (Morton 1987; W. Popenoe 1974), South

America, and much of the West Indies (Morton 1987). Some red mombin is exported from the West Indies to the United Kingdom as "June plum". The United States imported about 203 MT of frozen product in 1986 (Pearrow 1988).

Martin *et al.* (1987) feel there is potential to increase commercialization and production of red mombin in Central and South America. In juice form it may find appeal in mixed fruit blends.

YELLOW MOMBIN (*Spondias mombin L.; syn. S. lutea*)

Origin and Distribution

The yellow mombin is native to tropical America (Bailey 1950; Martin *et al.* 1987; Morton 1987) but is also found growing in the Caribbean, Asia (e.g., Indonesia, Malaysia) (Morton 1987), and Africa (Hutchinson and Dalziel 1948). Yellow mombin was introduced into the United States by the USDA on 3 separate occasions (1914, 1917, and 1921) and is found as a door-yard tree in extreme southern Florida (Campbell and Sauls 1980; Morton 1987; W. Popenoe 1974; Ruehle 1967).

Botanical Description

Yellow mombin trees have a medium (15 m) to large (20 m) height, with a spreading habit, and furrowed bark (Bailey 1950; Morton 1987; W. Popenoe 1974; Ruehle 1967). Leaves are pinnately compound, 20-30 cm long, with 7-21 ovate-lanceolate or lanceolate, 5-10 cm long, serrulate leaflets (Bailey 1950; Morton 1987; W Popenoe 1974). Trees are deciduous in areas with a distinct dry season (Campbell and Sauls 1980). The inflorescence is a loose terminal panicle, 15-30 cm long, with yellowish-white bisexual flowers (Bailey 1950; Campbell and Sauls 1980; W. Popenoe 1974; Morton 1987). Fruit is borne in clusters of 3-20 or more (Campbell and Sauls 1980; Morton 1987), is globular to ovoid in shape and 2.5-5 cm long, and 2.5 cm wide (Campbell and Sauls 1980; Morton 1987). Fruit is a drupe with a tough skin and scant flesh surrounding a large oblong seed (Bailey 1950; Campbell and Sauls 1980). The flesh is yellow, fibrous, soft, and juicy (Morton 1987; W. Popenoe 1974), with an acid to sub-acid, pungent flavor (W. Popenoe 1974; Sturz 1985).

Adaptation and Cultural Considerations

The yellow mombin is well adapted to hot tropical lowlands that are high in rainfall or possess a marked dry season (Martin *et al.* 1987). Trees will grow in warm subtropical areas which experience only occasional mild frosts (Campbell and Sauls 1980). Trees grow best in fertile, well-drained soils, but can be grown in poor soils if adequately fertilized (Campbell and Sauls 1980). Yellow mombin is successfully grown in south Florida along the coasts.

Yellow mombin may be propagated by seed, however, superior selections should be vegetatively propagated since seedlings are not true-to-type (Campbell and Sauls 1980; Martin *et al.* 1987). In the tropics, yellow mombin is commonly propagated by taking large stem cuttings (8-10 cm in diameter) and rooting them in the ground wherever desired (Popenoe 1979). Veneer grafting and shield or chip budding are also successful (Campbell and Sauls 1980). Trees begin bearing in 4-6 years from seed and 2-3 years from cuttings (Martin *et al.* 1987). In south Florida, yellow mombin trees flower in March-April and fruit mature during September-October (Martin *et al.* 1987).

Current Importance and Future Prospects

The yellow mombin or hog plum is not commercially important and seldom appears in local markets of Central and South America and the Caribbean (Morton 1987; W. Popenoe 1974). The fruit is considered inferior to other *Spondias* spp. and processing the fruit even on a local scale is reported to be limited (Morton 1987). Because of this, the future prospects for expanded culture of this fruit are limited.

AMBARELLA (OTAHEITE APPLE) (*Spondias cytherea* Sonn.; syn. *S. dulcis* Parkins)

Origin and Distribution

Ambarella is native to the South Pacific Islands (Martin *et al.* 1987; Morton 1987; Ruehle 1967) but has been introduced into tropical areas of Indochina, Australia, Central and South America, and the Caribbean (Morton 1987). It is grown to a small extent in the coastal areas of southern Florida (Campbell and Sauls 1980; Morton 1987; Ruehle 1967) and in Hawaii (Morton 1987).

Botanical Description

The ambarella is a medium to large, erect tree reaching 10-25 m in height (Bailey 1950; Martin *et al.* 1987; Ochse *et al.* 1961). Leaves are deciduous (Morton 1987), alternate, pinnately compound, and 15-60 cm long (Morton 1987; Ochse *et al.* 1961). Leaves possess 9-25 leaflets which are oppositely or slightly alternately arranged on the rachis (Morton 1987; Ochse *et al.* 1961; W. Popenoe 1974); each leaflet is 5.5-10 cm long, oval to oblong shaped, acuminate, and has entire or indistinctly crenate margins (Morton 1987; Ochse *et al.* 1961; Ruehle 1967). Small whitish flowers are borne on 20-50 cm long terminal panicles. Flowers are bisexual and possess 5 sepals and petals, 10 stamens and a 5-celled ovary (Ochse *et al.* 1961). The fruit is a drupe, borne in clusters of 3-20 or more (Campbell and Sauls 1980; Morton 1987), globular to oval (irregular) in shape, 5-10 cm long and 4-8 cm in diameter (Ochse *et al.* 1961), with a tough yellow to orange-yellow skin. The flesh is pale yellow, juicy, and varies from sweet to acid in flavor (Ochse *et al.* 1961; Ruehle 1967). A single seed is enclosed in a hard endocarp that often has fibers projecting into the pulp (Campbell and Sauls 1980; Ochse *et al.* 1961; Ruehle 1967). Flesh sweetness and thickness vary from tree to tree (Popenoe 1979).

Adaptations and Cultural Conditions

Ambarella trees are best adapted to hot tropical regions with a monsoon climate (Martin *et al.* 1987), but can be grown in humid subtropical areas which experience only occasional frosts (Campbell and Sauls 1980; Ochse *et al.* 1961; Popenoe 1979). Trees are generally adaptable to different soil conditions as long as they are well drained (Campbell and Sauls 1980; Martin *et al.* 1987; Ochse *et al.* 1961).

The ambarella may be propagated by seed, however, superior selections should be vegetatively propagated by hardwood cuttings, veneer grafting or shield or chip budding because they do not come true from seed (Campbell and Sauls 1980; Martin *et al.* 1987). Fruit production begins 4-5 years from seed and 2-3 years from cuttings (Martin *et al.* 1987). Flowering may occur from March to May in Florida and fruit mature in 200 or more days (Martin *et al.* 1987).

Current Importance and Future Prospects

Ambarella fruit are not important on the world market but are consumed fresh, pickled, and cooked (Morton 1987) to a limited extent throughout the tropics. Production and trade figures are not available. Some authors (Martin *et al.* 1987; Morton 1987) feel ambarella is an underutilized fruit and could be more important in some tropical regions.

AMRA (HOG PLUM, AMBRA) (*Spondias pinnata* Kurz; syn. *S. mangifera* Willd.)

Origin and Distribution

Amra is native to tropical Asia (Bailey 1950; Martin *et al.* 1987; Morton 1987) and can be found growing from the Himalayas of India to the Andaman Islands located near Burma and in Southeast Asia and

Malaysia. Trees also grow and fruit well in south Florida (Campbell and Sauls 1980).

Botanical Description

Amra is a small tree (to 8 m) with compound leaves 30-40 cm long (Bailey 1950; Martin *et al.* 1987). Leaflets number 5-11 and are 10-15 cm long, 3.8-7.6 cm broad, ovate-oblong or elliptic-oblong in shape, acuminate, and entire (Bailey 1950). Panicles are 25-36 cm long with whitish flowers (Bailey 1950). Ten to 15 fruit are borne in terminal clusters (Campbell and Sauls 1980). The fruit is a drupe, 4-5 cm long (Campbell and Sauls 1980), ovoid shaped, yellow-green and contains an oblong-elliptically shaped, smooth seed (Bailey 1950). The pulp has an acid, bitter flavor (Popenoe 1979).

Adaptation and Cultural Considerations

Trees are adapted to hot, tropical lowland conditions (Martin *et al.* 1987), but do well in the mild subtropical marine climate along the coast of south Florida (Campbell and Sauls 1980). Amra can be propagated by seed or cuttings (Martin *et al.* 1987).

Current Importance and Future Prospects

The amra is not grown commercially for its fruit but is consumed by some local people in tropical Asia (Campbell and Sauls 1980). No production, export and import figures are available. Amra fruit does not appear to have much potential for commercial exploitation because of the poor quality of the fruit (Martin *et al.* 1987).

ANNONACEAE

The Annonaceae consists of 46 genera and 500 to 600 species of trees or shrubs mostly in tropical regions of Asia, Africa, and America (Bailey 1950). Characteristics of the family include sepals and petals arranged in threes, numerous spirally arranged stamens, usually separate carpels, and ruminate endosperm (Bailey 1950).

Several species including sugar apple (*Annona squamosa* L.), atemoya (*A. squamosa* x *A. cherimola*), guanabana (*A. muricata* L.) and biriba (*Rollinia deliciosa* Saff.) are grown commercially for their fruit (Bailey 1950; Ochse *et al.* 1961). Others like the pawpaw (*Asimina triloba* Dunal) and climbing ylang-ylang (*Artabotrys hexapetalus* (L.F) Bhand.) are valued as ornamentals (Bailey and Bailey 1976; Ochse *et al.* 1961).

Ylang-ylang (*Cananga odorata* Hook. f.) has commerical value as an ingredient of perfumes.

CHERIMOYA (*Annona cherimola* Mill.)

Origin and Distribution

The cherimoya is thought to be native to the Andes of South America (i.e., Ecuador, Peru, and Bolivia) (Morton 1987; Ochse *et al.* 1961; W. Popenoe 1974; Ruehle 1967). However, it has become naturalized in parts of Central America and Mexico and has been introduced to the Malay Archipelago (Ochse *et al.* 1961), the Caribbean (Morton 1987; W. Popenoe 1974), India, the Mediterranean area, Australia (W. Popenoe 1974), the Middle East (Morton 1987), and parts of the United States (i.e., southern California, Florida, and Hawaii) (Morton 1987; W. Popenoe 1974; Ruehle 1967).

Botanical Description

The cherimoya is a small (5-9 m), erect or spreading tree (Bailey 1950; Ochse *et al.* 1961; W. Popenoe 1974). Leaves are simple, alternate, 2-ranked, 4-9 cm wide, ovate to ovate-lanceolate or obovate shaped, 10-25 cm long, sparsely pubescent above, velvety-tomentose beneath, and deciduous (Bailey 1950; Morton 1987; Ochse *et al.* 1961). Flowers are protogynous and fragrant; about 2.5 cm long, solitary or sometimes 2 or 3 together, on short peduncles, opposite leaves on young shoots and at old leaf scars on older wood (Ruehle 1967). Flowers possess 3 exterior petals which are oblong-linear in form and have a greenish color outside and pale-yellow or whitish color within; 3 inner petals are very small, scale-like, and ovate or triangular in shape (Bailey 1950; W. Popenoe 1974). Stamens and pistils are numerous; pistils are crowded together on a conical fleshy receptacle and stamens are arranged around the base of the receptacle (W. Popenoe 1974). The fruit is a syncarp (made up of the fusion of numerous carpels and receptacle tissue), variable in shape (e.g., conoid, spheriod, heart-shaped) with or without protuberances corresponding to individual carpels (Bailey 1950; Popenoe 1974). The fruit skin is thin, green when immature and greenish-yellow at maturity (Morton 1987; W. Popenoe 1974). Fruit weigh between 200-2000 g (Martin *et al.* 1987). Fruit flesh is creamy white, melting, juicy, sweet, and subacid with a good to excellent flavor (Bailey 1950; Ochse *et al.* 1961). Seeds are hard, numerous, easily separated from pulp, brown or black, obovate, somewhat compressed, and about 1 cm long (Bailey 1950; Morton 1987).

Adaptations and Cultural Considerations

The cherimoya grows best in tropical highlands (1000 m and higher) or cool subtropical areas and will not produce well in the hot, humid lowland tropics (Martin *et al.* 1987; Morton 1987; W. Popenoe 1974). Areas with a relatively dry climate but well-distributed rainfall are conducive to good production (Martin *et al.* 1987; W. Popenoe 1974) and irrigation in California benefits tree growth and fruit production (Brown 1989). Mature cherimoya trees can withstand cold temperatures down to about –1.1° to –3.3°C (Morton 1987; W. Popenoe 1974). Cherimoya fruit production has not been successful in south Florida because of the hot, humid climate and occasional freezing weather (W. Popenoe 1974; J. Popenoe 1974; Ruehle 1967). George (1985) reports young trees are killed by temperatures –3°C or below in Australia.

Cherimoya trees appear to grow on many different soil types as long as they are well drained (George 1985). Supplemental micronutrient fertilizer may be required for culture on some soils to prevent nutrient deficiencies.

Cherimoya flowers are protogynous, suggesting cross pollination is required for fruit set (Morton 1987; Ruehle 1967). Cherimoya fruit production in California and Israel appears to be limited by insufficient pollination (Nagel *et al.* 1989; W. Popenoe 1974; Ruehle 1967). This may be due to a lack of sufficient numbers and/or appropriate pollinating insects and the protogynous behavior of the flowers (Nagel *et al.* 1989; Ruehle 1967). High relative humidity during flowering may also markedly improve fruit set (George 1985).

Commercial growers in California successfully hand pollinate to increase fruit set, quality, and production (Nagel *et al.* 1989). Use of gibberellic acid has been tried to improve fruit set in Chile with little success (Saavedra 1979).

There are many cherimoya selections and named cultivars (Morton 1987; W. Popenoe 1974; Ruehle 1967). Many have been developed in South America (Morton 1987) and in the United States (i.e., California) (Morton 1987; W. Popenoe 1974). Three of the most popular cultivars in California are White, Bays, and Booth (Fig. 1.3) (Arpaia 1989).

Pest and disease problems of cherimoyas vary from region to region. California's major pest problem is infestation of the fruit by mealy bugs and the ants that attend them (Brown 1989). Several pests have been reported in Florida including a rust fungus (*Phakospora cherimoliae* (Lagh.) Cumm.) which causes premature leaf drop, ambrosia beetle (*Xyleborus* spp.) which may attack branches and twigs, and the annona seed borer (*Brephratelloides cubensis* (Ashmead) which oviposit into the

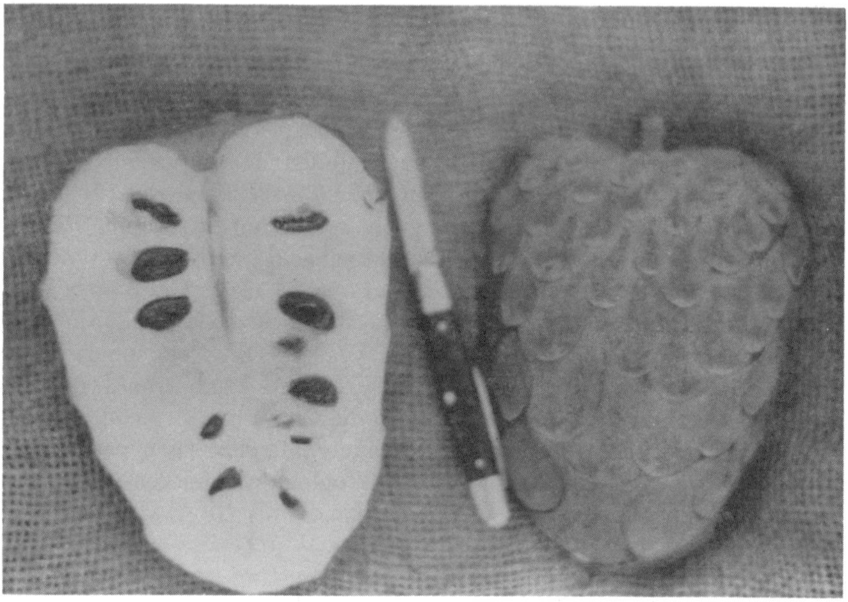

FIG. 1.3. BOOTH CHERIMOYA

seeds of young fruit (Ruehle 1967). In South America, various lepidopterous, scale, and seed and wood boring insects attack trees and fruit (Morton 1987).

Cherimoya may be propagated by seed, grafting, and budding (Martin *et al.* 1987; Morton 1987; W. Popenoe 1974). Several annona species have been used successfully as rootstocks, e.g., *A. reticulata, A. squamosa,* and *A. glabra* (Morton 1987; W. Popenoe 1974), although seedling cherimoya are generally used as rootstocks in California (Arpaia 1989). Cherimoya is also being evaluated as a rootstock for other annona species in Australia (George 1985). Trees come into production in about 5-6 years from seed and 3-4 years from grafts (Martin *et al.* 1987).

In California orchards, cherimoya trees are typically spaced at 6 m x 6 m or 5.5 m x 5.5 m (268 to 330 trees per hectare) (Brown 1989). Morton (1987) reports trees are spaced 7.5-9.0 m each way in Colombian orchards and 5 m apart in Spain.

In California, cherimoya trees may be trained to improve structure, fruit production and quality, and to control tree size (Brown 1989). In Australia, wind breaks, pruning, and defoliation are used during young tree development to encourage fruit bearing lateral shoots (George 1985).

Current Importance and Future Prospects

Commercial cherimoya production occurs in Spain, South America (e.g., Chile, Peru), Central America, Mexico, Israel, and California in the United States. Crop acreage and production figures are not readily available for production outside the United States. California has 101 to 121 ha (Arpaia 1989) and is estimated to produce about 453 MT in the 1989-90 season (Brown 1989). About 1.6 MT of frozen cherimoya was imported into the United States in 1985 from Thailand and New Zealand (Wright 1985) and 6.5 MT from Costa Rica, Dominican Republic and Guatemala in 1986 (Pearrow 1988). Spain produced 2190 MT of cherimoya in 1987 (Nunez 1988).

The cherimoya is considered an excellent fruit and can be eaten fresh or in processed form (Martin *et al.* 1987; Morton 1987). The prospects for increased consumption of cherimoya are encouraging in North America and Europe. Hectarage in a number of countries (e.g., New Zealand, Australia, Thailand, United States) will most likely increase.

SUGAR APPLE (SWEETSOP) (*Annona squamosa* L.)

Origin and Distribution

The sugar apple is native to tropical America and is grown throughout the tropics of Central and South America, the Caribbean, Indonesia, southern China, Africa, Asia, Australia, the United States (Florida, Hawaii) and India (Martin *et al.* 1987; Morton 1987; W. Popenoe 1974). It is also grown in subtropical regions of the Middle East (Martin *et al.* 1987; Morton 1987; W. Popenoe 1974). South Florida currently has about 19.8 ha of commercial sugar apple (Crane 1989).

Botanical Description

Sugar apple trees are small (3-6 m), deciduous and may branch near the base (Bailey 1950; Ochse *et al.* 1961; W. Popenoe 1974). Leaves are petioled, alternately arranged, lanceolate or oblong-lanceolate shaped, 5-17 cm long, 2-7 cm wide, acute at the apex, cuneate at the base, and pale green (Bailey 1950; Ochse *et al.* 1961). Inflorescences are produced singly or in clusters of 2-4 and are extra-axillary on drooping peduncles (Bailey 1950; Ochse *et al.* 1961; Phillips and Campbell 1980). Flowers possess 3 greenish-yellow flesh outer petals that are 2.5-3.8 cm long; and 3 minute ovate shaped inner petals (Bailey 1950; Ochse *et al.* 1961; W. Popenoe 1974). Flowers have many stamens and pistils; pistils crowded together on a conical receptacle and stamens arranged in whorls around the base of the receptacle (Ochse *et al.* 1961). The fruit is a syncarp, heart-shaped, spherical, ovate or conical in shape, 5-10 cm in diameter, and yellowish-green with a white or bluish bloom (Phillips and Campbell 1980; Morton 1987; W. Popenoe 1974) (Fig. 1.4). Carpellary segments are easily distinguished and separate when fruit is ripe (Phillips and Campbell 1980; W. Popenoe 1974; Ochse *et al.* 1961). The pulp is white, custard-like, sweet and slightly acidulous in flavor (Phillips and Campbell 1980; W. Popenoe 1974; Morton 1987). Seeds are numerous (exception seedless forms), of brown or black, 1-1.5 cm long, slightly compressed, and correspond to carpellary segments (Phillips and Campbell 1980; Ochse *et al.* 1961; W. Popenoe 1974).

Adaptations and Cultural Considerations

Sugar apple trees are adapted to tropical and warm subtropical areas that do not experience prolonged freezing weather (Phillips and Campbell 1980; Martin *et al.* 1987; Morton 1987). Mature trees have been observed to withstand temperatures of –1.7° to –2.2°C without severe injury, while young trees may be killed at temperatures of —1.1°C or below (Phillips and Campbell 1980; W. Popenoe 1974). Trees are drought tolerant; however, leaves abscise and fruit production is reduced under dry conditions (Phillips and Campbell 1980; Morton 1987).

FIG 1.4. SUGAR APPLE

Sugar apple trees grow best in fertile, well-drained soils but grow well in less fertile soil types with supplemental fertilizers (Phillips and Campbell 1980; W. Popenoe 1974). Supplemental fertilizer and irrigation are recommended for formal plantings to improve production and fruit quality (Phillips and Campbell 1980; W. Popenoe 1974). Trees have been observed to be intolerant of wet or flooded conditions (Phillips and Campbell 1980).

Orchard plant spacing varies among and within production areas (Morton 1987; Phillips and Campbell 1980). Morton (1987) reports spacings of 3 m x 3 m, 3 m x 5 m and 5 m x 5 m (400 to 1,111 trees per ha) as common in the Middle East. In south Florida, plant spacings vary from 3.1 m x 4.6 m to 9.1 m x 9.1 m (120 to 701 trees per ha) (Crane 1989).

Sugar apple trees suffer from several insect pests including ambrosia beetles (*Xyleborus* spp.) (Pena *et al.* 1984; Phillips and Campbell 1980) and Philephedra scale (*Philephedra tuberculosa* Nakahara and Gill) (Pena *et al.* 1984, 1987) which attack branches and twigs. Mealybugs (*Psuedococcus maritimus* (Erkon) and *P. nipae* (Mask.)) (Pena 1986) and the annona seed borer (*Brephratelloides cubensis* (Ashmead) attack the fruit (Pena *et al.* 1984; Phillips and Campbell 1980).

In south Florida, several diseases attack sugar apple fruit including anthracnose (*Colletotrichum gloeosporioides* Penz) and *Botryodiplodia theobromae* Pat. which cause fruit rot and mummification (McMillan 1986). Premature defoliation also occurs due to leaf infection by a rust fungus (*Phokospora cherimoliae* Cumm.) (McMillan 1986).

Sugar apple trees may be propagated by seed, grafting, budding (Ogden *et al.* 1981; Phillips and Campbell 1980) and inarching (Morton 1987). In south Florida, pregraft preparation of the budwood permits sugar apple grafting most of the year except during the winter when trees are dormant (Ogden *et al.* 1981). Rootstocks suitable for sugar apple include *A. reticulata* (Phillips and Campbell 1980; Morton 1987), and sugar apple seedlings (Phillips and Campbell 1980).

There are a moderate number of sugar apple selections and cultivars (Morton 1987; Phillips and Campbell 1980) which differ in size, peel and flesh color, season of maturity and seediness (Morton 1987). However, seedless forms are not used commercially because of cracking and splitting problems (Phillips and Campbell 1980; Whitman 1972). Most commercial plantings in Florida consist of seedling trees of known and unknown parentage (Martin *et al.* 1987) and little variation exists among seedling trees (Phillips and Campbell 1980).

Trees come into bearing after 3-5 years from seed and 2-3 years from grafting (Martin *et al.* 1987). Sugar apple trees flower in April-June in south Florida and fruit mature in 180 to 200 days (harvested in July-December) (Martin *et al.* 1987).

Current Importance and Future Prospects.

Sugar apples are grown commercially in the Middle East (Morton 1987), India, Malaysia, the West Indies (George 1985), Thailand (Wright 1985), Central and South America (W. Popenoe 1974), Dominican Republic (Pearrow 1988), and Florida (Campbell 1986). Production figures are not available for most areas. The fruit is important on a local scale in tropical Asia and America. Florida currently has about 20 ha (Crane 1989) and production is estimated to be 45 to 68 MT per year and increasing. Thailand exported about 4.0 MT of frozen sugar apple to the United States in 1985 (Wright 1985).

The potential for expanding the sugar apple market and, hence, production may be good. This is especially true of frozen or processed product (e.g., ice cream). Expansion of the fresh fruit market outside areas of production will most likely be limited until selections are found that have better storage characteristics (e.g., nonsplitting, slower ripening).

BIRIBA (ROLLINIA) (*Rollinia deliciosa* Saff.; syn. *R. pulchrinervis*)

Origin and Distribution

The biriba is native to Brazil but has been introduced into other South and Central American countries, the Caribbean, the United States (Morton 1987; W. Popenoe 1974), India (Ochse *et al.* 1961), and the Philippines (Morton 1987).

Botanical Description

The biriba is a rapidly growing medium sized (4-15 m) tree (Morton 1987). Leaves are obovate-oblong or elliptically shaped, round at the base, acuminate at the apex, 20-28 cm long, 8-10 cm broad, and glabrous above and beneath except for reddish-brown hairs on mid-rib and primary veins (Bailey 1950; W. Popenoe 1974). Flowers are leaf-opposed, solitary or rarely in threes and 2.5-4.0 cm long (Bailey 1950; Morton 1987). Flowers have 3 pubescent sepals; 3 laterally compressed, widely diverging and decurved outer petals; and 3 rudimentary inner petals (Bailey 1950; Morton 1987). Stamens and pistils are numerous; stamens crowded and arranged around the base of a raised receptacle while pistils are crowded on a conical receptacle (Bailey 1950). The fruit (Fig. 1.5) is a conical to heart-shaped or oblate syncarp, 8-15 cm diameter with aeroles distinctly outlined and terminating obtusely (Bailey 1950). The rind is yellow while the pulp is white or cream-colored, juicy, mucilaginous, and subacid to sweet flavored (Bailey 1950; Martin *et al.* 1987; W.

FIG. 1.5. BIRIBA

Popenoe 1974). Fruit possess many dark-brown compressed seeds, 1.6-2.0 cm long and 0.8-1.0 cm broad (Bailey 1950; Morton 1987).

Adaptations and Cultural Considerations

Biriba trees are adapted to hot tropical lowlands (Martin *et al.* 1987) but may survive in subtropical areas not frequented by cold temperatures or frost (Morton 1987; W. Popenoe 1974). Mature trees have been killed at –3.1°C in southern Florida (W. Popenoe 1974) and young trees by temperatures –3°C in Australia (George 1985).

The biriba is generally propagated by seed (Martin *et al.* 1987), however, it may be propagated by budding and grafting. Biriba trees flower in April-May in Florida and fruit mature in 90 or more days (Martin *et al.* 1987) while trees in Brazil flower and fruit all year long (Morton 1987).

Current Importance and Future Prospects

The biriba is commercially important to a limited extent in local Brazilian markets (Morton 1987). There are few if any formal commercial plantings and most of the fruit sold is harvested from natural stands (Morton 1987).

The potential for development of a biriba industry is doubtful. This is because there are many other superior annonaceous fruit; fresh biriba fruit damage easily and the pulp has a mucilaginous texture that is unappealing to some people.

APOCYNACEAE

The Apocynaceae includes 130 genera and about 1,000 species of herbs, shrubs or trees (Bailey 1950). Important family characteristics include sagittate anthers, absence of corona, stylar ring, milky sap, and usually separate ovaries but connate styles and stigmas (Bailey 1950).

Edible fruit of the Apocynaceae include karanda (*Carissa carandas* L.), natal plum (*C. grandiflora* A.DC.), Egyptian carissa (*C. edulis* Vahl), mangaba (*Hancornia speciosa* Gomez) and saba (*Saba senegalensis* (A.DC.) Pichon) (Martin *et al.* 1987).

MANGABA (*Hancornia speciosa* Gomez)

Origin and Distribution

Mangaba is native to Brazil and can be found growing only in some areas of South America (Martin *et al.* 1987). Little information is available concerning its distribution outside Brazil (see Chapter 8).

Botanical Description

Mangaba plants are shrubs or trees to 7 m in height (Martin *et al.* 1987). Leaves are opposite, lanceolate, and coriaceous. Fruit is a drupe, 3-6 cm in diameter with yellow skin striated with red (Martin *et al.* 1987). A more complete description can be found in Chapter 8.

Adaptations and Cultural Considerations

The natural habitats for mangaba are the hot, dry tropical areas of northeastern South America (Martin *et al.* 1987). Sandy soils are preferred (Martin *et al.* 1987). Mangaba is propagated by seed (Martin *et al.* 1987) and no selection or breeding for superior types has been reported (see Chapter 8).

Current Importance and Future Prospects

Mangaba culture is limited to the northeastern area of Brazil (Martin *et al.* 1987) and production is estimated at over 1,200 MT annually. The greatest potential for mangaba appears as a processed fruit. In Brazil it is used in ice cream, sherbet, syrups, and wines (Martin *et al.* 1987).

CARICACEAE

The Caricaceae is a small family consisting of 2 genera and 27 species indigenous to tropical and subtropical America (Bailey 1950). Characteristically, members of this family are single trunked arborescent herbs with terminal, alternate, palmately-lobed leaves, and latex (Bailey 1950; W. Popenoe 1974).

Several species are of economic importance including the papaya (*Carica papaya* L.) which is grown throughout the tropics and subtropics for its melon-like fruit and the mountain papaya (*C. candamarcensis* Hook.) grown on a small scale in some areas of the tropical highlands and cool subtropical areas of the world (Martin *et al.* 1987; Morton 1987; W. Popenoe 1974).

MOUNTAIN PAPAYA (*Carica candamarcensis* Hook.; syn. *C. pubescens* Koch)

Origin and Distribution

The mountain papaya is native to the Andean regions of South America (Morton 1987; W. Popenoe 1974) and is also grown to a limited extent in the tropical highlands and cool subtropical areas of Sri Lanka, India, New Zealand, and the Caribbean (Martin *et al.* 1987; Morton 1987; W. Popenoe 1974).

Botanical Description

Mountain papaya plants are arborescent herbs 5 m or more in height (Martin *et al.* 1987) with terminal, alternate, dark green above and pale green beneath, pinnately 5-lobed leaves up to 38 cm wide (Bailey 1950; W. Popenoe 1974). Flowers are green and pubescent (Bailey 1950). Fruit (Fig. 1.6) are 6-20 cm long, 5-angled, conical, yellow to orange outside, and yellow inside (Bailey 1950; Martin *et al.* 1987; Morton 1987). Fruit are aromatic and have a sour to sweet flavor (Martin *et al.* 1987; Morton 1987; W. Popenoe 1974).

Adaptations and Cultural Considerations

Mountain papaya are native to tropical highlands (1,800-3,000 m) and cool subtropical lowlands with well distributed rainfall (Martin *et al.* 1987; Morton 1987; W. Popenoe 1974). Plants are reported to withstand temperatures as low as –2.2°C without serious injury (W. Popenoe 1974). Plants are propagated by seed and begin fruit production in 2 years (Martin *et al.* 1987).

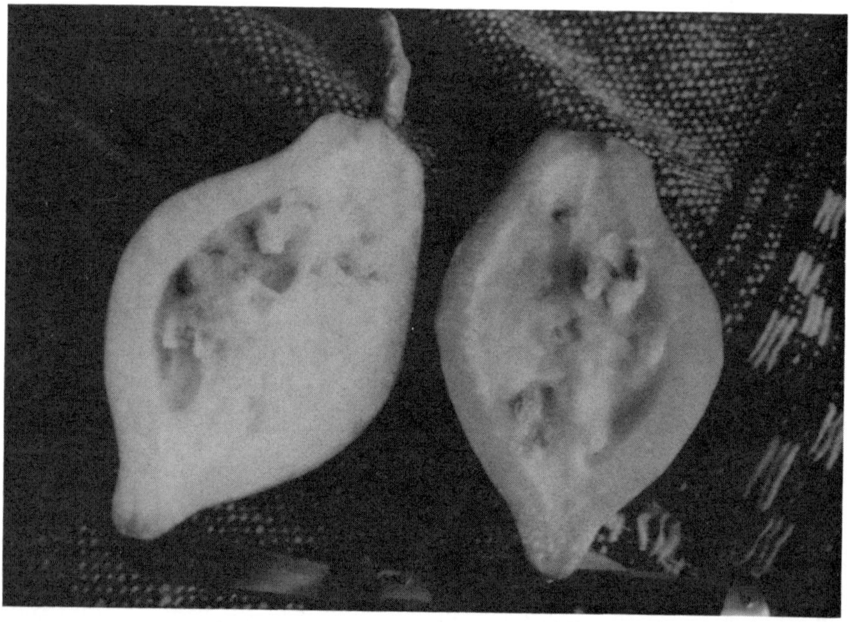

FIG. 1.6. MOUNTAIN PAPAYA

Current Importance and Future Prospects

Mountain papaya is produced commercially to a limited extent in Chile where it is eaten after cooking (Morton 1987). The pulp of the mountain papaya has a very firm texture and the fruit is not commonly eaten fresh. The fruit appears to have only limited potential for wider production and marketing. In New Zealand fruit are used to make preserves and sauces.

GUTTIFERAE

The Guttiferae is a tropical family consisting of about 35 genera and nearly 400 species (Bailey 1950; W. Popenoe 1974) of trees and shrubs usually possessing polygamous flowers (Bailey 1950). Many modern taxonomists classify this family as Clusiaceae.

There are several Guttiferae species in limited cultivation. These include the mangosteen (*Garcinia mangostana* L.) and mamey apple (*Mammea americana* L.) (Morton 1987; W. Popenoe 1974).

BACURI (BACUR) (*Platonia insignis* Mart.)

Origin and Distribution

Native to the Amazon region of Brazil and Guyana (Martin *et al.* 1987), the bacuri is also found growing in Paraguay and Colombia (Morton 1987).

Botanical Description

Bacuri are large trees to 25 m (Martin *et al.* 1987) with leaves that are coriaceous, oblong, and acute (Bailey 1950). Flowers are solitary and terminal with 5 sepals and petals; petals are rose colored outside and white inside (Bailey 1950; W. Popenoe 1974). Fruit is an oval berry, 10-12 cm in diameter, yellow to brown outside and white inside with 1-5 seeds (Bailey 1950; Martin *et al.* 1987). The fruit has a pleasant subacid flavor and matures in 200-300 days (Martin *et al.* 1987) (see Chapter 10).

Adaptations and Cultural Considerations

Bacuri trees are adapted to hot, humid tropical lowland areas and tolerant of various soil conditions, including poorly drained sites (Martin *et al.* 1987). Trees are propagated by seed (Martin *et al.* 1987) but form root suckers easily after the parent tree is cut down (W. Popenoe 1974) (see Chapter 10).

Current Importance and Future Prospects

Bacuri is sold on a limited commercial scale in Brazil where canned sections and processed products are made (Morton 1987). Fruit is harvested from wild stands.

The potential for an increase in production and utilization of bacuri has not been thoroughly evaluated (Martin *et al.* 1987).

MORACEAE

The Moraceae contains 55 genera and about 950 species (Bailey 1950). Plants in this family characteristically possess a milky sap. The family is diverse, possessing species indigenous to tropical and temperate climates (Bailey 1950). Important genera include Morus (mulberry spp.), Ficus (rubber and edible fig spp.), Cannabis (hemp), and Artocarpus (breadfruit, jackfruit) (Bailey 1950).

Breadfruit (*Artocarpus altilis* Fosb.) and jackfruit (*A. heterophyllus* Lamb) are found growing throughout the tropics and have been selected and cultivated for centuries in Southeast Asia and Polynesia (Martin *et al.* 1987).

BREADFRUIT (*Artocarpus altilis* Fosb.)

Origin and Distribution

Breadfruit is native to Southeast Asia and Polynesia (Martin *et al.* 1987; W. Popenoe 1974) and is widely distributed throughout the tropics including the West Indies, Sri Lanka, India, Central America, and northern South America (Ochse *et al.* 1961; W. Popenoe 1974).

There are seedy and seedless forms of breadfruit (Ochse *et al.* 1961; W. Popenoe 1974) and many selections and cultivars have been named (Morton 1987; W. Popenoe 1974).

Botanical Description

Breadfruit trees attain 20-30 m or more in height, possess a straight trunk buttressed at the base, and have thick scaffolding limbs (Morton 1987; Ochse *et al.* 1961). Leaves are evergreen (Bailey 1950), clustered near the ends of branchlets and are alternately arranged (Ochse *et al.* 1961; W. Popenoe 1974). Leaves are 40-90 cm long, 20-50 cm wide (Morton 1987), thick, leathery, ovately outlined (3-11 lobed at the apical end) with entire bases (Morton 1987; Ochse *et al.* 1961; W. Popenoe 1974). The main veins are pubescent and colored yellow and laminae are bright green above and dull yellow-green below (Morton 1987; Ochse *et al.*

1961). The tree is monoecious, with axillary inflorescences (Ochse *et al.* 1961). Staminate flowers are on club-like spikes, 15-30 cm long, 2.5-3.8 cm wide, yellow or greenish-yellow, and with 1 stamen per flower (Morton 1987; Ochse *et al.* 1961). Pistillate inflorescence are globose-ovoid shaped, 12-30 cm long, 3.8 cm wide, and have many small flowers (Morton 1987; Ochse *et al.* 1961). Fruit (Fig. 1.7) are 20-30 cm diameter, spherical to ovoid syncarps produced near the ends of short branches (Martin *et al.* 1987; Ochse *et al.* 1961). Fruit are green when immature and turn yellow to brown at maturity (Morton 1987). There are seedy and seedless forms of breadfruit (Martin *et al.* 1987; Ochse *et al.* 1961). The fruit rind is patterned with 4-6 sided irregularly shaped faces, level or conically projecting from the surface (Morton 1987). Pulp is whitish in immature fruit and yellowish in fully ripe fruit (Ochse *et al.* 1961; W. Popenoe 1974). Seeds if present, are oval, compressed, 2.0-2.5 cm long, and dull brown with dark brown stripes (Morton 1987).

Adaptations and Cultural Considerations

Breadfruit is strictly adapted to hot, humid tropical lowland areas and will not tolerate temperatures below 5°C (Martin *et al.* 1987; Ochse

FIG. 1.7. BREADFRUIT - TREE, LEAVES AND FRUIT

et al. 1961; Sedgley 1985A). Breadfruit trees appear to be tolerant of various soil types if they are well drained, and grow best in deep, rich, well drained sandy loams (Ochse *et al.* 1961).

Seedless breadfruit is propagated by root cuttings (Martin *et al.* 1987; Ochse *et al.* 1961; W. Popenoe 1974), marcottage (Martin *et al.* 1987; Ochse *et al.* 1961), and budding (Ochse *et al.* 1961) while seedy breadfruit is propagated by seed (Martin *et al.* 1987; W. Popenoe 1974). Vegetatively propagated trees begin bearing in about 6 years while seedlings take about 8 years (Sedgley 1985A).

The time and duration of the breadfruit crop varies with geographic location (Martin *et al.* 1987; Morton 1987) and plant selection (i.e., both clones and seedlings differ) (Sedgley 1985A). In some areas (e.g., South Pacific) some breadfruit is available all year long (Morton 1987) while in others 1-2 crops of varying duration may be produced (Martin *et al.* 1987; Morton 1987). Sedgley (1985A) reports 10-year-old trees produce 500-700 fruit per year, each fruit weighing between 2-5 kg.

Breadfruit trees are subject to attack by a number of fungal species including *Phytophthora*, *Phomopsis*, *Fusarium*, and *Pythium* and insect pests like mealy bugs and scales (Morton 1987).

Current Importance and Future Prospects

Breadfruit is not produced in formal orchards but rather harvested from door-yard trees and groves. It may be packed and exported fresh (Wright 1985), processed into flour or dried (Morton 1987), frozen and then exported (Wright 1985) or cooked and consumed locally (Morton 1987).

Although breadfruit is grown and consumed throughout the tropics of the east and west, production figures are not available. The United States imported 438.3 MT of fresh breadfruit from the Caribbean and 13.0 MT of frozen breadfruit from Asia in 1986 (Pearrow 1988).

The potential for increased demand and production of breadfruit is good. The fruit is versatile, many cultivars exist, and small scale processing technology could be expanded (Morton 1987).

JACKFRUIT (*Artocarpus heterophyllus* Lam.)

Origin and Distribution

Jackfruit is indigenous to tropical Asia (Martin *et al.* 1987) with India (Ochse *et al.* 1961; W. Popenoe 1974) and Malaysia (Ruehle 1967) cited as possible centers of origin. Nevertheless, jackfruit is widely distributed throughout the tropical areas of the world (e.g., Africa, Australia, Sri Lanka, and East Indies) (Martin *et al.* 1987; Morton 1965).

Botanical Description

Jackfruit trees are evergreen, possess viscous white latex (Morton 1965; W. Popenoe 1974), are monoecious, and may reach 25 m in height (Martin *et al.* 1987; Ochse *et al.* 1961; Ruehle 1967). Leaves are 10-20 cm long, 5-10 cm wide, alternate, stipulate, and petioled (Ochse *et al.* 1961); oblong, oval or elliptic in form (Purseglove 1969); coriaceous, stiff, dark green and shiny above and pale green below (Ochse *et al.* 1961).

Male inflorescences may arise in leaf axils, along the trunk, and along larger branches above female inflorescences. The male inflorescences are 5-10 cm long and 2-3 cm in diameter; oblong or oblong clavate in shape and have long peduncles (Purseglove 1969). Male flowers are small, numerous, have a 2-lobed calyx and a single stamen (Ochse *et al.* 1961; Purseglove 1969).

Female inflorescences are 4-15 cm long, solitary or geminate, and arise from the axils of primary and secondary branches and the main trunk (Ochse *et al.* 1961). The inflorescences are borne on short, stout peduncles with a green fleshy ring at the apex (Purseglove 1969). Female flowers are small, pubescent, and tubularly shaped; the ovary is oblong and compressed; the style is terminal and obliquely inserted; and the stigma club-shaped (Bailey 1950; Purseglove 1969).

The fruit (Fig. 1.8) is a gigantic yellowish syncarp, 30-90 cm long and 25-50 cm in diameter covered with short triangular to hexagonal, pyramidal, acuminate spines (Bailey 1950; Ochse *et al.* 1961). The fruit interior contains 100-500 seeds (achenes), 2-4 cm long and 1-2 cm thick (Morton 1987). The pericarp is white to yellow, coriaceous, and juicy (Morton 1987; Ochse *et al.* 1961; Purseglove 1969). Fruit are highly aromatic (Morton 1987) and with a sweet flavor (Martin *et al.* 1987).

Adaptation and Cultural Considerations

Jackfruit grow best in hot, humid tropical climates (Martin *et al.* 1987), and will tolerate occasional cold (Ochse *et al.* 1961) but may be severely damaged by frost (Martin *et al.* 1987).

Jackfruit are tolerant of a variety of well-drained soil types but grow best in rich, deep, sandy-loams (Martin *et al.* 1987; Ochse *et al.* 1961; W. Popenoe 1974). Trees thrive best with good soil moisture but do not tolerate flooded conditions (Morton 1965; Ochse *et al.* 1961).

There are numerous jackfruit varieties ranging from sweet, juicy, soft-fleshed types to more dry, coriaceous, acid types (Morton 1965; Ochse *et al.* 1961; W. Popenoe 1974). NS-1 and Tabuey are popular cultivars in south Florida. Precocity and season of bearing also vary among cultivars (Morton 1965). Jackfruit trees generally flower in June-July

FIG. 1.8. JACKFRUIT

and December-February in south Florida and fruit mature in 180-200 days (Martin *et al.* 1987).

Jackfruit trees can be propagated by seed, grafting (Martin *et al.* 1987) and cuttings (Ochse *et al.* 1961; Sedgley 1985A). Inarching is commonly practiced in Florida. Seedling trees take 4-14 years to begin bearing fruit (Morton 1965) while grafted trees usually begin bearing in 4-6 years.

In India, the jackfruit tree and fruit are attacked by a number of insect pests including a stem and fruit borer (*Margaronia caecalis*), a moth larvae (*Perina nuda*) and several disease organisms (Morton 1987). Fruit rot caused by *Rhizopus artocarpi* occurs in south Florida (McMillan 1974) and in the native range of the species.

Morton (1987) reports jackfruit trees are spaced 12 m apart in Indian orchards while the small amount of jackfruit planted in south Florida (less than 1 ha) is spaced 5.5 to 7.6 m apart within rows and 6.1 to 7.6 m apart between rows. Sedgley (1985A) suggests 6-12 m spacings be used in Australia.

Current Importance and Future Prospects

Jackfruit is grown commercially in India, Asia (Morton 1965), Africa (Morton 1987), and south Florida (Anonymous 1988; Campbell 1986). India and Sri Lanka have an estimated 26,000 ha and 4,452 ha of jackfruit, respectively (Morton 1987), while south Florida has less than 1 ha. Production, export, and import figures are not available.

Fresh jackfruit and its many processed forms (Morton 1965) are sold wherever it is grown. Several authors (Martin *et al.* 1987; Morton 1987) feel the potential for increased demand and production of jackfruit is good. In south Florida, production and acreage may increase slightly to meet the demand of the local Asian population.

MYRTACEAE

The Myrtaceae is a large family (75 genera and about 2,750 species) of shrubs and trees characterized by flowers with numerous stamens and pellucid-dotted leaves (Bailey 1950). Most species are indigenous to tropical and subtropical areas; however, America and Australia are the main centers of origin (Bailey 1950; Purseglove 1969).

Numerous plants in the Myrtaceae are of economic importane. These include cloves (*Eugenia caryophyllus* (Sprengel) Bullock and Harrison), allspice (*Pimenta dioica* (L.) Merrill) (Purseglove, 1969), guava (*Psidium guajava* L.), myrica oil (*Pimenta acris* Kostel.), cajeput (*Melaleuca leucadendron Linn.)* and eucalyptus (*Eucalyptus* spp.) (Bailey 1950).

Others like araza (*Eugenia stipitata* subsp. *sororia* McVaugh) (Clement 1983), jaboticaba (*Myrciaria cauliflora* (DC.) Berg), feijoa (*Feijoa sellowiana* Berg) and Malay apple (*Syzygium malaccensis* Merr. et Perry) are less known but are grown commercially for their fruit in some areas (Martin *et al.* 1987).

JABOTICABA (*Myrciaria cauliflora* (DC.) Berg)

Origin and Distribution

Native to Brazil, jaboticaba has been introduced into tropical and subtropical areas of South America (Martin *et al.* 1987; Morton 1987), the United States (Florida and Hawaii) (Ochse *et al.* 1961), and Australia (Maggs 1985A).

Botanical Description

Jaboticaba is a small tree or large shrub to 12 m in height (Martin *et al.* 1987; Ochse *et al.* 1961) with a symmetrically dome-shaped canopy (Campbell 1977; Ochse *et al.* 1961). The bark is attractive with varicolored light and dark patches (W. Popenoe 1974; Phillips and Goldweber 1984).

Jaboticaba is an evergreen tree with light to dark green lanceolate to elliptically shaped leaves (Phillips and Goldweber 1984; Ochse *et al.* 1961; W. Popenoe 1974). Leaves are opposite (Bailey 1950), glabrous, 2.5-10 cm long and 1 cm wide, acute to acuminate shaped at the apex and rounded at the base; margins are entire (Bailey 1950; Ochse *et al.* 1961; W. Popenoe 1974). The small flowers are white with four petals and many stamens (W. Popenoe 1974) and are borne singly or in clusters directly on the bark of the trunk and limbs (Bailey 1950; Ochse *et al.* 1961; Phillips and Goldweber 1984). The fruit (Fig. 1.9) is nearly sessile or borne on short pedicels (Bailey 1950; Campbell 1977); globose shaped, dark purple to black, 2.0-3.5 cm in diameter, with a tough skin and sweet, juicy, whitish pulp (Campbell 1977; Ochse *et al.* 1961; Phillips and Goldweber 1984; Ruehle 1967). There are 1-4 laterally compressed oval to round seeds (Bailey 1950; W. Popenoe 1974; Ochse *et al.* 1961; Ruehle 1967) measuring 0.6-1.0 cm long (Bailey 1950).

Adaptations and Cultural Considerations

Jaboticaba trees are best adapted to cool tropical and warm subtropical climates with medium to high rainfall (Martin *et al.* 1987). Best flowering and fruiting occur in areas with a period of low winter temperatures (Wiltbank *et al.* 1983). Mature trees have been reported to tolerate minimum temperatures as low as –2° to –3°C, although some defoliation and branch death occurred (Phillips and Goldweber 1984; W. Popenoe 1974).

Trees grow most vigorously in well-drained, fertile, and mildly acid soils (Martin *et al.* 1987; Phillips and Goldweber 1984). Trees are not well adapted to the highly calcareous soils of south Florida (Campbell 1977;

FIG. 1.9. JABOTICABA FRUIT GROWING SINGLY OR IN CLUSTERS ON LIMBS

Phillips and Goldweber 1984). Addition of rotted organic matter to the soil at planting and foliar fertilization with micronutrients improves plant growth (Phillips and Goldweber 1984).

Jaboticaba trees can be propagated by seed, budding, grafting, marcottage, and inarching (Ochse *et al.* 1961; Wiltbank *et al.* 1983). Common selections are propagated from seed because jaboticaba seeds are polyembryonic and plants come true to seed (Campbell 1977; Phillips and Goldweber 1984). However, seedling trees take 6-10 years or more to begin bearing, whereas grafted trees take only 4-5 years (Bailey 1950; Martin *et al.* 1987; Wiltbank *et al.* 1983). Grafting jaboticaba onto more vigorously growing genera to induce precocity and scion vigor has not been successful (Ogden and Campbell 1982).

Jaboticaba trees may bear 5 to 6 crops per year (Phillips and Goldweber 1984); however, temperatures and rainfall influence the frequency and time of flowering (Wiltbank *et al.* 1983). Fruit mature in 30-40 days from bloom (Wiltbank *et al.* 1983; Martin *et al.* 1987).

Jaboticaba trees respond positively to fertilizer applications. In Florida, monthly applications of N-P-K-Mg are recommended (Phillips and Goldweber 1984). Periodic foliar sprays of boron, copper, manganese and zinc, and soil drenches with chelated iron are also suggested (Phillips and Goldweber 1984). Mature trees are less frequently fertilized. In Brazil, high rates of N fertilization are reported to reduce fruit production (Wiltbank *et al.* 1983).

There are numerous jaboticaba cultivars in Brazil (Morton 1987; Wiltbank *et al.* 1983) with varying fruit and bearing characteristics. Several cultivars of importance include Paulista, Rajada, Bronca, Ponhema, and Sabara (Morton 1987; Wiltbank *et al.* 1983). Planting distance recommendations vary with cultivar and range from 4.5 m x 4.5 m (Morton 1987) to 10 m x 10 m (W. Popenoe 1974; Wiltbank *et al.* 1983).

In Brazil jaboticaba trees are occasionally attacked by aphids, scales, weevils, and beetles (Wiltbank *et al.* 1983), and flowers and fruit may be infected by a rust (*Puccinia psidii* Wint.) during periods of warm weather and high rainfall (Wiltbank *et al.* 1983).

Current Importance and Future Prospects

Jaboticaba fruit are eaten fresh and processed into wine, jelly, marmalade, and juice (Morton 1987). Clement (1983) reports large acreages in southern Brazil; trees are also common in the home landscape (Wiltbank *et al.* 1983).

There are no readily available production or trade figures for jaboticaba fruit although it is sold fresh and in processed form in Brazil (Morton 1987). Several authors (Martin *et al.* 1987; Wiltbank *et al.* 1983) suggest jaboticaba has good commercial potential as a fresh and processed fruit product because of its excellent flavor.

ARAZA (ARACA BOI) (*Eugenia stipitata* subsp. *sororia* McVaugh)

Origin and Distribution

The araza is native to western Amazonia—Peru, Bolivia (Donadio 1983), and Colombia (Morton 1987). Araza has not been widely introduced outside its native range.

Botanical Description

Araza plants are large shrubs or small trees (Fig. 1.10) (Clement 1983; Swift and Prentice 1983) to 3 m in height (Morton 1987) with dense dark green foliage (Clement 1983). Leaves are elliptically shaped, 9-12.5 cm long, and 2.5-4.5 cm wide (Morton 1987). Flowers are small (about 1.25 cm diameter) and possess 75 stamens (Morton 1987) (see Chapter 12 for more details). The oblately shaped fruit (Fig. 1.11) is a small to large berry up to 800 g (Clement 1983) with bright yellow skin (exocarp) and pulp (mesocarp) (Clement 1983). Seeds are generally numerous, kid-

Courtesy of Charles R. Clement

FIG. 1.10. ARAZA TREE

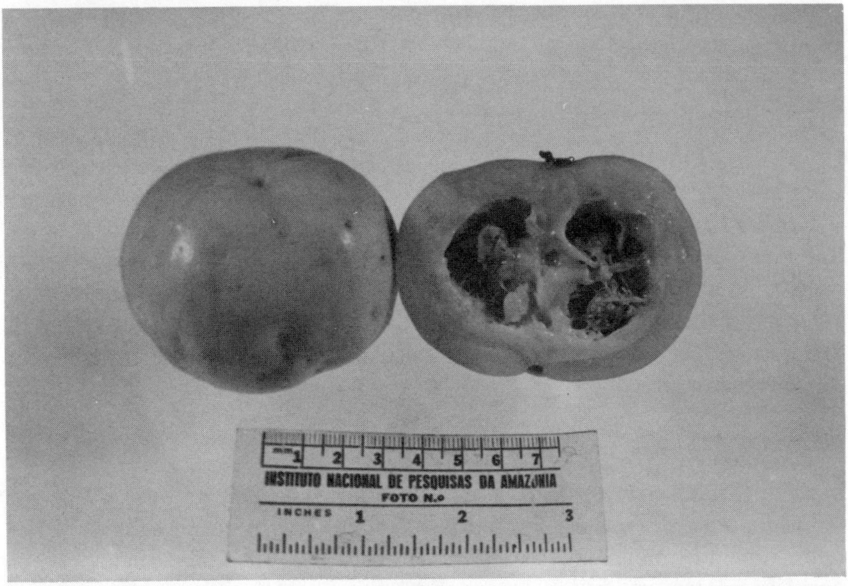

Courtesy of Charles R. Clement

FIG. 1.11. ARAZA FRUIT

ney shaped and 3-7 mm long (Morton 1987). The fruit is aromatic and the pulp is acid (Clement 1983).

Adaptations and Cultural Considerations

The araza appears well adapted to the hot lowland tropics of its native habitat as an understory plant and in more open formal plantings (Swift and Prentice 1983; see Chapter 12). Trees are precocious and begin to bear within 2-3 years after planting (Swift and Prentice 1983).

Research has suggested that availability of water improves fruit production dramatically (Clement 1983). Trees are propagated by seed (Swift and Prentice 1983), and seedlings vary in fruit color (Swift and Prentice 1983), size (Clement 1983) and seediness.

In Ecuador, araza trees are planted 3 m within rows and 5-6 m between rows (555-666 trees/ha) (Swift and Prentice 1983), while in Peru a 3 m x 4 m spacing (833 trees/ha) has been used (Clement 1983).

Araza trees appear to bear up to 4 crops per year although some fruit is available all year (Swift and Prentice 1983). Fruit is currently picked when yellow and soft, which makes handling for the fresh market difficult (Swift and Prentice 1983). This is because green fruit do not ripen properly. Fruit are eaten fresh, juiced and added to ice cream (Donadio 1983).

Araza trees encounter several pests including a leaf spot (*Cercospora myrticola*) and the Mediterranean fruit fly (*Ceratitis capitata* (Wied.) (Swift and Prentice 1983).

Current Importance and Future Prospects

There is limited commercial production of araza in Ecuador, Peru (Swift and Prentice 1983) and Brazil (Clement 1983). Several authors (Clement 1983; Donadio 1983; Swift and Prentice 1983) suggest that araza has good processing potential as a commercial crop. This is due to its high productivity (Clement 1983; Swift and Prentice 1983), attractive color and aroma, high protein and moderate vitamin C content (Clement 1983), and poor handling characteristics for the fresh fruit market (Swift and Prentice 1983).

OXALIDACEAE

The Oxalidaceae contains about 7 genera and 230 species, mostly of tropical and subtropical distribution (Bailey 1950). Characteristics of the family include moderate to high concentrations of oxalic acid in plant parts (e.g., leaves, fruit), compound leaves with pulvini at the base of leaflets, dimorphic styles or stamens, a superior 5-celled ovary with a persistent central column, and fruit that is a dehiscent capsule or berry that does not break into separate carpels (Bailey 1950).

Oxalic acid extracted from various species is used as a tarnish and rust removal agent on semi-precious metals (Bailey 1950; Morton 1987) and as a mordant in the dyeing industry (Morton 1987). Other species like carambola (*Averrhoa carambola* L.) and bilimbi (*Averrhoa bilimbi* L.) are eaten fresh or in processed form.

CARAMBOLA (*Averrhoa carambola* L.)

Origin and Distribution

Native to Southeast Asia (Campbell and Malo 1981; FAO 1982; Martin *et al.* 1987), the carambola is found cultivated or naturalized throughout most tropical and warm subtropical areas of the world (Knight 1983).

Botanical Description

The carambola is a small (7-10 m high) spreading (6-7.6 m wide) (Campbell and Malo 1981) single or multi-trunked tree with grayish-brown to dark gray smooth bark (Ochse *et al.* 1961). Leaves are alternate, imparipinnate, with the rachis thicker at the base than apex (Bailey 1950). Five to eleven leaflets increase in size toward the leaf tip (Ruehle 1967) and are alternately (nearly oppositely) arranged along the rachis on short petiolules (Ochse *et al.* 1961). Leaflets are 1.5-9 cm long, 1-4.5 cm wide, ovate to elliptic-oblong shaped, rounded at the base and acuminate at the apex, with dark to yellowish green color (Ochse *et al.* 1961). Flowers are borne in panicles on twigs and small diameter branches in the leaf axils of present or fallen leaves (Ochse *et al.* 1961). Flowers are small, perfect, about 1 cm in diameter (Campbell and Malo 1981) and 0.8 cm long (Purseglove 1969), with five sepals and petals (Ochse *et al.* 1961); pink to lavender (Campbell and Malo 1981). The flowers have five nonfunctional epipetalous stamens (staminodes) and five fertile episepalous stamens (Ochse *et al.* 1961). The ovary is greenish white and 4-5 celled. Trees possess flowers with either 4-5 long (0.2 cm) or short (0.05-0.075 cm) styles (heterostyle) (Ochse *et al.* 1961). The fruit is a fleshy 4-5 celled berry, 5-12.5 cm long, 3-6 cm in diameter with five (occasionally 4-6 or more) longitudinal ribs (Ochse *et al.* 1961) and is star shaped in cross section (Campbell and Malo 1981; Ochse *et al.* 1961) (Fig. 1.12). Fruit range in weight from 100-110 g or more (Campbell and Koch 1989). The skin (exocarp) is thin, smooth, light yellow to dark orange-yellow with a waxy cuticle (Campbell and Malo 1981; Ochse *et al.* 1961). The flesh is yellow, crisp and juicy with a mild to strongly sour or sweet flavor.

FIG. 1.12. ARKIN CARAMBOLA

There are 10-12 (sometimes none) ovoid shaped, 0.6-1.3 cm long, compressed, light brown seeds enclosed by a gelatinous aril (Campbell and Malo 1981; Ochse *et al.* 1961).

Adaptation and Cultural Considerations

The carambola is best adapted to hot, wet tropical lowland areas (Campbell and Malo 1981; Martin *et al.* 1967), but grows well in warm subtropical areas that experience only occasional freezing temperatures (Campbell *et al.* 1985). In Florida, air temperatures of –1.1° to 0°C may kill immature leaves. Young trees and the mature leaves and twigs of older trees may be killed at –1.7° to –2.8°C (Campbell *et al.* 1985). Damage to small branches on mature trees has been observed after –1.7° to –3.9°C temperatures, whereas –4.4° to –6.7°C temperatures may kill large branches and trees (Campbell *et al.* 1985). Sprinkler irrigation has been observed to be effective in protecting trees and fruit from prolonged freezing temperatures.

Carambola cultivars in Florida vary in wind tolerance. Symptoms of wind damage include defoliation, desiccation, twig dieback, stunted growth, and damaged fruit (wind scar). Golden Star trees (Campbell

1965) have been observed to tolerate growing in open and windy sites better than Arkin. The best sites for Arkin in south Florida are sites adjacent to natural pine lands and mature groves, and some growers have constructed artificial wind breaks. Arkin groves with good wind protection grow more vigorously and begin bearing before groves without wind protection. Wind damage (e.g., fruit drop, limb breakage) may be a problem in some coastal areas of Malaysia (Wahab *et al.* 1989) and Hawaii (Hamilton and Hamilton 1989).

Trees are tolerant of a seasonal dry period (Martin *et al.* 1987) but grow and produce better with well distributed rainfall and/or periodic irrigation during dry periods (Campbell *et al.* 1985; Campbell and Malo 1981). In Malaysia observations indicate prolonged dry periods (i.e., greater than two months) reduce flowering while periods of rainfall stimulate vegetative growth and flowering (Wahab *et al.* 1989). Carambola trees tolerate a variety of soil types as long as they are well drained (Campbell and Malo 1981; Martin *et al.* 1987). However, recent studies by Joyner and Schaffer (1989) suggest carambola trees are moderately flood tolerant. Young Golden Star carambola trees survived continuous flooding up to 18 weeks and intermittent flooding during an 18-week period. Carbon assimilation, transpiration, and CO_2 stomatal conductance declined significantly after five weeks of continuous flooding, however, recovered to nonflooded values after plants were returned to a nonflooded condition. Biomass accumulation of roots and shoots was reduced for continuously and intermittently flooded trees compared to nonflood controls. This would be a problem under commercial conditions.

Carambola trees grow best in mildly acidic to neutral pH (5.5-7.0) soils (Campbell *et al.* 1985). Trees growing in the calcareous limestone soils of south Florida with a pH of 7.5-8.5 often have iron, zinc, and manganese deficiencies. These can be prevented by applying chelated iron as a soil drench, and soil (chelated forms) or foliar applications of zinc and manganese 2-6 times per year (Campbell *et al.* 1985; Campbell and Malo 1981). Four to six dry material applications of nitrogen, phosphorus, potassium, and magnesium at 567 kg per ha are also recommended in south Florida (Campbell *et al.* 1985). Preliminary field work has suggested that use of controlled release fertilizers may reduce fertilizer application frequency for young trees (Ferguson *et al.* 1988). Long term studies are needed to determine if application rates can also be reduced.

There are two basic forms of carambola trees, those that produce sweet fruit and those that produce tart fruit. Sweet types produce mature fruit with greater sugar to acid ratios (Campbell *et al.* 1985; Wagner *et al.* 1975). In addition to sugar and acid content, cultivars and selections differ in fruit size, weight, number of seeds (Wagner *et al.* 1975), rib size and color (Campbell *et al.* 1985).

Carambola trees possess heterostyly and Knight (1965) has demonstrated some self- and cross-incompatibility. Highest percent fruit set (2.79-2.98%) resulted from crossing short-styled flowers with pollen from long-styled flowers and vice-versa. Fruit set was reduced for long-styled crossed with long-styled (0.47%) and long-styled selfed (0.62%). Short-styled selfed or crossed with short-styled resulted in no fruit set. However, this cross incompatibility appears incomplete because commercially acceptable yields of isolated blocks of Golden Star (Campbell *et al.* 1985) and Arkin are common. The Malaysian cultivars B2, B10 and B17 require cross pollination for good fruit set (Wahab *et al.* 1989).

Carambola trees produce some fruit sporadically throughout the year (Campbell *et al.* 1985; Campbell and Malo 1981). However, the time of major production varies with growing area. The major crop in Taiwan is between June and March (Green 1987). Fruit are available year-round in Malaysia (4-6 production peaks occurring per year) (Wahab *et al.* 1989) and Guyana (Ramsammy 1989). In Florida, two major flowering and fruiting periods have been observed. The first major bloom occurs from April to June and fruit is harvested from August to October while the second main bloom is September to October and fruit is harvested from December to February (Campbell *et al.* 1985). In Australia, the main flowering period is May to June, and the major cropping season is late winter (Sedgley 1985B). Fruit development from flowering to horticultural maturity has been shown to vary among cultivars; requiring about 61 days for Golden Star and 65 days for Arkin (Campbell and Koch 1989).

Carambola trees are precocious bearers with fruit production generally beginning within 18-24 months after planting and increasing rapidly as trees mature. In Taiwan and Malaysia, even extensively thinned trees produce 136 kg and 200 kg per tree per year, respectively (Green 1987). Donadio (1989) reports Brazilian production ranges from 20-100 kg per tree per year but probably averages 30 kg. In south Florida, yields of 45 kg per tree per year are not uncommon for 5-year-old trees, with 8-10 year-old trees producing up to 136 kg per year, and 12-13 year-old trees producing 136-141 kg per tree per year (Campbell *et al.* 1985).

Numerous cultivars have been selected in Taiwan, Malaysia, and the United States (Florida, Hawaii) where active selection and breeding programs exist. Knight (1989) has described a carambola improvement program in the United States. The major sweet cultivars in Malaysia are B2, B10 and B17; in Taiwan the major sweet cultivars include Meeshi, Er-lin, Soft sih, and Cheng-Tsey (Green 1987). In Florida, the major sweet cultivar is Arkin and the tart cultivar is Golden Star (Crane *et al.*

1989). In Hawaii, the major cultivars are Sri Kembangan, Kari (1-11), B10, B16 and Kajang (Hamilton and Hamilton 1989).

Carambola rootstock selection and development has been limited. However, field trials by Knight (1982) demonstrated seedling plants grown from selfed Golden Star and open pollinated M-18960 performed better in the highly alkaline (limestone) soils of south Florida than seedlings derived from Asian cultures.

Carambola trees can be propagated by seed, grafting, budding, and air layering (Ochse *et al.* 1961; Campbell *et al.* 1985; Campbell and Malo 1981). Topworking is also successful (Campbell *et al.* 1985). In Florida, veneer grafting and chip budding of actively growing young seedlings is practiced (Campbell *et al.* 1985). Preparing the budwood 3-4 days ahead of time by removing leaves improves the proportion of successful grafts. Success with air layering in Florida has been limited due to poor root development and field establishment. Tissue culture propagation has met with only limited success because differentiation of shoots but not roots has been obtained (Litz and Conover 1980; Litz and Griffis 1989).

Plant spacing varies with growing region and cultural practices. In Taiwan many carambolas are grown on 1.8 m high trellises, and trees are spaced 5.5-6.4 m apart (244-331 trees per ha) within and between rows (Green 1987). Trees planted on hilly terrain in Malaysia are planted at 3.4-3.7 m centers (730-865 trees per ha) (Green 1987) while other plantings utilize 5 m x 5 m (400 trees per ha) and 6 m x 6 m (277 trees per ha) (Wahab *et al.* 1989). Florida growers use many different plant spacings, however, most (>60%) plantings are spaced 4.6-6.1 m within rows and 6.1-7.6 m between rows (215-356 trees per ha) (Crane *et al.* 1989). Brazilian groves are typically spaced 4 m x 4 m (625 trees per ha) and 6 m x 6 m (277 trees per ha) (Donadio 1989). Hawaiian growers currently space trees at 3.8 m x 6.7 m (392 trees per ha) and 3.8 m x 7.6 m (346 trees per ha) (Hamilton and Hamilton 1989).

Tree training is practiced extensively in Taiwan (Green 1987) and Malaysia (Green 1987; Wahab *et al.* 1989) while in Florida and Guyana it is not. Many Taiwanese (Green 1987) plantings are trellised which involves extensive pruning and limb bending. In contrast, some Florida growers periodically hedge and top and/or selectively prune trees when crowding occurs (Campbell *et al.* 1985).

Fruit thinning is extensively practiced in Taiwan (Green 1987) and Malaysia (Green 1987; Wahab *et al.* 1989). This increases fruit size and uniformity. Thinning is not practiced in Florida.

Carambola trees and fruit are attacked by insects and diseases. Taiwanese (Green 1987) and Malaysian (Wahab *et al.* 1989) growers bag their fruit to prevent infestation by fruit fly and moth larvae. Leaf spotting caused by *Cercospera averrhoa*, *Phomopsis* sp., and *Phyllosticta* sp. occurs in Florida, although they are not considered major problems (Campbell *et al.* 1985). *Leptothyrium* sp. has been isolated from "sooty mold" fruit symptoms and *Phomopsis* sp. has also been associated with small circular reddish fruit spots (Simone and Kucharek 1988). Anthracnose (*Colletotrichum gleosporiodes* Penz.) also attacks ripe and injured fruit (McMillan 1986).

In Florida, several scale species attack leaves and twigs including plumose scale (*Morganella lingispina* (Morgan)) and *Philephedra tuberculosa* (Nakahara and Gill) (Pena 1986). Fruit damage by stink bugs, *Acanthocephala* sp., and a fruit blotch miner have also been observed (Pena 1986).

Harvest timing and postharvest handling are critical to successful marketing of fresh carambola fruit. Carambola is a nonclimacteric fruit (Oslund and Davenport 1983) although mature fruit (fruit harvested at "color break") will develop normal golden yellow color after picking (Campbell *et al.* 1987). Sugar content of the fruit does not increase after picking; therefore, fruit picked too early (green) is sour or insipid.

Commercial growers in Flordia (Campbell 1987) and Malaysia (Wahab *et al.* 1989) carefully harvest fruit by hand at color break—light green fruit with yellow color developing in the furrows of the ribs and transport them quickly to the packinghouse. At the packinghouse, fruit are graded, packed, and then stored at 5-10°C and 85-95% relative humidity (Campbell *et al.* 1987; Wahab *et al.* 1989). Research (Campbell *et al.* 1987) has shown fruit can be stored for up to 44 days at 5°C without chilling injury and then develop normal color when transferred to 23°C. This capacity to store the fruit greatly enhances the distance fruit can be shipped and the length of time it can be marketed.

Most carambola fruit is eaten fresh (Campbell *et al.* 1985; Green 1987) or used in salads, desserts, and cooked as star-shaped slices. Carambola can also be dried (Campbell and Campbell 1983), made into fruit leathers, canned, and frozen at home (Smith and Campbell 1983). Guyana has developed numerous processed products for commercial sale including wines, sauces, preserves, pickles, prepared dessert breads and cakes. A few entrepreneurs in the United States produce high quality pickles, sauces, relishes and jams.

Current Importance and Future Prospects

Commercial carambola production occurs in Taiwan, Malysia (Green 1987), southern China (W. Popenoe 1974), the Philippines (Pearrow 1988), Australia (Sedgley 1985B), the United States (Florida, Hawaii),

Brazil (Donadio 1989), Suriname (Groeizaam and Lewis 1989) and Guyana.

Hectarage figures for many countries are not available, however, Malaysia is estimated to have about 896 ha (Wahab *et al.* 1989), Taiwan 2,792 ha (Green 1987), Brazil 300 ha (Donadio 1989), Hawaii 30 ha (Hamilton and Hamilton 1989), south Florida 192 ha and Guyana 2,450 ha (Ramsammy 1989).

Carambola production in Taiwan in 1986 was estimated at 34,712 MT while Donadio (1989) estimated Brazil produces 300 MT annually. Malaysia produced about 24,000 MT during 1989 (Wahab *et al.* 1989). Estimated production for south Florida during the 1988-89 sason was 590 MT and for 1990-91, 907 MT or more.

Trade in carambola is expanding worldwide as people become more familiar with its unique shape and taste. Malaysia is reported to have exported 13,022 MT in 1988 (Wahab *et al.* 1989) to such countries as Hong Kong, Singapore, England, Australia and Germany. Taiwan exports small amounts to Japan, Canada, Hong Kong and Singapore (Green 1987). South Florida shippers only sporadically export small amounts to Europe because domestic markets have not been fully developed. The Philippines exported 2.4 MT of frozen carambola to the United States in 1986 (Pearrow 1988).

Carambola production and consumption appears on the rise throughout the world. Carambola production and consumption will most likely increase as selection and breeding for superior tasting cultivars with improved handling characteristics are found.

BILIMBI (*Averrhoa bilimbi* L.)

Origin and Distribution

The bilimbi is found throughout Southeast Asia (Martin *et al.* 1987; Morton 1987) and is thought to be native to Malaysia (W. Popenoe 1974; Purseglove 1969), but it is only known as a cultivated species (W. Popenoe 1974; Ruehle 1967). The bilimbi has been introduced into Australia (Sedgley 1985B), the Caribbean (Morton 1987), Central and South America (Morton 1987), Florida (Ruehle 1967) and Hawaii (Morton 1987).

Botanical Description

Bilimbi trees reach 18 m in height (Kennard and Winters 1960) but more commonly are 15 m or less (Purseglove 1969; Sedgley 1985B). Leaves are alternate, imparipinnate, 30-60 cm long (Morton 1987) with 11 to 37 alternate (nearly opposite) leaflets (Ruehle 1967). Leaflets are oblong, rounded at the base. Flowers, borne in panicles along the trunk and main branches (Kennard and Winters 1960; Ruehle 1967; Sedgley 1985B) are five petalled (Morton 1987) and 2 cm long (Purseglove 1969) with 10 perfect stamens (W. Popenoe 1974). The fruit is a 5-celled berry, cylindrical, 5-7.5 cm long, faintly 5-angled (Purseglove 1969) and with a persistent calyx (Morton 1987) (Fig. 1.13). It is greenish-yellow when mature (W. Popenoe 1974). The exocarp (skin) is thin, smooth, and glossy (Morton 1987) and the flesh (mesocarp) green, soft, juicy, and very acid with a few small seeds (Purseglove 1969). Sweet forms of bilimbi are reported to exist (Kennard and Winters 1960; Morton 1987).

Adaptations and Cultural Considerations

The bilimbi is best adapted to hot, wet tropical lowland areas (Martin *et al.* 1987) that experience little or no freezing temperatures. Young trees may be killed at temperatures of –1° to –2°C. Campbell *et al.* (1977) observed young trees were killed and older trees severely damaged in Florida after –4° to –7°C temperatures. Bilimbi is much more sensitive to cold than carambola.

Courtesy of Kendal and Julia Morton

FIG. 1.13. BILIMBI FRUIT, SEEDS AND LEAVES

Bilimbi trees are tolerant of various soil types and a seasonal dry period (Martin *et al.* 1987). Trees bloom and fruit several times per year (Martin *et al.* 1987) although some fruit is available just about all year (Morton 1987). Fruit mature from flowering in about 90 days (Martin *et al.* 1987).

The bilimbi is generally propagated by seed and begins bearing in 5-7 years (Martin *et al.* 1987). Trees can also be propagated by air layering (Sedgley 1985B; Kennard and Winters 1960). Budding and grafting has not been very successful (Morton 1987; W. Popenoe 1974). There are no named cultivars (W. Popenoe 1974).

Current Importance and Future Prospects

Bilimbi is not important as a fresh fruit on the world market but is made into curries, jellies, chutney, preserves and pickles on a local scale (Martin *et al.* 1987). It is also juiced (Morton 1987) and candied (Kennard and Winters 1960). Its sourness precludes its use as a fresh fruit to any large extent. No trade or production figures are available. The commercial potential of bilimbi appears limited due to the sour taste of the fresh fruit and the availability of many other processed fruits.

PALMAE

The Palmae has 128 genera and about 1,000 species of tropical and subtropical distribution (Bailey 1950). Characteristics of the family include coriaceous plicate leaves, woody flowers and inflorescences, three sepals and petals, usually six stamens, and three carpels, each with one seed (Bailey 1950).

Palm species of economic importance include the oil palm (*Elaeis quineensis* Jacq.), peach palm (*Bactris gasipaes* H.B.K.), coconut palm (*Cocos nucifera* L.), date palm (*Phoenix dactylifera* L.) (Martin *et al.* 1987), and numerous ornamental species.

PEJIBAYE (PEACH PALM) (*Bactris gasipaes* L.) (syn. *Guilielma gasipaes* (H.B.K.) L. H. Bailey)

Origin and Distribution

Native to Central America (Martin *et al.* 1987) the pejibaye has been cultivated and distributed throughout South America (Martin *et al.* 1987; Morton 1987), the Caribbean, Philippines, and southern Florida (Morton 1987).

Botanical Description

The pejibaye is a single or multi-trunked (caespitos) palm—suckering at the base to form a cluster of palm trunks (Kennard and Winters 1960; Martin *et al.* 1987) (Fig. 1.14). Each trunk may reach up to 18 m in height and be 15 cm in diameter (Ochse *et al.* 1961). The trunk is covered with rings of sharp, stiff, 5 cm long, black spines (Ochse *et al.* 1961) (Fig. 1.15). Some selections do not have spines. Leaves are pinnate, 2.5-3.7 m long and dark green (Ochse *et al.* 1961). Leaflets are up to 60 cm long and 3.2 cm wide with spines along some veins.

Pejibaye palm is monoecious with male and female flowers borne on the same raceme (Ochse *et al.* 1961). Racemes, at first enclosed by a spathe, are 45-60 cm long and are borne from the trunk below leaves or in the axils of lower leaves (Fig. 1.16). Staminate flowers have six stamens; the pistillate flowers have an annular calyx and small campanulate, three-toothed corolla with a trilocular ovary with three sessile stigmas (Ochse *et al.* 1961). Pendant racemes hold up to 300 fruit and weigh up to 11.3 kg (Ochse *et al.* 1961) (Fig. 1.17). Five to six racemes are produced per crop. Fruit are conical or ovoid in form, 2.5-5 cm long and nearly covered by the green calyx (Ochse *et al.* 1961). The skin (exocarp) is thin and light yellow to deep orange or brown (Ochse *et al.* 1961). The flesh (mesocarp) is light orange to yellow with a dry, mealy texture (Ochse *et al.* 1961). The single seed is about 1.9 cm long, black, with a hard shell surrounding a white kernel.

Adaptations and Cultural Considerations

Pejibaye palms require a hot, wet tropical climate (Martin *et al.* 1987; Maggs 1985B) with temperatures between 16-25°C, up to 250 cm of rain annually, and an altitude between sea level and 1,200 m (Maggs 1985B). Grown in protected areas, pejibaye palms have survived freezing temperatures in south Florida (Campbell *et al.* 1977).

Trees grow best in clay loams or loam soils (Ochse *et al.* 1961). Production is reduced in areas with dry periods (Martin *et al.* 1987) and at elevations of 1,500 m and above (Ochse *et al.* 1961).

Seedling pejibaye begin bearing in 6-8 years and remain productive for 50-75 years or more (Ochse *et al.* 1961). Morton (1987) reports cool climates may delay bearing until seedlings are 10-12 years old. The bearing season varies with climate, and multiple crops are common (Maggs 1985B; Morton 1987) (see Chapter 15 for more details). In Colombia, fruit are harvested in January and February (Morton 1987) while in Costa Rica trees bloom in March and fruit is available throughout the fall (Maggs 1985B).

Courtesy of Odoardo Brito

FIG. 1.14. A CLUSTER OF PEJIBAYE PALM TRUNKS

Pejibaye trees vary in size of fruit clusters, fruit color and size, length of the spines, and absence of spines (Ochse *et al.* 1961). Seedless forms also exist (Maggs 1985B; Ochse *et al.* 1961).

Courtesy of Odoardo Brito

FIG. 1.15. PEJIBAYE PALM TRUNK SHOWING RINGS OF BLACK SPINES

At present, several Central and South American countries including Costa Rica (Martin *et al.* 1987; Morton 1987), Panama, Colombia, and Brazil (Morton 1987) have germplasm collections for developing superior

Courtesy of Odoardo Brito

FIG. 1.16. RACEMES EXTENDING FROM THE TRUNK OF THE PEJIBAYE PALM

pejibaye palms. Clement (Chapter 15) describes eight races of pejibaye, each with distinct fruiting characteristics.

Trees are spaced 6 m apart for fruit production (Ochse *et al.* 1961) and 2-4 suckers are allowed to develop to maturity—the rest being removed and used as propagation material (Morton 1987). Trees planted for palm heart production are spaced more closely (1.5-3 m apart) and are harvested in 2.5-3 years after planting (Morton 1987).

Propagation is by seed and suckers. Suckers are cut from the parent plant after reaching about 1-1.5 m in height and 7-8 cm in diameter; roots should be present (Maggs 1985B). The foliage of the suckers is reduced and the plants shaded until they are established.

Yields of pejibaye fruit are estimated to be up to 115 kg per clump of trees and production per ha of 23 MT per year (Maggs 1985B) (see Chapter 15 for more details).

Courtesy of Odoardo Brito

FIG. 1.17. A RACEME OF PEJIBAYE PALM FRUIT

Current Importance and Future Prospects

Pejibaye palms produce edible fruit and terminal meristems. Fruit must be cooked (usually boiled) before consumption and can be fried, roasted, and ground into meal after boiling (Morton 1987). The pulp can also be fermented to make an alcoholic beverage. The palm hearts can be eaten fresh or cooked and added to salads and casseroles (Morton 1987).

At present, commercial production of pejibaye appears limited to Costa Rica (Ochse *et al.* 1961). Clement (Chapter 15) estimates Costa Rica has over 2,000 ha of palm heart production and about 300 ha in fruit production.

The pejibaye is considered by several authors (Martin *et al.* 1987; Maggs 1985B) to have considerable potential. Pejibaye palm is a productive crop with a diverse number of uses. Commercial acreage may increase in the future.

PASSIFLORACEAE

The Passifloraceae includes 500 species in 12 genera (Beal and Farlow 1985), most of them (400 species) (Martin and Nakasone 1970) are indigenous to tropical South America but some from Asia and the South Pacific Islands (Beal and Farlow 1985). Various *Passiflora* species have been introduced and cultivated in South and Central America, the Caribbean, Australia, Hawaii (Martin and Nakasone 1970) and Florida.

Characteristics of *Passiflora* species include herbaceous or woody, tendril-bearing vines or shrubs; 3-5 sepals and petals present or zero; a crown or many filaments (corona) between petals and stamens; superior ovary, raised on a gynophore; ovary 1-celled with 3-5 parietal placentae; 1-5 styles; fruit is a berry or capsule. *Passiflora* species have an aril surrounding the seeds (Bailey 1950).

Passiflora species of economic importance include purple (*Passiflora edulis* Sims.) and yellow (*P. edulis* Sims F. *flavicarpa* Degner) passion fruits and the giant granadilla (*P. quadrangularis* L.) (Martin and Nakasone 1970; Morton 1987). The sweet granadilla (*P. ligularis* Juss.), water lemon (*P. laurifolia* L.) and banana passion fruit (*P. mollissima* Bailey) are also produced extensively in some countries in tropical America (Martin and Nakasone 1970).

GRANADILLA (*Passiflora quadrangularis* L.) and (*P. ligularis* Juss.)

Origin and Distribution

The term granadilla is applied to both *P. quadrangularis* (also called the giant granadilla) and *P. ligularis* (also called the sweet granadilla) (Purseglove 1969). Both are members of the subgenus *Granadilla* (Martin and Nakasone 1970).

The giant granadilla (*P. quadrangularis*), native to tropical America, is pantropical in distribution (Martin *et al.* 1987; Martin and Nakasone 1970). In contrast, the sweet granadilla (*P. ligularis*), indigenous to Central and South America, is found mostly from Mexico to northern South America (Martin and Nakasone 1970; Martin *et al.* 1987). Both species have been introduced into new regions, including Australia (Beal and Farlow 1985), Hawaii (W. Popenoe 1974), and Florida (Howell 1976).

Botanical Description

P. quadrangularis—Giant granadilla are vigorously growing perennial vines from 5-50 m long with glabrous stems which become hollow when mature (Ochse *et al.* 1960). Stems are four-sided (quadrangular) and four-winged (Purseglove 1969) with axillary tendrils 22.5-35 cm long (Ochse *et al.* 1961). Leaves are alternate, stipulate, with ovate or elliptic lamina, 10-25 cm long, 8-18 cm wide, dark green above and pale green below with 10-12 pairs of lateral veins (Purseglove 1969). Flowers are large, 10-12 cm in diameter when fully open, axillary, solitary and pendulous; peduncle 1.5-2 cm long and angled (Ochse *et al.* 1961). Three leaf-like light green bracts subtend a five-lobed calyx (tubular at the base), yellow-green below, reddish above (Purseglove 1969). Five petals are 4-4.5 cm long, 1.5-2 cm wide and densely dotted with red (Ochse *et al.* 1961). The corona exhibits two outer layers of 5-7 cm long filaments, white and red at the base, and white and purple above (Purseglove 1969). Five stamens are connivent with gynophore; apical free parts obliquely erect, yellowish-green (Ochse *et al.* 1961), and hand downwards (Purseglove 1969). The ovary raised on a gynophore is glabrous, ellipsoidal, three-furrowed longitudinally, 1-1.5 cm long, one-locular with many parietal ovules (Purseglove 1969). The fruit is a fleshy ovate-oblong berry, 20-30 cm long, 10-15 cm in diameter with a yellowish-green external color (Purseglove 1969) (Fig. 1.18). Fruit weigh 225-450 g or more (Knight and Sauls 1986). The mesocarp is white, 2.5-4 cm thick, juicy and with an insipid taste (Purseglove 1969). The seed cavity is filled with dark brown, obovate to rhomboid, flat seeds surrounded by yellow arils (Ochse *et al.* 1961).

P. ligularis L.—Sweet granadilla is a vigorous perennial vine with broad-ovate, cordate, 10-20 cm long, entire leaves (Purseglove 1969); and is stipulate (Morton 1987). There are one to two flowers to a node (Morton 1987) and 7-10 cm in diameter; the sepals and petals are greenish-white and the corona is white with zones of purple (Purseglove 1969). Fruit is hard-shelled (Purseglove 1969), ovoid, 6-9 cm long, 4.5-5 cm in diameter; greenish, yellow, orange, or purplish and sometimes spotted or mottled

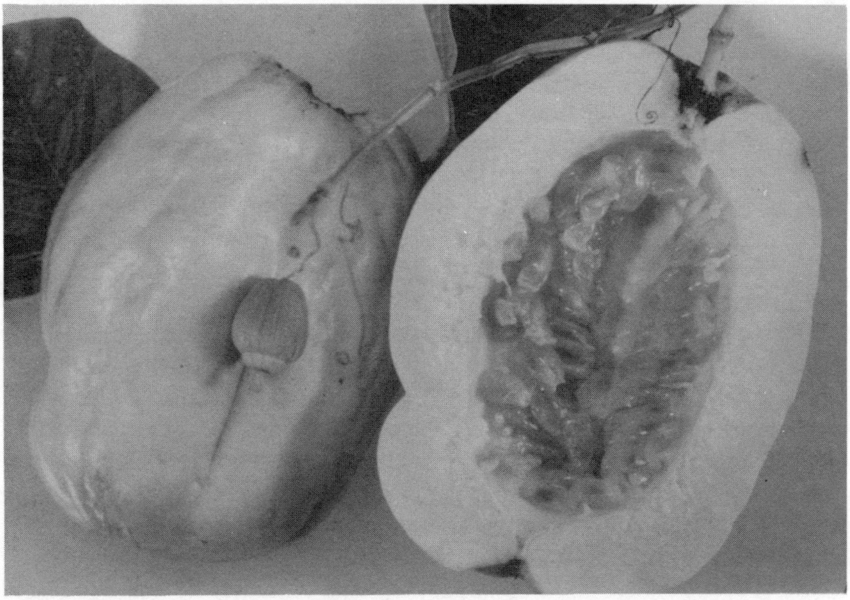

FIG. 1.18. GIANT GRANADILLA WITH ARILS ENCLOSING FLATTENED-OVAL SEEDS

(Martin and Nakasone 1970). The pulp is white and sweet with a distinctive flavor (Martin and Nakasone 1970).

Adaptations and Cultural Considerations

Giant granadilla (*P. quadrangularis* L.) grows best in hot, humid tropical lowland areas (Martin *et al.* 1987) but also grows well in protected sites in warm subtropical areas like south Florida (Knight and Sauls 1986). Vines generally survive light frost but may be severely damaged or killed after prolonged freezing temperatures in south Florida (Campbell *et al.* 1977). In contrast, sweet granadilla (*P. ligularis* L.) is a highland species that grows at elevations between 610-2,743 m (Martin and Nakasone 1970), and grows best in cool tropical climates with well distributed moderate to high rainfall (Martin *et al.* 1987). Vines do not grow well in hot tropical lowland areas (Martin *et al.* 1987).

Giant granadilla tolerates various soil types but grows best on fertile, well-drained loams (Beal and Farlow 1985; Ochse *et al.* 1961). Vines do not tolerate flooding and prefer a slightly acid (pH 6) to neutral pH (Beal and Farlow 1985). Morton (1987) suggests vines do not tolerate highly alkaline soils. Soil requirements and preferences for sweet granadilla are similar.

The giant (Beal and Farlow 1985; Purseglove 1969) and sweet (Morton 1987) granadilla are protandrous and both species set little fruit without the presence of adequate pollinators (Morton 1987). Hand pollination is recommended for adequate fruit set for giant (Purseglove 1969) and sweet (Morton 1987) granadilla. In addition, giant granadilla requires mild temperatures for normal fruiting and may flower but set no fruit during the hottest period of the summer (Knight and Sauls 1986).

Trellising of giant granadilla is practiced in Australia (Beal and Farlow 1985) and Indonesia (Morton 1987). Vines are spaced 2-3 m apart each way in Indonesia (Morton 1987).

Fertilizer practices of giant granadilla in Australia include 0.8-1.2 kg N, 0.3-0.5 kg P, and 1.1-1.5 kg K applied per year per vine in a split (2-3) application (Beal and Farlow 1985). Vines are also irrigated during dry periods. Cultural practices have not been described for sweet granadilla but are similar to the giant granadilla (Beal and Farlow 1985) and passion fruit (*P. edulis* L.) (Knight and Sauls 1986; W. Popenoe 1974).

The harvest season of giant granadilla varies with region and is in autumn and spring in Australia (Beal and Farlow 1985), June through October in Venezuela (Morton 1987), and summer in south Florida (Knight and Sauls 1986). Giant granadilla fruit are clipped by hand in Australia and yields range from 70-150 kg per vine per year (Beal and Farlow 1985).

Giant and sweet granadilla may be propagated by seed, cuttings (Beal and Farlow 1985; Martin *et al.* 1987), and air-layering (Beal and Farlow 1985).

Few insect and disease problems are specifically described for giant and sweet granadilla although several are described for related *Passiflora* species (Beal and Farlow 1985; Knight and Sauls 1986; Morton 1987). Rats, birds and bats may be a problem in some areas (Purseglove 1969).

Current Importance and Future Prospects

Giant and sweet granadilla are grown commercially in Central and South America, Indonesia (Morton 1987), Australia, Fiji, and Sri Lanka (Beal and Farlow 1985). Several authors feel giant granadilla production has moderate potential for expansion (Martin *et al.* 1987; Martin and Nakasone 1970). Most importantly, usefulness of giant and sweet granadilla in developing superior *Passiflora* hybrids is recognized and being exploited (Beal and Farlow 1985; Martin and Nakasone 1970).

The giant granadilla may be cooked and eaten as a vegetable (immature fruit) and the pulp eaten fresh (Martin *et al.* 1987). The arils may be eaten fresh, juiced, and added to desserts (Morton 1987). Sweet granadilla pulp is eaten fresh and the arils juiced and consumed in drinks and added to desserts (Morton 1987).

Specific production and trade figures for giant and sweet granadilla are not available. In some cases these two species may enter commerce as passion fruit and, therefore, it makes determination as to their trade and production difficult to assess.

Giant and sweet granadilla appear to have limited potential for increased production. However, both species are invaluable as germplasm sources in breeding programs designed to develop superior *Passiflora* hybrids. Increased production of hybrids from both species may be good because of their unique flavor and versatility.

PUNICACEAE

The Punicaceae consists of only one genus and two species (Bailey 1950), and is thought to be native to southern Asia (Kennard and Winters 1960; Ochse *et al.* 1961; Ruehle 1967) and/or Middle East (Martin *et al.* 1987; W. Popenoe 1974; Purseglove 1969). Characteristics of the family include bisexual flowers, a tubular or campanulate receptacle, one style and stigma, and the fruit is a berry (Bailey 1950).

Only the pomegranate (*Punica granatum* L.) is commercially important; grown for its fruit and as an ornamental tree (Purseglove 1969; Westwood 1978).

POMEGRANATE (*Punica granatum* L.)

Origin and Distribution

Native to southern Asia (Kennard and Winters 1960; Ochse *et al.* 1961; Ruehle 1967) and/or the Middle East (Martin *et al.* 1987; W. Popenoe 1974; Purseglove 1969), the pomegranate has been distributed throughout tropical, cool tropical, subtropical (Martin *et al.* 1987) and warm temperate regions of the world (Kennard and Winters 1960; Westwood 1978).

Botanical Description

The pomegranate is a shrub or small tree to 6 m in height (Martin *et al.* 1987) with spiny branches (Westwood 1978). Trees are deciduous in cooler climates (Kennard and Winters 1960; Purseglove 1969). Leaves are opposite, often crowded on small axillary twigs (Ochse *et al.* 1961), short petioled, glabrous, 2-8 cm long, and obovate to oblong shaped (Westwood 1978). One to five flowers may be borne on small axillary twigs near the apex; one flower terminal, the others solitary, arising from adjacent leaf axils (Ochse *et al.* 1961). Flowers, or short peduncles are red, bisexual, 4-6 cm long and 5-7 cm in diameter when expanded (Ochse *et al.* 1961). The ovary is inferior, many celled, surrounded by the receptacle and below the 5-8 sepals and petals; one stigma and style are present (Ochse *et al.* 1961). There are numerous stamens inserted irregularly on the receptacle disk (Ochse *et al.* 1961). The fruit is a false berry, subglobose in shape (Westwood 1978), shiny red, yellowish green or whitish, crowned with a persistent calyx, and has a coriaceous rind (Ochse *et al.* 1961) (Fig. 1.19). Internally, the fruit has few to many locules containing many seeds. The astringent, acid to sub-acid aril surrounding the seeds is edible (Ochse *et al.* 1961).

Adaptations and Cultural Considerations

Pomegranate trees grow and produce best in subtropical areas with long, dry summers, and cool winters (Martin *et al.* 1987; Ochse *et al.* 1961; Westwood 1978). Fruit produced in warm, humid areas like Florida is inferior (Ruehle 1967). Westwood (1978) reported that trees have a very short chilling requirement and can tolerate freezing temperatures of –9° to –12°C. Pomegranate trees in southern Florida withstood prolonged freezing temperatures with no damage (Campbell *et al.* 1977).

Courtesy of Seymour Goldweber

FIG. 1.19. POMEGRANATE

Trees are tolerant to a wide variety of soils (Martin *et al.* 1987; Ochse *et al.* 1961; Sheets and DuBois 1986). The pomegranate tolerates dry soil conditions (Martin *et al.* 1987) but does best with some irrigation during dry periods (Sheets and DuBois 1986). Trees also tolerate poorly drained soil conditions (Martin *et al.* 1987).

Orchards are spaced 4-5 m (Ochse *et al.* 1961) to 5-6 m apart between and within rows; trees may also be double-set (2-3 m apart) to form hedge rows (Sheets and DuBois 1986). Commercially, tree training and pruning are practiced in the United States (Childers 1983; Sheets and Dubois 1986) and India (Chapter 17). Fertilizer recommendations in Florida include 1 kg per tree of an 8-8-8 or similar material two times (November and March) per year (Sheets and Dubois 1986).

Pomegranate trees may be propagated by seed, cuttings, marcottage (Martin *et al.* 1987), grafting, and budding (Childers 1983). Seedling trees take 3-4 years to begin bearing (Martin *et al.* 1987) and produce fruit of variable quality (Sheets and DuBois 1986). Fruit may be produced year-round in south Florida (Sheets and DuBois 1986) but mainly is harvested in late summer and fall in northern Florida (Ruehle 1967).

There are numerous pomegranate cultivars which must be propagated vegetatively (Kennard and Winters 1960). The major cultivar in California is Wonderful while various cultivars are used in the Middle East and Asia (see Chapter 17 for descriptions).

Pomegranates appear to be relatively pest-free although a fruit borer is reported to invade the fruit in India and Egypt, and the Mediterranean fruit fly infests the fruit in Hawaii (W. Popenoe 1974). Scales and mites are occasional problems in Florida (Sheets and DuBois 1986).

Current Importance and Future Prospects

Pomegranates are commercially produced to a varying degree in numerous regions (e.g., Middle East, South America, Asia). Demand for the fruit is greater in the Middle East and India than in the Western Hemisphere. Mohan Kumar (Chapter 17) reports that demand in India currently outstrips production. The United States (California) produced about 3 MT annually from 1969-1973 (Westwood 1978).

The pomegranate can be eaten fresh and the juice extracted form the aril for beverages and syrup (Ochse *et al.* 1961). The juice has a sweet to sub-acid flavor. The prospects for expanded production in some regions (e.g., India) appears good while in others it is limited (e.g., United States). This may be because of the seediness and cartilaginous nature of the fruit.

RUTACEAE

The Rutaceae consists of over 100 genera and about 900 species of trees, shrubs, and herbaceous plants mostly of tropical distribution. However, some species extend into subtropical and temperate zones (Bailey 1950). Plants of the Rutaceae have been introduced throughout the world.

Characteristics of the Rutaceae include flowers with 4-5 sepals and petals, a 2-5 lobed and celled superior ovary (Bailey 1950), ovary raised on a nectary disc, fruit and leaves with oil glands, and pellucid dotted leaves.

The most important sub-family of the Rutaceae is the Aurantioideae. Two important characteristics of this sub-family include mostly polyembryonic seeds and a specialized berry called a hesperidium.

Many species in the Rutaceae (Aurantioideae) are of major economic importance. These include the orange (*Citrus sinensis* (L.) Osbeck), grapefruit (*C. paradisi* Macf.), mandarin (including tangerine) (*C. reticulata* Blanco), lemon (*C. limon* (L.) Burm.), lime (*C.* x Tahiti), and a large number of hybrids.

CALAMANSI (CALAMONDIN) (*Citrus madurensis* Lour. syn. *C. mitis* Blanco)

Origin and Distribution

Native to the Philippines (Bailey 1950; Hume 1957; Jackson 1990), calamansi has been introduced and distributed throughout the tropical and subtropical world. Calamansi trees are grown for their fruit and their ornamental value in the home landscape and as an indoor potted plant.

Botanical Description

The trees are evergreen, 2-7.5 m in height (Morton 1987) with upright branches and a dense canopy (Bailey 1950; Hume 1957). Leaves are alternate, 4-7.5 cm long (Morton 1987), broadly oval, dark green above and pale green beneath with narrowly winged, articulated petioles (Hume 1957). Trees are nearly thornless (Hodgson 1967). Flowers are borne singly or in clusters of 2-3 terminally or in leaf axils near branch tips (Morton 1987); these are usually 20 stamens (Hume 1957). Fruit is a hesperidium, small, 2.9-3.2 cm in diameter (Hume 1957), oblate, with an orange-red color; the exocarp is thin, smooth and finely pitted (Hodgson 1967) (Fig. 1.20). The fruit has 7-10 segments that separate easily from each other (Hume 1957); the pulp is orange, juicy, and acid tasting

Courtesy of Kendal and Julia Morton

FIG. 1.20. CALAMANSI (CALAMONDIN)

(Hodgson 1967). Fruit usually contain 2-5 small, polyembronic seeds with green cotyledons (Hodgson 1967).

Adaptations and Cultural Considerations

Calamansi trees may be grown under a wide range of climatic conditions in tropical (e.g., Philippines, Central America) (Morton 1987) and subtropical (e.g., Japan, Florida) (Hodgson 1967) areas. Trees are relatively cold hardy (Hume 1957) and have survived prolonged freezing temperatures in south Florida with little to no damage (Campbell *et al.* 1977).

The calamansi is tolerant of various soil types provided they are well drained. In Florida, trees grow well in acid sandy and alkaline limestone soils.

Calamansi may be propagated by seed (Morton 1987; Jackson 1990), budding and cuttings (Morton 1987). Calamansi seedlings are used as rootstocks for calamansi in the Philippines (Morton 1987) and as a rootstock for kumquat in the United States (Hodgson 1967). Trees bear some fruit year-round but the peak season in Florida is winter (Jackson 1990) and in mid-August to October in the Philippines (Morton 1987).

There are a number of systemic diseases which attack calamansi trees; these include exocortis, xyloporosis, psorosis, tristeza, and leaf mottle-yellow (Wallace 1978). Fruits are also susceptible to attack by fruit flies (Morton 1987).

Current Importance and Future Prospects

Calamansi fruit juice is used as a condiment in cooking, and whole fruits and rind are made into marmalade and candied (Morton 1987). The calamansi fruit is produced on a commercial scale in the Philippines (Morton 1987), whereas in Florida, trees are mostly grown for their ornamental value.

The potential for expanded calamansi production does not appear to be great because there are many other citrus fruits with identical uses that are already produced commercially.

SAPOTACEAE

The Sapotaceae consists of 31 genera and about 400 species of trees and shrubs of tropical distribution (a few temperate zone species also exist) (Bailey 1950). Characteristics of the family include milky sap, a superior ovary and a berry fruit (Bailey 1950).

Some plants in the Sapotaceae are grown commercially, these include the Shea butter tree (*Butyrospermum paradoxum* Gaertn. f. Hepper), sapodilla (*Manilkara zapota* van Royen), mamey sapote (*Calocarpum sapota* Merr.), caimito (*Chrysophyllum cainito* L.) (Martin *et al.* 1987; Purseglove 1969), canistel (*Pouteria campechiana* Baehni), abiu (*Pouteria caimito* Rodlk) (W. Popenoe 1974), and lucuma (*Pouteria lucuma* Ruiz and Pavon).

LUCUMA (LUCUMO) (*Pouteria lucuma* (Ruiz and Pavon); syn, *Lucuma obovata* H.B.K.)

Origin and Distribution

Native to South America, Ecuador, Peru, northern Chile (Martin *et al.* 1987; Scholefield 1985), lucuma has been introduced to other areas of tropical America (Martin *et al.* 1987). Lucuma has also been introduced into south Florida but is not well adapted there (Morton 1987).

Botanical Description

Lucuma trees are evergreen (Purseglove 1969), range up to 12 m in height (Martin *et al.* 1987), and have a dense rounded canopy (W.

Popenoe 1974). Leaves are clustered near the ends of branches, elliptic-ovate in shape (W. Popenoe 1974), 12.5-25 cm long, dark green on the upper surface and pale green or brown-hairy underneath (Morton 1987). The fruit is rounded (ovate, obovate), 7.5-10 cm in diameter with a brownish-green exocarp (Morton 1987) (Fig. 1.21). Pulp is yellow, dry, mealy, and sweet (Morton 1987; Purseglove 1969). The fruit contains 1-5 dark brown, round to broadly oval seeds (Morton 1987).

Adaptations and Cultural Considerations

Lucuma is best adapted to cool subtropical climates with moderate rainfall (Martin *et al.* 1987). Trees can tolerate a somewhat dry condition (Scholefield 1985) and are not well adapted to hot tropical lowland areas (Martin *et al.* 1987) (see Chapter 19 for more details). Moderate to severe damage occurred to mature lucuma trees after exposure to prolonged freezing temperatures in south Florida (Campbell *et al.* 1977).

The lucuma tolerates various soil types provided they are well drained (Scholefield 1985). Trees flower in June-August in Florida and fruit mature in about 180 days (Martin *et al.* 1987). In Chile, fruit are harvested from April to December (see Chapter 19 for more details). Propagated

FIG. 1.21. LUCUMA FRUIT WITH EXPOSED SEED

by seed grafting (Martin *et al.* 1987; Scholefield 1985), lucuma seedlings take up to eight years to start bearing fruit (Scholefield 1985).

Current Importance and Fruit Prospects

Lucuma is commercially produced in Chile, Bolivia (Morton 1987) and Peru on a small scale. The fruit is eaten fresh and added to desserts and beverages (Martin *et al.* 1987).

At present, lucuma production and trade appear to be limited to local areas (Scholefield 1985) and whether the fruit has potential for expanded production outside its current area has not been determined. Lizana *et al.* (1986) suggest lucuma has increased potential for local markets while Martin *et al.* (1987) suggest the fruit has fair potential for increased commercialization.

STERCULIACEAE

The 48 genera and 750 species in the Sterculiaceae are mostly indigenous to the tropics (Bailey 1950). The family consists of trees, shrubs, herbs, and vines characterized by alternate leaves, flowers with superior ovaries, and milky sap (Bailey 1950).

Several species are of economic importance, including cacoa (*Theobroma cacao* L.) and the cola nut (*Cola acuminata* Schott et Endl.) (Bailey 1950).

CUPUACU (CUPUASSU) (*Theobroma grandiflorum* K. Schum.)

Origin and Distribution

The cupuassu is native to Brazil and is common in the Amazon region (Martin *et al.* 1987). Cupuassu has been introduced into other tropical regions only to a small extent.

Botanical Description

Cupuassu trees are variable in height (5-33 m) (Clement 1983; Purseglove 1969) with dimorphic branching (Purseglove 1969). Whorls of 3 branches at 1.5-3.0 m intervals are produced along the trunk (Clement 1983). The bark is rough and russet brown (Clement 1983). Leaves are simple, stipulate, petiolate with pulvini (Purseglove 1969), and longer than they are wide (Clement 1983). Inflorescences are solitary or in cymes (Purseglove 1969) borne along main and secondary branches (Clement 1983). Flower parts are in fives (Purseglove 1969). Fruit is a drupe (Purseglove 1969) that varies in shape (cylindrical to 5-sided) and size

(10-60 cm long and 10-20 cm wide) (Clement 1983). The exocarp is thick (1 cm), russet brown, and surrounds a central cavity containing 5 rows of pulp-covered seeds (Clement 1983). The pulp surrounding each seed varies from 0.5-1.5 cm thick and is white to yellowish white (Clement 1983; Martin *et al.* 1987). The pulp is fibrous and somewhat juicy.

Adaptations and Cultural Considerations

Cupuassu is adapted to a hot, humid tropical climate (Martin *et al.* 1987). The cupuassu industry is in its infancy and little or no plant selection has occurred. Cultural practices have not been determined (see Chapter 10). Trees are reported to respond well to fertilizer and pruning (Clement 1983). Cupuassu is propagated by seed (Martin *et al.* 1987) and grafting (Clement 1983).

Current Importance and Future Prospects

The cupuassu is produced commercially only in northern Brazil (Donadio 1983). The fruit is harvested from wild stands, and some small orchards (Clement 1983). Clement (Chapter 10) estimates 2.8-4.7 MT are produced per hectare annually.

Cupuassu fruit pulp has a strong aroma and agreeable flavor that is used for making juices, preserves, and flavoring ice creams (Clement 1983; Donadio 1983). The pulp is also eaten fresh locally and can be frozen (Clement 1983).

Cupuassu's excellent flavor and processing potential suggest there will be an increase in demand and production of this fruit (Clement 1983; Donadio 1983; Martin *et al* 1987). Some processed product is currently exported to Europe (Clement 1983).

REFERENCES

ANONYMOUS. 1988. Marketing Florida tropical fruits and vegetables - summary 1986-87. Federal-State Market News Service, Winter Park, FL.

ARPAIA, M. 1989. Personal communication. University of California - Riverside, Riverside, CA.

BAILEY, L. H. 1950. The Standard Cyclopedia of Horticulture. The Macmillan Co., N.Y.

BAILEY, L. H. and BAILEY, E. Z. 1976. Hortus Third. Macmillan Publishing Co., Inc., New York.

BEAL, P. R. and FARLOW, P. J. 1985. Passifloraceae. *In* Tropical Tree Fruits of Australia. P. E. Page (Editor). Queensland Dept. of Primary Industries, Brisbane, Australia.

BROKAW, R. 1989. Personal communication. California Cherimoya Association. Saticoy, CA.

BROWN, J. 1989. Personal communication. California Tropics. Carpenteria, CA.

CAMPBELL, B. A. and CAMPBELL, C. W. 1983. Preservation of tropical fruits by drying. Proc. Fla. State Hortic. Soc. 96, 229-231.

CAMPBELL, C. A. 1987. Carambola fruit development and storage in Florida. M.S. Thesis. University of Florida, Gainesville.

CAMPBELL, C. A. and KOCH, K. E. 1989. Weight, color, and composition of development of sweet and tart carambola fruit. J. Am. Soc. Hortic. Sci. 114, 455-457.

CAMPBELL, C. A., KOCH, K. E. and HUBER, D. J. 1987. Postharvest response of carambolas to storage at low temperatures. Proc. Fla. State Hortic. Soc. 100, 272-275.

CAMPBELL, C. W. 1965. The 'Golden Star' carambola. Circ. S-173. Coop. Ext. Serv., University of Florida, Gainesville.

CAMPBELL, C. W. 1977. Cultivation of tropical fruits of the Myrtaceae in southern Florida. Proc. Am. Soc. Hortic. Sci. - Trop. Reg. 21, 3-7.

CAMPBELL, C. W. 1986. Tropical fruits crops in Florida - a rapidly changing situation. Proc. Fla. State Hortic. Soc. 99, 217-219.

CAMPBELL, C. W., KNIGHT, JR., R. J. and OLSZACK, R. 1985. Carambola production in Florida. Proc. Fla. State Hortic. Soc. 98, 145-149.

CAMPBELL, C. W., KNIGHT, JR., R. J. and ZARESKI, N. L. 1977. Freeze damage to tropical fruits in southern Florida. Proc. Fla. State Hortic. Soc. 90, 254-257.

CAMPBELL, C. W. and MALO, S.E. 1981. The carambola fruit. Fruit Crops Fact Sheet FC-12. Coop. Ext. Serv., University of Florida, Gainesville.

CAMPBELL, C. W. and SAULS, J. W. 1980. Spondias in Florida. Fruit Crops Fact Sheet FC-63. Coop. Ext. Serv., University of Florida, Gainesville.

CHILDERS, N. F. 1983. Modern Fruit Science. Horticultural Publications, Gainesville, FL.

CLEMENT, C. R. 1983. Underexploited Amazonian fruits. Proc. Am. Soc. Hortic. Sci.-Trop. Reg. 27(A) 117-141.

CRANE, J. H. 1989. Current statistics and plant density of commercial carambola, mamey sapote, lychee, longan, sugar apple, atemoya and passion fruit plantings in south Florida. Proc. Fla. State Hortic. Soc. 102 (in press).

CRANE, J. H., CAMPBELL, C. W. and OLSZACK, R. 1989. Current statistics for commercial carambola groves in south Florida. Proc. Interamer. Soc. Trop. Hortic. 33 (in press).

DONADIO, L. C. 1983. A review of South American fruits. Proc. Am. Soc. Hortic. Sci. - Trop. Reg. 27(A) 21-41.

DONADIO, L. C. 1989. Carambola growing in Brazil. Proc. Interamer. Soc. Trop. Hortic. 33 (in press).

FAO. 1982. Fruit-bearing forest trees - technical notes. FAO Forestry Paper No. 34. FAO - United Nations, Rome.

FERGUSON, J. J., CRANE, J. H. and OLSZAK, R. 1988. Growth of young carambola trees using standard and controlled-release fertilizers. Proc. Interamer. Soc. Trop. Hortic. 32, 20-24.

GARNER, R. J. and CHAUDHRI, S. E. 1976. The propagation of tropical fruit trees. Anthony Rowe Ltd., Wiltshire, Great Britain.

GEORGE, A. P. 1985. Annonaceae. *In* Tropical Tree Fruits for Australia. P. E. Page (Editor). Queensland Dept. of Primary Industries, Bribane, Australia.

GREEN, J. 1987. Carambola production in Malaysia and Taiwan. Proc. Fla. State Hortic. Soc. 100, 275-278.

GROEIZAAM, M. and LEWIS, D. 1989. The cultivation and utilization of carambola in Suriname. Proc. Interamer. Soc. trop. Hortic. 33 (in press).

HAMILTON, B. and HAMILTON, S. 1989. Personal communication. Kurtistown, HI.

HODGSON, R. W. 1967. Horticultural varieties of citrus. *In* The Citrus Industry, Vol. 1. W. Reuther, H. J. Webber, and L. D. Batchelor (Editors). University of California Press, Berkeley.

HOWELL, C. W. 1976. Edible fruited *Passiflora* adapted to south Florida growing conditions. Proc. Fla. State Hortic. Soc. 89, 236-238.

HUME, H. H. 1957. Citrus Fruits. Macmillan Co., New York.

HUTCHINSON, J. and DALZIEL, J. M. 1948. The useful plants of West Africa. Watmoughs Ltd., London.

JACKSON, L. K. 1990. Citrus Growing in Florida. The University Presses of Florida, Gainesville.

JOSEPH, K. T. 1979. The cashew flower. Indian Cashew J. 12, 7-9.

JOYNER, M. E. B. and SCHAFFER, B. 1989. Flooding tolerance of 'Golden Star' carambola trees. Proc. Fla. State Hortic. Soc. 102 (in press).

KENNARD, W. C. and WINTERS, H. F. 1960. Some fruits and nuts for the tropics. USDA-ARS Misc. Publ. 801.

KNIGHT, JR., R. J. 1965. Heterostyly and pollination in carambola. Proc. Fla. State Hortic. Soc. 78, 375-378.

KNIGHT, JR., R. J. 1980. Origin and world importance of tropical and subtropical fruit crops. *In* Tropical and Subtropical Fruits. S. Nagy and P. E. Shaw (Editors). AVI Publishing Co., Westport, CT.

KNIGHT, JR., R. J. 1982. Response of carambola seedling populations to Dade County's oolitic limestone soil. Proc. Fla. State Hortic. Soc. 95, 121-122.

KNIGHT, JR., R. J. 1983. Tropical fruit of Asia with potential for expanded world production. Proc. Am. Soc. Hortic. Sci. - Trop. Reg. 27(A) 71-93.

KNIGHT, JR., R. J. 1989. Carambola cultivars and improvement programs. Proc. Interamer. Soc. Trop. Hortic. 33 (in press).

KNIGHT, JR., R. J. and SAULS, J. W. 1986. The passion fruit. Fruit Crops Fact Sheet FC-60. Coop. Ext. Serv., University of Florida, Gainesville.

LITZ, R. E. and CONOVER, R. A. 1980. Partial organogenesis in tissue culture of *Averrhoa carambola*. HortScience 15, 735.

LITZ, R. E. and GRIFFIS, JR., J. L. 1989. (*Averrhoa carambola* L.). Biotech. Agric. and Forestry, Vol. 5. Y.P.S. Bajaj (Editor). Springer-Verlag, Berlin.

LIZANA, L. A., REGINATO, G. and LAVANDEROS, J. C. 1986. Maturity characterization of lucuma. Proc. Interamer. Soc. Trop. Horti. 30, 105-120.

MAGGS, D. H. 1985A. Myrtaceae. *In* Tropical Tree Fruit for Australia. P. E. Page (Editor). Queensland Dept. of Primary Industries, Brisbane, Australia.

MAGGS, D. H. 1985B. Palmae, subfamily Cocoinae. *In*. Tropical Tree Fruits for Australia. P. E. Page (Editor). Queensland Dept. of Primary Industries, Brisbane, Australia.

MARLER, T. 1989. Personal communication. Fairchild Tropical Garden, Miami, FL.

MARTIN, F. W., CAMPBELL, C. W. and RUBERTE, R. M. 1987. Perennial edible fruits of the tropics: An inventory. USDA-ARS Agric. Hdbk. 642.

MARTIN, F. W. and NAKASONE, H. Y. 1970. The edible species of *Passiflora*. Econ. Bot. 24, 333-343.

MCMILLAN, R. T. 1974. *Rhizopus artocarpi* rot of jackfruit (*A. heterophyllus*). Proc. Fla. State Hortic. Soc. 87, 392-393.

MCMILLAN, R. T. 1986. Serious diseases of tropical fruits in Florida. Proc. Fla. State Hortic. Soc. 99, 224-227.

MORTON, J. F. 1965. The jackfruit (*Artocarpus heterophyllus* Kam.): Its culture, varieties, and utilization. Proc. Fla. State Hortic. Soc. 78, 336-344.

MORTON, J. F. 1987. Fruits of Warm Climates. J. F. Morton Publisher, Miami, FL.

MUKHERJEE, S. K. 1956. Variations in cashew nut (*Anacardium occidentale* Linn.). Indian J. Hortic. 13, 12-14.

NAGEL, J., PENA, J. E., and HABECK, D. 1989. Insect pollination of the atemoya in Florida. Fla. Entomol. 72, 207-211.

NAMBIAR, M. C. 1977. Cashew. *In* Ecophysiology of Tropical Crops. P. T. Alvim and T. T. Kozlowski (Editors). Academic Press, New York.

NUNEZ, J. V. 1988. National markets. Comercializacion Hortofruticola 10, 30-31. (Spanish)

OCHSE, J. J., SOULE, JR., M. J., DIJKMAN, M. J. and WEHLBURG, C. 1961. Tropical and Subtropical Agriculture. The Macmillan Co., New York.

OGDEN, M. H. and CAMPBELL, C. W. 1982. Intergeneric and interspecific rootstock trials for jaboticaba (*Myrciaria cauliflora* (Mart.) Berg. Proc. Fla. State Hortic. Soc. 95, 119-121.

OGDEN, M. H., CAMPBELL, C. W. and LARA, S. P. 1981. Grafting annonas in southern Florida. Proc. Fla. State Hortic. Soc. 94, 355-358.

OSLUND, C. R. and DAVENPORT, T. L. 1983. Ethylene and carbon dioxide in ripening fruit of *Averrhoa carambola* HortScience 18, 229-230.

PEARROW, J. 1988. U.S. imports of fruits and vegetables under plant quarrantine regulations, Fiscal year 1986. ERS Staff Report No. AGES881107. CED/ERS/USDA, Washington, D.C.

PENA, J. E. 1986. Status of pests of minor tropical fruit crops in south Florida. Proc. Fla. State Hortic. Soc. 99, 227-230.

PENA, J. E., BARANOWSKI, R. M. and LITZ, R. E. 1987. Life history, behavior, and natural enemies of *Philephedra tuberculosa* (Homoptera:Coccidae). Fla. Entomol. 70, 423-427.

PENA, J. E., GLENN, H. and BARANOWSKI, R. M. 1984. Important insect pests of *Annona* spp. in Florida.

PHILLIPS, R. L. and CAMPBELL, C. W. 1980. The sugar apple. Fruit Crops Fact Sheet FC-38. Coop. Ext. Serv., University of Florida, Gainesville.

PHILLIPS, R. L. and GOLDWEBER, S. 1984. The jaboticaba. Fruit Crops Fact Sheet FC-39. Coop. Ext. Serv., University of Florida, Gainesville.

POPENOE, J. 1974. Status of annona culture in south Florida. Proc. Fla. State Hortic. Soc. 87, 342-344.

POPENOE, J. 1979. The Genus Spondias in Florida. Proc. Fla. State Hortic. Soc. 92, 277-279.

POPENOE, W. 1974. Manual of Tropical and Subtropical Fruits. Hafner Press, N.Y.

PURSEGLOVE, J. W. 1969. Tropical Crops: Dicotyledons 1 and 2. Longmans, Green, and Co., Ltd., London.

RAMSAMMY, P. 1989. The carambola in Guyana. Proc. Interamer. Soc. Trop. Hortic. 33 (in press).

RUEHLE, G. D. 1967. Miscellaneous tropical and subtropical Florida fruits. Agric. Ext. Serv., IFAS, University of Florida. Bull 156A.

SAAVEDRA, E. 1979. Set and growth of *Annona cherimola* Mill. fruit obtained by hand pollination and chemical treatments. J. Am. Soc. Hortic. Sci. 104, 668-673.

SCHOLFIELD, P. B. 1985. Sapotaceae. *In* Tropical Tree Fruits for Australia. P. E. Page (Editor). Queensland Dept. of Primary Industries, Brisbane, Australia.

SEDGLEY, M. 1985A. Moraceae. *In* Tropical Tree Fruits for Australia. P. E. Page (Editor). Queensland Dept. of Primary Industries, Brisbane, Australia.

SEDGLEY, M. 1985B. Oxalidaceae. *In* Tropical Tree Fruits for Australia. P. E. Page (Editor). Queensland Dept. of Primary Industries, Brisbane, Australia.

SHEETS, M. D. and DUBOIS, M. L. 1986. The pomegranate. Fruit Crops Fact Sheet FC-44. Coop. Ext. Serv., University of Florida, Gainesville.

SIMONE, G. and KUCHAREK, T. 1988. Extension plant pathology report. Coop. Ext. Serv., University of Florida, Gainesville.

SMITH, J. and CAMPBELL, B. 1983. Methods of home preservation of fruits. Proc. Am. Soc. Hortic. Sci. Trop. Reg. 27(A) 179-183.

STURROCK, D. 1940. Tropical Fruits of Southern Florida and Cuba and Their Uses. The Arnold Arboretum of Harvard University, Jamacia Plain, MA.

STURZ, J. D. 1985. Anacardiaceae. *In* Tropical Tree Fruits for Australia. P. E. Page (Editor). Queensland Dept. of Primary Industries, Brisbane, Australia.

SWIFT, J. F. and PRENTICE, W. E. 1983. Native fruits of the Ecuadorian Amazon: Production techniques and processing requirements. Proc. Am. Soc. Hortic Sci. - Trop. Reg. 27(A) 95-110.

WAGNER, C. J., BRYAN, W. L., BERRY, R. E. and KNIGHT, JR., R. J. 1975. Carambola selection for commercial production. Proc. Fla. State Hortic. Soc. 88, 466-469.

WAHAB, N. A., AHMAD, I. and HASSAN, A. 1989. Carambola production, processing, and marketing in Malasia. Proc. Interamer. Soc. Trop. Hortic. 33 (in press).

WALLACE, J. M. 1978. Virus and viruslike diseases. *In* The Citrus Industry, Vol. 1. W. Reuther, H. J. Webber, and L. D. Batchelor (Editors). University of California Press, Berkeley.

WESTWOOD, M. N. 1978. Temperate-Zone Pomology. W. H. Freeman and Co., San Francisco, CA.

WHITMAN, W. F. 1972. Two decades of tropical fruit introduction. Proc. Fla. State Hortic. Soc. 99, 280-287.

WILTBANK, W. J., CHALFUN, H. N. J. and ANDERSEN, O. 1983. The jaboticaba in Brazil. Proc. Am. Soc. Hortic. Sci. - Trop. Reg. 27(A) 57-69.

WRIGHT, M. L. 1985. United States imports of fruits and vegetables under plant quarantine regulations. Fiscal year 1985. IED/ERS/USDA. ERS Staff Report No. AGE860304.

2

Cashew Apple and Nut

Wilfred F. Wardowski
M. Joseph Ahrens

The cashew, *Anacardium occidentale* L., is a member of the Anacardiaceae family. It is indigenous to South America and was introduced as a soil binding tree to the west coast of India by Portuguese traders in the 16th century. The Portuguese and Spanish spread the cashew throughout the tropics from South America and India, but it did not become important in international trade until the early 1920's. Five countries were reported to produce 84% of the world production: India, Brazil, Mozambique, Tanzania, and Kenya (190, 85, 75, 72, 15 MT, respectively; Bittenbender 1984; Swaminathan 1984). In most Spanish speaking countries the cashew is known as maranon, in Venezuela as merey, and caju or cajueiro in Portuguese (Morton 1987).

Early in its development as a commercial crop, it was recognized that cashew did best in sandy soils and did not tolerate frost (Paul and Canagaratnam 1937; Stephens 1935). However, it was reported to be well suited to all manner of soil types and conditions, not being fastidious in its requirements (Stephens 1935) and thriving in soils unfit for other crops (Paul and Canagaratnam 1937). Typically, the trees were sown by seed and allowed to grow unattended, without supplemental fertilizer or irrigation (General de Agricultura, Guatemala 1942). By the 1950's, though, it became recognized that the cashew did benefit if grown within some broad nutritional and cultural guidelines. Tropical regions with distinct wet and dry seasons, which receive 76 to 127 cm of rain, and having medium fertility, sandy soils were recommended as candidates for prime cashew production areas (Haarer 1953). Thereafter, magnesium deficiency was discovered to occur in cashew (Hewitt and Bull 1956). It was not until ten years later that the first fertilizer guidelines appeared (Galli 1964). However, it is still recognized that cashew is a hardy plant, suited to many areas of the tropics. It grows well from near sea level to 700 m. It can be established in marginal lands and infertile soils; is very drought tolerant (Alexander *et al.* 1982), but is especially sensitive to cold damage. The literature abounds with work on the propagation and culture of cashews (Baskara Rao and Hameed Kahan 1984).

The cashew produces both male and bisexual flowers with male flowers predominating the first month. Pollination is by insects. Flowering varies with climate, but usually lasts 3-4 months. The fruit (containing the edible nut) develops for 6-8 weeks after pollination, with the apple (pseudofruit) developing during the last two weeks. The fruit (nut) and apple drop together over a period of 6-8 weeks (Wait and Jamieson 1986).

The true cashew fruit is the popular kidney-shaped nut including its shell. The shell is honeycombed and contains a caustic phenolic resin which makes processing difficult. The nut is marketed worldwide, and is very popular in Australia, Europe, Hong Kong, Japan, North America and the USSR. The list of products which can be derived from the shell resin is lengthy, including adhesives used in particle boards, bactericides, insecticides, and varnishes (Chittenden and Paddon 1973; Gulati *et al.* 1964; Irvine 1955).

The nut has been discarded or little used in many Latin American countries where the apple is harvested. Brazil and India developed nut processing industries, but have not used vast tonnages of cashew apples (Morton 1987). Morton (1970) suggested that cashew nut production and processing could be expanded and improved in South America.

After the nut develops, the receptacle (elongated peduncle) swells to become a pseudofruit known as the cashew apple. The cashew apple is fleshy, pear or round shaped; yellow, red, or red and yellow, with a yellow interior. It may reach a size of 10 cm long and 5 cm wide. The cashew apple grows very slowly until the true fruit (nut) matures, at which time the apple grows very rapidly and ripens (Fig. 2.1).

Removal of the nut initiated rapid growth and ripening of the apple in the Philippines (Pratt and Mendoza 1980). In another study, removal of nuts at 15 days after anthesis resulted in withered and dropped fruit. Applications of 100 and 1000 ppm gibberellic acid prevented fruit drop with no further growth, and 100 or 1000 ppm 2,4-D or 2,4,5-T resulted in faster fruit development than with nuts attached (Bose 1964).

The cashew apple was reported to be a nonclimacteric fruit after harvest as measured by respiration rate (50 ml/kg-hr) and steady ethylene production (200-400 ml/kg-hr) (Pratt and Mondoza 1980). Biale

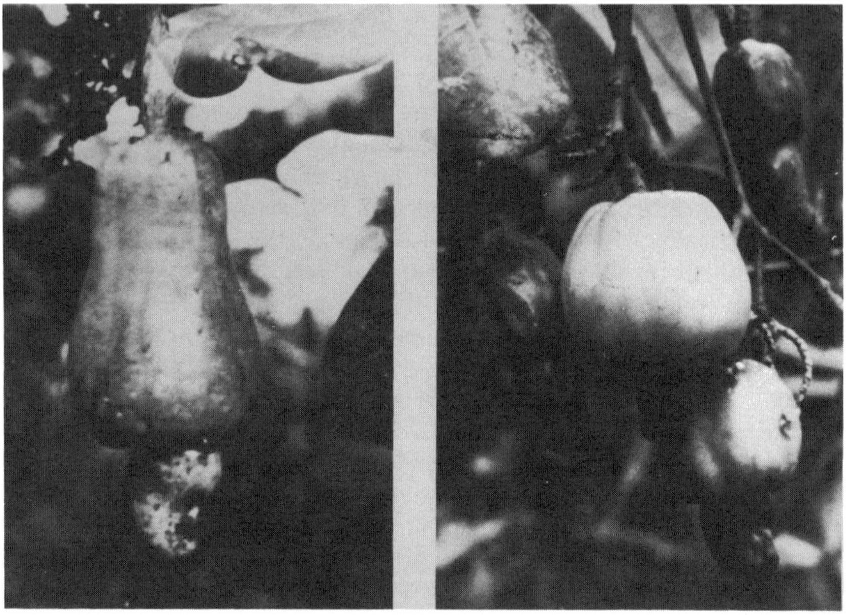

Courtesy of Dr. A. B. Graf

FIG. 2.1. LARGE CASHEW APPLES (PESUDOFRUIT) WITH PEAR AND ROUND SHAPES ABOVE THE TRUE FRUIT NUT AND SHELL.

and Barcus (1970) reported that the nonclimacteric Brazilian cashew fruit had a respiration rate of 74-76 ml O_2 and 62-72 ml CO_2 per mg-hr.

HARVESTING

Cashew apples and nuts fall together and are collected from the ground two to four times over six to eight weeks if the weather is dry. Rain during nut fall presents problems with decay of the apples and germination of the nuts. If the apples are used, they must be collected immediately when they fall, or be hand harvested because they spoil quickly (Wait and Jamieson 1986).

APPLE STORAGE

Cashew apples stored for five weeks at 0-2°C and 85-90% relative humidity were reported to suffer freezing injury (McGregor 1987). Pruthi *et al.* (1963) reported that cashew apples spoilage occurred within 24 hr of harvest and was caused by *Canadia krusei*. Storage time could be increased to three weeks storage after steeping apples in 0.25% citric acid with 400 to 500 ppm SO_2. Quick spoilage is still cited as the most common limiting factor to the use of large amounts of cashew apple for numerous products.

COMPOSITION OF CASHEW APPLE

Cashew Apple Pulp and Peel

The yield of the pseudofruit, cashew apple, is four to five times that of cashew nuts. However, in the major producing countries of Brazil and India the apple is little used. The apples contain astringent and acrid chemicals which produce an unpleasant taste. The apples are also subject to injury and rapid decay. Although cashew apples have good nutritional value, the problems of distasteful ingredients and poor keeping quality limit their use.

Approximately 30 to 40% of the fruit by weight remains after the juice is extracted. This pulp has been determined on a dry weight basis (Table 2.1) to contain 10.7% pectin, 9.0% protein and 8.5% fiber (Nanjundaswamy *et al.* 1984).

The ascorbic acid content of cashew apple is reported to be much greater than that for oranges: four and one-half times (Bittenbender 1984); six times (Cecchi and Rodriguez-Amaya 1981); and five to seven times (Lopes 1972). While the cashew apple is rich in vitamin C, it is a poor source of carotene (0.32 mg/100 g). Blends of the pulp with carotene rich pulps from mango or papaya were reported to store well for five months at ambient temperatures (Sondhi 1962). Cecchi and Rodriguez-

TABLE 2.1. COMPOSITION OF CASHEW APPLE RESIDUE

	Dry weight basis
	----- % -----
Total ash	1.8
Total pectin as Ca pectinate	10.7
Total tannins	5.0
Proteins (N x 6.25)	9.0
Ether extractives	4.4
Crude fiber	8.5
	--- mg % ---
Ca	20.5
Fe	35.6
P	152.1

Source: Nanjundaswamy *et al.* (1984).

Amaya (1981) identified eight pigments (ß-carotene highest amount) present in both red and yellow cashew apples, and found that color differences reflected the amount but not the type of carotenoid present (Table 2.2). These authors also (1981) converted the provitamin A active carotenoids to vitamin A values of 0.37 to 1.07 μg/g cashew apples (Table 2.2). Aurochrome and auroxanthin were detected by chromatographic means, but values were not reported.

Fatty acid contents of cashew apple pulp and peel were reported by Maia *et al.* (1975). The most common fatty acids found were oleic and palmitic, accounting for over 80% of the fatty acids present (Table 2.3). Additionally, they presented the composition of fatty acids for red apple immature pulp and peel, and yellow apple immature pulp.

Xavier Filho *et al.* (1962) reported on the amino acid contents of five Brazilian fruits; cashew apple, bacuri (*Platonia insignis* Mart.), genipap (*Genipa americana* L.), imbu (*Spondias tuberosa* Arr. Cam.) and sapodilla (*Achras sapota* L.) (Table 2.4). Their two dimensional chromatography results were evaluated as relative amounts of amino acids from none, trace and 1 to 5. Amino acids present in the greatest abundance (values 2- 5) in the cashew apple were α-alanine, α-aminobutyric acid, aspartic acid, glutamic acid, glycine, proline, and the leucines. Arginine, lysine, ornithine, cystine/cysteine, glutamine, phenylalanine, serine, threonine, tryptophane and tyrosine appeared in smaller amounts (trace-2) (Table 2.4).

TABLE 2.2. CAROTENOID CONTENT AND VITAMIN A VALUE OF CASHEW APPLE FRUITS AND PASTEURIZED JUICE IN BRAZIL

	Yellow or red apple (Lot #)[1]				Juice brand[2]	
	Y (1)	R (2)	Y (3)	R (4)	A	B
	μg/g					
α-carotene	tr	0.3	0.1	tr	tr	tr
β-carotene	0.6	1.8	0.5	1.6	0.7[3]	0.7[3]
ζ-carotene	tr	tr	tr	tr	tr	tr
Cis-β-carotene	0.5	0.5	0.1	0.4	0.1[3]	0.1[4]
Cryptoxanthin	0.5	1.0	0.5	1.1	0.5[4]	0.5[3]
Cryptochrome	tr	tr	tr	tr	0.6[3]	0.5[3]
Total	1.6	3.6	1.2	3.1	1.9[3]	1.7[3]
Vitamin A Value	0.42	1.02	0.37	1.07	0.47	0.47

Source: Cecchi and Rodriguez-Amaya (1981).

[1]8-10 fruit/lot, each lot from different area.

[2]Each value the mean of 3 lots, each lot a mix of 2 bottles of commercial pasturized juice.

[3]± 0.1

[4]± 0.0

TABLE 2.3. PERCENTAGE COMPOSTION OF FATTY ACIDS OF CASHEW APPLE MATURE PULP AND PEEL, JUICE, TESTA AND KERNEL

	Red		Yellow				
Fatty acid	Pulp	Peel	Pulp	Peel	Juice	Testa	Kernel
Capric	tr	tr	6.6	1.2	—	—	—
Lauric	0.8	0.1	0.2	0.5	23.7	0.2	—
Myristic	0.5	0.4	0.3	0.5	0.4	0.3	—
Myristoleic	0.2	0.4	0.1	1.3	1.3	—	—
Palmitic	18.5	21.3	15.0	17.6	14.9	16.4	7.5
Palmitoleic	1.3	1.5	1.5	2.0	0.7	1.1	—
Hexadecadienoic	0.3	0.2	0.2	0.2	0.4	1.4	—
iso-Stearic	0.3	tr	tr	tr	—	—	—
Stearic	1.2	3.0	0.8	1.5	0.9	6.4	4.5
Oleic	64.4	65.4	65.8	68.3	50.6	35.3	73.7
Linoleic	2.5	3.2	2.2	3.2	1.7	30.4	14.3
Arachidic	tr	tr	0.6	0.5	—	—	—
Linolenic	6.0	3.3	3.8	2.4	4.7	5.8	tr
Gadoleic	4.0	1.0	2.9	1.2	2.0	1.6	—
Eicosadecadienoic	—	—	—	—	—	0.8	—

Source: Maia *et al.* (1975), except Barroso *et al.* (1973) for kernel.

A detailed study of the flavor and aroma components of cashew apples shipped by air from Venezuela was conducted in England by MacLeod and de Troconis (1982). They determined that the levels of aroma compounds (3.6 ug/100 kg fresh fruit) were low when compared to those of other tropical fruits. Of 46 identified components, it is interesting to note that car-3-ene with a fruity, sickly odor represented 24.3% of the total amount. Even more interesting, car-3-ene had not previously been reported in cashew apple from India. It has been reported in tomato, currant, passion fruit and citrus fruits (MacLeod and de Troconis 1982).

Cashew Apple Juice

Juice extraction remains mostly a small scale local operation due to the inability to store and transport apples for mass production. A hand-operated, home-constructed fruit juice expeller developed in the Philippines processed 138 kg fruit per hr and yielded 109 l juice per hr (Farrales *et al.* 1958).

Nanjundaswamy *et al.* (1984) reported the characteristics of red and yellow cashew apple juice from two regions in India (Table 2.5). The values determined ranged within ± 5% of 71% for percent juice extracted, and ± 5% for pH, total soluble solids, and total sugars. Wide

TABLE 2.4. RELATIVE AMOUNTS OF FREE AMINO ACIDS IN FIVE BRAZILIAN FRUITS[1]

	Cashew Apple	Bacuri	Genipap	Imbu	Sapodilla
α-alanine	5	4	3	2	1
β-alanine	–	1	–	–	–
Arginine	2	5	3	2	5
Asparagine	–	1	1	5	T
Aspartic acid	3	5	5	2	3
α-aminobutyric acid	4	4	3	1	3
Cystine/Cysteine	1	–	–	–	–
Glycine	3	1	1	T	T
Glutamic Acid	5	5	4	1	2
Glutamine	T	4	T	T	T
Histidine	–	T	–	T	T
Leucines	3	–	–	–	–
Lysine	1	2	1	T	2
Ornithine	T	T	T	T	T
Phenylalanine	1	–	–	–	–
Pipecolic acid	–	T	–	T	–
Proline	3	T	3	1	3
Serine	1	3	3	T	2
Threonine	2	–	–	–	–
Tyrosine	T	–	–	–	–

Source: Xavier Filho *et al.* (1962).
[1]Relative amounts T = trace, 1 = low, 5 = high.

TABLE 2.5. COMPOSITION OF JUICE FROM RED AND YELLOW CASHEW APPLE FROM TWO REGIONS IN INDIA

	Kerala		Karnataka	
	Red	Yellow	Red	Yellow
Fruit weight (g)	50.2	50.5	26.1	46.2
pH	3.9	4.0	3.9	4.1
True ascorbic acid (mg %)	117	126	200	153
	%			
Juice	71.2	72.4	69.8	72.2
Total soluble solids	12.6	11.8	12.3	12.2
Acidity as malic acid	0.19	0.20	0.35	0.32
Total sugars	8.7	8.3	8.4	8.0
Total tannins	0.34	0.35	0.55	0.45

Source: Nanjundaswamy *et al.* (1984).

variability was noted for fruit weight (26.1-50.5 g), acidity as malic acid (0.19-0.35%), ascorbic acid content (117-200 mg/100 g), and total tannins present (0.34-0.55%). Sugars in Nigerian juice varied from 8.4 to 21.6% with no relationship to fruit color or shape (Falade 1981) (Table 2.6).

Much higher ascorbic acid levels (170-306 mg/100 ml) were found by Sastry *et al.* (1962) for red and yellow cashew apple juice from three

TABLE 2.6. COMPOSITION OF YELLOW AND RED, ELONGATED AND ROUND CASHEW APPLES FROM NIGERIA EXPRESSED AS MEAN AND RANGE

	Yellow		Red	
	Elongated	Round	Elongated	Round
	%			
Pectin	0.226 0.04-0.27	0.09 0.03-0.22	0.10 0.05-0.18	0.10 0.02-0.22
Carbohydrates	0.28 0.12-0.43	0.35 0.11-0.87	0.37 0.16-0.67	0.34 0.11-0.94
Sugars	14.8 9.4-19.8	14.8 8.4-21.6	15.5 13.0-19.6	14.7 10.3-21.0
Amino acids	1.04 0.80-1.16	1.04 0.68-1.18	1.09 0.70-1.32	1.04 0.74-1.21
Protein	0.24 0.16-0.35	0.21 0.09-0.31	0.23 0.14-0.52	0.32 0.11-0.47
	mg %			
Vitamin C	318 156-429	305 156-416	351 234-429	338 162-455
Na	3.4 2.5-5.3	3.8 3.1-5.5	5.0 3.5-6.4	3.6 2.7-6.4
K	6.2 4.8-7.5	5.9 4.7-8.5	5.4 5.1-5.9	5.7 4.5-6.6
Ca	11.8 6.0-20.0	12.3 4.0-26.0	15.0 8.0-20.0	13.4 8.0-28.0
Mg	6.2 4.3-8.2	7.3 4.3-9.4	7.0 5.3-8.2	7.0 4.3-9.8
Cu	0.08 0.02-0.12	0.09 0.02-0.45	0.05 0.02-0.07	0.07 0.02-0.11
Fe	0.83 0.19-0.98	0.46 0.19-0.98	0.54 0.39-0.59	0.48 0.19-0.78
Mn	0.22 0.11-0.32	0.16 0.05-0.32	0.11 0.05-0.16	0.16 0.05-0.32
S	22.3 4.0-40.0	28.5 4.0-60.0	14.4 4.0-24.0	26.0 4.0-60.0
P	11.2 7.5-14.5	10.1 7.0-14.0	7.1 4.2-10.0	9.0 7.0-11.0

Source: Falade (1981).

different regions in India (Table 2.7). They also reported a wider range of total tannins (0.1-0.7 g/100 ml) and a true tannin range of 0.01-0.5 g/100 ml. The differences among fruit from different regions were large.

Nigerian apple juice had a wide variation of vitamin C, 156 to 455 mg/100 ml (Falade 1981). Higher values were observed for red than for yellow types and for elongated than for round fruits (Table 2.6). Lopes (1972) reported that red and yellow fruit from 24 trees from Mozambique averaged 286 mg per 100 ml (range 232-371) ascorbic acid, 0.52% (038-0.94%) acidity expressed as malic acid, and 0.11% (0.6-0.22%) tannins. Vitamin A of pasteurized cashew apple juice (Table 2.2) was calculated, and loss during processing discussed by Cecchi and Rodriguez-Amaya (1981). The calculated loss was based on that expected from the mix of red and yellow fruit.

The astringent and acrid taste of cashew fruit has been traced to the tannin content and an unidentified oily substance. Two successful methods for removing objectionable flavors are to steam for five min at 0.4 kg/cm^2 plus subsequent cold water rinsing, and to cook the fruit in a 2% common salt solution for 3-4 min (Jain *et al.* 1951). The effects of steam and cooking in salt solution were reported by Nanjundaswamy *et al.* (1984) (Table 2.8). In each treatment the total soluble solids were reduced from 14.2% to 13.4% and the acidity was increased by one third from 0.3 to 0.4%. Tannins were drastically reduced from 0.3% to 0.07% by the use of steam leaving a slightly astringent taste, and to 0.04% for fruit cooked in brine resulting in total freedom from astringency, but with a slight salty taste.

A third astringency removal method, tannin precipitation with gelatin added to juice, has also been successful (Table 2.9). With this procedure the juice is clarified by filtration following the gelatin treatment. Total soluble solids are reported to be reduced from 12.50 to 10.20%, and acidity lowered from 0.46 to 0.33%. Ascorbic acid was lowered by 17% from 264 to 219 mg/100 ml and tannins were reduced from 0.32 to 0.03% (Nanjundaswamy *et al.* 1984; Chakrabarthy 1961).

TABLE 2.7. REGIONAL VARIATION IN TANNIN AND ASCORBIC ACID CONTENTS OF RED AND YELLOW CASHEW APPLES

Region	No. of Samples	Tannins (g/100 ml)		Ascorbic acid (mg/100 ml)
		Total	True	
Mysore	28	0.4-0.7	0.30-0.50	180-250
Ullal	20	0.32-0.69	0.12-0.50	192-306
Trichur	18	0.10-0.28	0.01-0.10	170-250

Source: Sastry *et al.* (1962).

TABLE 2.8. EFFECT OF STEAMING AND COOKING CASHEW FRUIT IN SALT SOLUTION ON THE QUALITY OF JUICE

	Juice source		
	Fresh	Steamed	Brine cooked
Taste and aroma	Astringent	Slightly	Nonastrigent
	%		
Total soluble solids	14.2	13.4	13.4
Acidity as malic acid	0.3	0.4	0.4
Tannins	0.3	0.07	0.04

Source: Nanjundaswamy *et al.* (1984).

Like the pulp, the fatty content of cashew apple juice is comprised largely of oleic and palmitic acids. In addition, however, there is also a relatively high (23.7) percentage of lauric acid; these fatty acids account for nearly 90% of those present (Maia *et al.* 1975) (Table 2.3).

Vilasa Chandran and Damodaran (1984) reported on the characteristics of apples from 16 varieties selected for high nut yields. Their detailed results on each variety are reduced to the data in Table 2.10. Compared to the data presented in Tables 2.5, 2.7, 2.8 and 2.9, their selected varieties averaged less juice (possibly a factor of a different extraction method), higher acidity, higher sugar, equal to higher tannins, and higher ascorbic acid. Higher sugar and ascorbic acid are desirable for cashew apples to be made into juice products.

Cashew Apple Juice Concentrate—Unclarified and clarified cashew apple juices were concentrated (Pruthi *et al.* 1963) using a forced circulation evaporator which produced a superior product compared to open pan concentration (Table 2.11). Ascorbic acid retention was slightly higher (85%) for unclarified juice than for clarified juice (78%). Clarified juice

TABLE 2.9. CHANGES IN CASHEW APPLE JUICE BY GELATIN TREATMENT

	Juice	Gelatin and clarification
pH	4.25	4.20
Ascorbic acid (mg %)	264	219
	%	
Total soluble solids	12.50	10.50
Acidity	0.46	0.33
Tannins	0.32	0.03

Source: Nanjundaswamy *et al.* (1984).

TABLE 2.10. CHARACTERISTICS OF APPLES AND JUICE FROM 16 SELECTED HIGH YIELDING CASHEW NUT VARIETIES

	Max	Min	Average
Apples			
Fruit weight (g)	75.4	29.2	41.5
		%	
Juice	64.0	44.9	56.9
Protein	0.92	0.42	0.67
Ether extracts	0.76	0.39	0.52
Carbohydrates	11.53	8.13	9.96
Crude fiber	0.91	0.62	0.75
		mg %	
Ca	15	5	10
P	27	12	19
Fe	1.6	0.8	1.2
Apple Juice		%	
Total soluble solids	14.67	11.06	12.68
Reducing sugars	13.37	9.88	11.37
Acidity as malic acid	0.58	0.36	0.44
Total tannins	0.74	0.33	0.49
True tannins	0.36	0.20	0.27
		mg %	
Ascorbic acid	321	224	256

Source: Vilasa Chandran and Damodaran (1984).

concentrate was judged to have a better color and flavor. Viscosity of concentrate from unclarified juice was four times that from clarified juice. The addition of 1000 ppm sulphur dioxide reduced browning and allowed successful storage at ambient temperature (24-30°C) for 16 weeks.

Alcoholic Beverages

The characteristics of cashew apple juice lend to the production of alcoholic beverages. Subba Rao (1984) noted that the juice contained 10.7% reducing sugar (fructose and glucose), 0.3% acid, 0.3% protein and had a pH of 4.0. As presented in Table 2.12, the average values are in agreement, but have wide ranges (Shondhi 1962).

Patwardhan (1970) also reported on fermented cashew apple juice in India (Table 2.13) which contained 16.4 g/100 l absolute alcohol and had an astringent flavor. The yield of wine in large scale trials (Subba Rao 1984) averaged over 60% V/W (liters wine/kg cashew apple) with an alcohol content of approximately 2.5% W/W of the fruit and 3.17% alcohol

TABLE 2.11. CHANGES DURING EVAPORATIVE CONCENTRATION OF CASHEW APPLE JUICE

	Unclarified		Clarified	
	Juice	Concentrate	Juice	Concentrate
pH	4.1	4.0	4.3	4.1
Viscosity (CP)	20.0	1820	20.0	450
Ascorbic acid (mg %)	250	1423	193	1108
Color	Light yellow	Dark brown	Light yellow	Brown
Flavor	Normal	Mild	Normal	Mild
	%			
Solids	12.0	77.0	9.0	71.0
Acidity as malic acid	0.4	2.6	0.4	2.1
Reductones as ascorbic acid	25.0	14.5	17.6	11.5
Retention of ascorbic acid	—	85.0	—	78.0

Source: Pruthi *et al.* (1963).

in wine. However, wine made in India has not always been well controlled, and sanitation quality control can always be improved in cottage level production. Feni is a beverage with about 38-40% alcohol made as a traditional liquor in Goa, India by multiple distillation. Brandy, made as a distillate of wine under more controlled conditions, retained a cashew apple flavor without astringency (Table 2.14).

TABLE 2.12. CHARACTERISTICS OF CASHEW APPLE JUICE

	Max	Min	Average
Fruit weight (g)	140.7	13.0	37.9
Brix	18.3	7.2	12.4
pH	4.6	3.7	4.0
True ascorbic acid (mg %)	265.5	17.2	105.4
Protein (N x 6.25)	0.3	0.2	0.3
	%		
Juice	84.3	46.9	67.1
Acidity as malic acid	0.7	0.1	0.3
Total reducing sugars	17.7	5.3	10.7
Glucose	11.6	5.1	7.9
Fructose	6.5	0.2	2.6
True tannin (as tannic acid)	0.7	0.0	0.2

Source: Shondhi (1962).

TABLE 2.13. COMPOSITION OF FERMENTED CASHEW APPLE JUICE (WINE)

	Value
Brix	3.00
pH	4.00
Higher alcohol[1]	16.40
	-- % --
Acidity as malic acid	0.63
Volatile acid as acetic acid	0.03
Ethyl alcohol	5.51
Total reducing sugars	0.38
Furfural	1.25
Cu	Nil
	mg %
Tannin as tannic acid	145
Esters as ethyl acetate	0.023
Aldehydes as acetaldehyde	0.002

Source: Patwardhan (1970).
[1]g/100 l absolute alcohol

NUT PROCESSING

Cashew nut processing includes drying, humidifying, sizing, roasting, shelling, husking, grading and packing/storage. The nuts are sometimes cleaned prior to storage by simply removing empty or deformed nuts, pieces of wood, small stones, and other foreign objects.

Preparation for Shelling

Nuts may be soaked in water or humidified in piles, and then held in a humid atmosphere so that the shell has a set moisture content from 15 to 45% depending on the methods. This procedure helps rupture the cells

TABLE 2.14. COMPOSITION OF CASHEW APPLE BRANDY

Constituent	g/100 l of Absolute alcohol
Ethyl alcohol	42.85
Volatile acids as acetic acid	12.28
Esters as ethyl acetate	55.97
Aldehyde as acetaldehyde	18.28
Higher alcohol	149.70
Furfural	3.22
Cu	1.04

Source: Subba Rao (1984).

and extract the cashew nut shell liquid (CNSL, also known as cashew nut shell oil). Increased moisture in the kernel makes the kernel less brittle during cracking. When the nuts are soaked in water, the temperature must be less than 30°C to avoid releasing tannins (Ohler 1966).

The nuts are then sized to insure uniform roasting, a procedure necessary to drive off the caustic nut shell oil. The irregular shape of nuts makes sizing difficult, however the principle of two inclined slightly diverging outward turning cylinders is reported to be a satisfactory method (Ohler 1966).

Shelling

Once the shell oil has been volatilized, the nut may be shelled without discoloration of the kernels. This may be done in a variety of ways. A foot operated mechanical aid to cashew nut shell removal improved the yield of whole kernels in Nigeria, although the single person 2.5 nut rate per min was slower than 7 nuts per min by hand cracking (Hall 1966). In contrast, Joubert and Thomas (1965) reported a rate of 3 nuts per min (200 nuts per hr for a ten hr day) by experienced women in India using wooden mallets and a hooked wire to hold the nuts, and Esteves (1962) reported 2.5 nuts per min by hand in Brazil.

Esteves described two machines in which A) the nuts were held and pressed between a plunger and short spikes (28-35 kg per 8 hr) (1962) and B) the nuts were held in a clamp and pressed (67 raw nuts per hr) between two metal cutting plates yielding 17.4 kg nuts and 9.3 kg whole nuts per hr (1963). Ohler (1966) reported on several mechanical cracking methods: A) sized nuts cracked between two inward turning cylinders (300 kg per 8-hr day with a 20% yield of whole kernels increasing to 36.5% after manual cracking of uncracked nuts; B) rotating in conical abrasive surfaces in a vertical rotor until the kernel and shell separate (1000 kg nuts per hr); C) centrifugal cracking at 1200 to 1800 rpm and throwing at 250 km per hr against vertical knives (870 to 1200 kg nuts per hr with a yield of 67% whole kernels); and D) cutting or sawing in not fully mechanized systems (22 kg nuts per day with 74% yield of whole nuts, and another system yielding 400 kg per day with no stated whole nut yield).

Husking

Husking must be done very carefully to remove the testa or papery seed coat from the very brittle kernels. Kernel bluish-black patchy discoloration by the testa limits marketing although the taste is not affected. Mathew and Parpia (1970) reported that the dark pigment may be removed by a dilute acid wash and concluded that the darkening of cashew nuts during processing is probably due to a polyphenol-iron complex. Ohler (1966) noted that husking is done after drying kernels at 70°C for 4 to 8 hr and that it is labor intensive with one worker husking 4 to 12 kg per day.

Grading

Intact and broken kernels may be sorted into as many as six grades, and the market is poor for broken or badly roasted whole kernels (Joubert and Thomas 1965). Grading was reported to be done by hand and be based on count per lb (Ohler 1966). Morton (1970) suggested that with low fat and high protein content, a market could be found for broken nuts because of high food value.

Packing and Storage

Export grades are packed under vacuum or carbon dioxide to prolong keeping quality of the kernels (Joubert and Thomas 1965). Roasted nuts may be stored for up to two years packed in carbon dioxide and one year packed under a vacuum (Alexander *et al.* 1982).

Well dried nuts can be stored for up to two years in the shells. Nuts establish an equilibrium moisture with the surrounding air, and the shell moisture content can be used to predict the moisture content of the whole fruit and the kernel (Okwelogu and Mackay 1969). They noted that nuts stored in moving air absorbed moisture faster than nuts stored in still air. The maximum safe moisture content for raw cashew nuts was estimated to be 8.9 to 9.2% in equilibrium with 70% relative humidity at 27°C. At 27°C with over 75% relative humidity air, the nuts were invaded by *Aspergillus*, *Fusarium*, *Paecilomyces*, *Penicillium* and *Rhizopus* (Okwelogu and Mackay 1969).

COMPOSITION OF CASHEW NUT

Kernel

Early work on the composition of cashew nut (Damodaran and Sivaswamy 1936) isolated 17 to 18% of the fat-free seed as anacardein, a newly identified protein. Nonroasted cashew nuts from Brazil were reported (Barroso *et al.* 1973) to contain four measurable fatty acids, namely oleic (73.4%), linoleic (14.3%), palmitic (7.5%) and stearic (4.5%) (Table 2.3). Three kernel sizes varied in fatty acid contents, but there were no consistent trends. The total kernel oil averaged 45.6%. Nagaraja (1987) reported that 96% of the lipids of high yielding cashew nuts were neutral lipids, with the remaining 4% being glycolipids and phospholipids. Further, Nagaraja indicated that the triglycerides were rich in unsaturated fatty acids (oleic and linoleic), and glycolipids were rich in saturated fatty acids (lauric and myristic).

Murthy and Yadava (1972) presented the oil and carbohydrate contents of 24 Indian and exotic cashew nut varieties; the values varied widely among varieties (Table 2.15).

Kuppuswamy *et al.* (1958) detailed the amino acid composition of cashew kernel gobulins, and reported that arginine represented about

TABLE 2.15. OIL AND CARBOHYDRATE CONTENTS OF 24 CASHEW NUT VARIETIES

				Kernel carbohydrate			
No.[1]	Origin	Shell oil	Kernel oil	Reducing sugar	Non-reducing sugar	Total sugar	Starch
		%					
1	Kenya	20.0	42.5	3.0	4.4	7.6	10.0
2	Ex-Tanganika	28.8	42.3	2.4	3.6	6.2	9.5
3	Singapore	24.2	34.8	2.9	3.8	6.9	10.4
4	Chirala	24.6	43.4	1.7	1.7	3.5	5.5
5	Rangium	28.4	42.7	1.2	3.5	4.9	5.4
6	Kattupalli	23.4	41.3	1.3	4.0	5.6	6.7
7	Gangadevan-kuppam	21.9	39.8	1.2	5.0	6.5	7.4
8	Peppariyan-kuppam	20.2	37.0	1.5	2.4	4.0	5.1
9	Edayanchavadi	20.6	40.0	3.2	2.9	6.2	11.2
10	Chinglepet	21.7	36.7	1.1	2.0	3.6	5.0
11	South Arcot	31.1	36.0	2.9	5.5	8.7	10.5
12	Ramapuram	26.6	34.5	2.6	3.9	6.7	8.9
13	Ramapuram (irrigated)	19.2	42.2	2.6	3.1	5.8	9.5
14	Vazhisodanai-palayam	17.4	40.3	1.0	5.8	7.1	8.6
15	Badiyadka	16.6	38.1	1.5	4.0	5.7	6.7
16	Vittal	18.9	40.9	2.2	4.3	6.7	9.7
17	Belthangady	24.3	40.8	1.6	3.2	4.9	8.3
18	Agumbe	19.2	40.9	1.1	1.3	2.4	5.8
19	Tirthahalli	32.9	38.8	1.7	3.7	5.6	7.3
20	Bhatkal	23.5	40.7	1.3	2.6	4.0	10.9
21	Kumta	25.2	46.7	1.8	1.7	3.7	5.1
22	Mapusa	30.3	43.6	2.3	3.7	6.2	9.0
23	Pillerna	25.3	43.8	0.9	3.9	5.1	4.8
24	Vengurla	29.1	41.2	1.5	2.6	3.8	4.7

Source: Murthy and Yadava (1972).
[1]1-3 Exotic varieties collected from Cashew Research Station, Vriddhachalam, Tamil Nadu; 4 Andhra Pradesh; 5-14 Kerala; 16-21 Mysore State; 22-23 Goa; 24 Maharashtra.

10% of the amino acid composition of the kernel globulins. Phenylalanine and valine were nearly equal (4.5%) (Table 2.16).

Meal from whole kernels were reported by Piva *et al.* to contain *ca.* 42% protein, 1% lipid, 0.5% calcium and 0.2% phosphorus (Table 2.17). Meal from cashew nut kernels were found to contain proteins including ca. 4.6% lysine, 1.3% tryptophan, 2% cystine and 1.6% methionine (Table 2.18). Rats gained more weight on a diet of cashew nut meal than on soybean or casein meals (Piva *et al.* 1971).

Shell

The cashew nut shell is grey-brown, smooth, ca. 3 mm thick with a soft honeycomb structure containing CNSL. The edible nut or kernel is protected from the CNSL by a thin testa which is easily removed. CNSL is a dark brown viscous liquid reported to be 15 to 20% by weight of the unshelled nut in Africa, 25 to 30% of the weight in India (Cornelius 1966) and *ca.* 25% overall (Ohler 1966). The oil contents of shells from 24 varieties were reported by Murthy and Yadava (1972) to be far less than that found in nuts (Table 2.15).

Four methods reported to extract CNSL (Cornelius 1966) are A) roasting nuts and collecting expelled liquid (Indian native method, ca. 50% yield), B) extract with hot CNSL without charing the kernels (yield 85 to 90%), C) super heated steam treatment and collect condensate (used to improve yields of B), and D) solvent extraction. These methods are also described by Ohler (1966).

Several reports indicate that CNSL contains ca. 90% anacardic acid, which forms cardanol by decarboxylation on heating, and ca. 10% cardol (Fig. 2.2) (Cornelius 1966, Ohler 1966). Various levels of anacardic acid (80 to 88%) are given in Table 2.19 for CSNL from different locations

TABLE 2.16. AMINO ACID CONTENT OF CASHEW KERNEL GLOBULINS

Amino Acid	% of protein
Arginine	10.3
Histidine	1.8
Lysine	3.3
Tyrosine	3.2
Cystine	1.0
Phenylalanine	4.4
Methionine	1.3
Threonine	2.8
Valine	4.5

Source: Kuppuswamy *et al.* (1958).

TABLE 2.17. COMPOSITION OF CASHEW KERNEL AND EXTRACTED MEAL FROM WHOLE KERNEL

	Kernel	Meal
	----- % of dry matter -----	
Moisture	6.92	9.10
Crude protein	21.52	41.55
Crude fiber	1.13	1.55
Crude lipids	46.09	0.95
Ash	3.38	5.53
N-free extractives	27.88	50.42
Ca	0.22	0.51
P	0.70	0.15

Source: Piva *et al.* (1971).

(Tyman and Morris 1967). Later studies found only 71.7% anacardic acid, but 18.7% cardol, 4.7% cardanol, 2.7% novel phenol and 2.2% of two unidentified ingredients (Tyman and Morris 1967). Rocha Gonsalves and Santos Costa (1975) reported 70% anacardic acid as being extremely stable even in the presence of strong acids. Compounds related to those found in CNSL are present in poison ivy, Japanese lac tree and maidenhair tree (Cornelius 1966).

TABLE 2.18. AMINO ACID COMPOSITON OF CASHEW EXTRACTION MEAL

	% of dry matter	% of protein
Alanine	1.70	4.10
Arginine	4.11	9.90
Aspartic acid	3.82	9.20
Cystine	0.78	1.90
Glutamic acid	9.47	22.80
Glycine	1.82	4.40
Histidine	0.81	1.95
Isoleucine	1.70	4.10
Leucine	2.99	7.20
Lysine	1.91	4.60
Methionine	0.66	1.60
Phenylalanine	1.86	4.50
Proline	1.49	3.60
Serine	1.99	4.80
Threonine	1.41	3.40
Tryptophan	0.52	1.25
Tyrosine	1.22	2.95
Valine	2.32	5.60

Source: Piva *et al.* (1971).

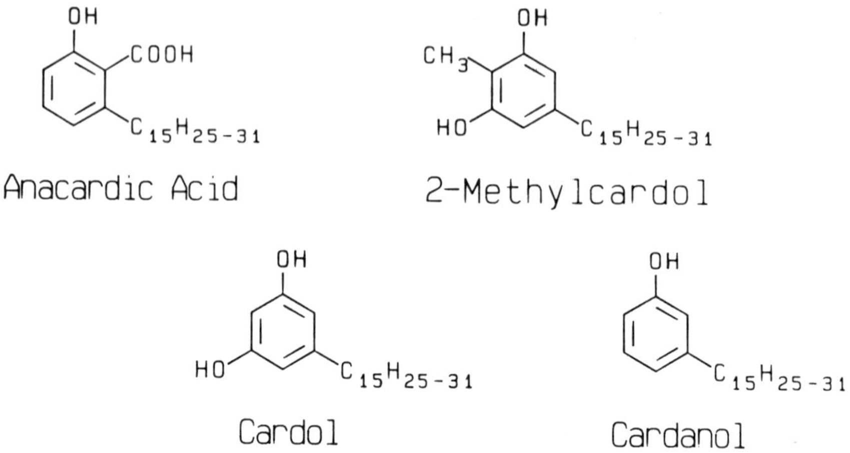

FIG. 2.2. CHEMICAL STRUCTURES OF PRINCIPAL CASHEW NUT SHELL LIQUID INGREDIENTS.

OTHER PRODUCTS

In addition to the more common products discussed above, some additional uses of cashew are worth noting.

The production of cashew apple 'pasa' or date-like carmel in Costa Rica is a small scale sun dried process resulting in a variable quality product with mold and yeast contamination (Ortez *et al.* 1982). Improved product was obtained by osmotic drying followed by forced air drying to *ca.* 16% moisture. Other uses of cashew apple noted in the literature include jams, marmalade and enriched flours.

The phenolic raw material of CNSL has been used in the manufacture of brake linings, industrial belting and clutches, reinforcing synthetic rubber, for oil or acid resistance, in lacquers, in electrical insulation material, as a metal anti-corrosive material, for waterproofing and as adhesives (Cornelius 1966). Cashew nut shell liquid has been used as an adhesive to produce medium density particle board which met British standards (Chittenden and Padden 1973). However, they concluded that due

TABLE 2.19. COMPOSITION OF CASHEW NUT SHELL LIQUID FROM DIFFERENT LOCATIONS

	% Composition			
Location	Anacardic acid	Cardol	2-Methylcardol	Cardanol
India	87.8	13.7	2.5	2.1
Mozambique	79.9	13.9	2.8	3.4
Brazil	79.8	13.6	2.9	3.7

Source: Tyman & Morris (1967).

to the requirement of large amounts of solvent it would be more economical to import synthetic resins into developing countries. Chittenden and Padden did suggest that other nonsolvent uses could be made of CNSL.

Pereira *et al.* (1963) carefully analyzed the ingredients of the seed tegument versus contamination from the shell oil and suggested that tannins could be made from lipids and proteins could be used in animal feeds.

REFERENCES

ALEXANDER, D. McE., SCHOLEFIEL P. B., and FRODSHAM, A. 1982. Some Tree Fruits for Tropical Australia. CSIRO (Commonwealth Scientific and Industrial Research Organization), Melbourne.

BARROSO, M. A. T., WHITING, F. M., BROWN, W. H. and STULL, J. W. 1973. Fatty acids of Brazilian cashew kernels. HortScience 8(2) 99.

BHASKARA RAO, E. V. and HAMEED KAHAN, H. (Editors). 1984. Cashew Research and Development. Indian Society for Plantation Crops, Kerala, India.

BIALE, J. B. and BARCUS, D. E. 1970. Respiratory patterns in tropical fruits of the Amazon Basin. Trop. Sci. XII(2) 93-104.

BITTENBENDER, H. C. 1984. Handbook of Tropical Fruits and Spices. Dept. of Hortic., Mich. State Univ., East Lansing.

BOSE, T. K. 1964. Growth of receptacle in cashew as influenced by growth substances. Curr. Sci. 33, 120.

CECCHI, H. M. and RODRIGUEZ-AMAYA, D. B. 1981. Carotenoid composition and vitamin A value of fresh and pasteurized cashew-apple (*Anacardium occidental* L.) juice. J. Food Sci. 46(1) 147-149.

CHAKRABARTHY, R. N. 1961. Associateship thesis. Central Food Technological Research Institute, Mysore, India.

CHITTENDEN, A. E. and PADDON, A. R. 1973. Cashew nut shell liquid as an adhesive in particle board manufacture. Trop. Sci. 15(4) 329-352.

DAMODARAN, M. and SIVASWAMY, T. G. 1936. A new globulin from the cashew nut (*Anacardium occidentale*). Biochem. J. 30, 604-608.

DIRECCION GENERAL DE AGRICULTURA, GUATEMALA. 1942. Cultivation of Cashew. Rev. Agric. Guatemala 19, 61-62. (Spanish)

ESTEVES, A. B. 1962. Henry cashew nut shelling machine. Estud. Agron. (Lisbona) 3(1) 11-20. (Portuguese)

ESTEVES, A. B. 1963. Sima cashew nut shelling machine. Estud. Agron. (Lisbona) 4(4) 141-162. (Portuguese)

FALADE, J. A. 1981. Vitamin C and other chemical substances in cashew apple. J. Hortic. Sci. 56(2) 177-179.

FARRALES, J. F., JOCSON, A. T. and ANDRES, B. C. 1958. Progress report on the design and development of a cashew juice expeller. Philip. J. Agric. 23, 71-75.

GALLI, O. 1964. Fertilization of cashew. Agromonico. 16, 6. (Portuguese)

GULATI, A. S., KRISHNAMACHAR, V. S. and RAO, B. C. S. 1964. Quaternary nitrogen germicides derived from the monophenolic components of cashewnut shell liquid. Indian J. Chem. 2, 114-118.

HAARER, A. E. 1953. Usefulness of the cashew tree and nuts. New Commonw. 25. 500-501.

HALL, F. J. 1966. Cashew nut processing—Part II: trials of equipment in Western Nigeria, May-June 1965. Trop. Sci. 8, 160-171.

HARIHARAN, M., LEELAVATHY, K. M. and UNNIKRISHAN, K. 1973. Changes in the free amino-acid content of cashew nut kernel during development. Sci. Cult. 39(9) 407-408.

HEWITT, E. J. and BULL, R. A. 1956. A preliminary survey of deficiency symptoms in crop plants in the oil palm belt of Nigeria. J. W. African Inst. Oil Palm Res. 4, 83-105.

IRVINE, F. R. 1955. West African insecticides. Colon. Plant Anim. Prod. 5, 34-38.

JAIN, N. L., BHATIA, B. S., ANAND, J. C. and GIRDHARILAL. 1951. Cashew apple products. J. Sci. Indian Res. lOA, 209-210.

JOUBERT, A. J. and THOMAS, D. des S. 1965. The cashew nut. Farming S. Afr. 40(11) 6-7.

KUPPUSWAMY, S., SREEVIVASAN, M. and SUBRAMANYAN, V. 1958. Special Report Series No. 33, Indian Council of Medical Research, New Deli, India.

LERCKER, G. and PALLOTTA, U. 1984. Cashew seed and oil composition. *In* Cashew Research and Development. E. V. Bhaskara Rao and H. Hameed Kahan (Editors). Indian Society for Plantation Crops, Kerala, India.

LOPES, M. H. C. 1972. Chemical composition and exploitation of the cashew apple in Mozambique. Agron. Mocamb. 6(2) 119-131. (Portuguese)

MacLEOD, A. J. and de TROCONIS, N. G. 1982. Volatile flavour components of cashew 'apple' (*Anacardium occidentale*). Phytochemistry 21(10) 2527-2530.

MAIA, G. A., BROWN, W. H., WHITING, F. M. and STULL, J. W. 1975. Cashew fatty acids. HortScience 10(3) 233-234.

MANDAL, R. C., DHANDAR, D. G. and DHUME, C. V. 1984. Cashew apple liquor industry in Goa. *In* Cashew Research and Development. E. V. Bhaskara Rao and H. Hameed Kahan (Editors). Indian Society for Plantation Crops, Kerala, India.

MATHEW, A. G. and PARPIA, H. A. B. 1970. Polyphenols of cashew kernel testa. J. Food Sci. 35, 140-143.

McGREGOR, B. M. 1987. Tropical Products Transport Handbook, USDA Handb. 668.

MORTON, J. M. 1970. Prospects for cashew advancement in Colombia and Venezula. Proc. Fla. State Hortic. Soc. 83, 405-410.

MORTON, J. L. 1987. Fruits of Warm Climates. J. L. Morton, Miami, FL.

MURTHY, B. G. K. and SIVASAMBAN, M. A. 1984. Recent trends in CNSL utilization. *In* Cashew Research and Development. E. V. Bhaskara Rao and H. Hameed Kahan (Editors). Indian Society for Plantation Crops, Kerala, India.

MURTHY, K. N. and YADAVA, R. B. R. 1972. Note on the oil and carbohydrate contents of varieties of cashewnut (*Anacardium occidentale* L.). Indian J. Agric. Sci. 42(10) 960-961.

NAGARAJA, K. V. 1987. Lipids of high-yielding varieties of cashew (*Anacardium occidentale* L.). Plant Foods Hum. Nutr. 37(4) 307-311.

NANJUNDASWAMY, A. M., RADHAKRISHNIAH SETTY, G. and PATWARDHAN, M. V. 1984. Utilization of cashew apples for the development of processed products. *In* Cashew Research and Development. E. V. Bhaskara Rao and H. Hameed Kahan (Editors). Indian Society for Plantation Crops, Kerala, India.

OHLER, J. G. 1966. Cashew nut processing. Trop. Abstr. 21(9) 549-554.

OKWELOGU, T. N. and MACKAY, P. J. 1969. Cashew moisture relations. J. Sci. Food Agric. 20, 697-702.

ORTIZ, A. J. COOKE, R. D. and QUIROS M., R. A. 1982. The processing of a date-like carmel from cashew apple. Trop. Sci. 24(1) 29-38.

PATWARDHAN, V. M. 1970. A dissertation on "cashew apple fruit" and an investigation on "fermented cashew apple beverage". Thesis submitted for M.Sc., Mysore Univ. Cited by Subba Rao. 1984.

PAUL, W. R. C. and CANAGARATNAM, C. 1937. The production of cashew nuts in Mannar. Trop Agriculturist 88, 6-11.

PEREIRA JR., A., GRACA, J. B. DA, DODINHO, L. S. and PEREIRA, M. M. 1963. Chemical composition of the cashew nut: 3. Contribution to the study of *Anacardium occidentale* L. seds tegument. Estud. Agron. (Lisboa) 4, 115-132. (Portuguese)

PIVA, G. and SANTI, E. 1984. On nutritional value of the protein in the cashew seed. *In* Cashew Research and Development. E. V. Bhaskara Rao and H. Hameed Kahan (Editors). Indian Society for Plantation Crops, Kerala, India.

PIVA, G., SANTI, E. and EKPENYONG, T. 1971. Nutritive value of cashewnut extraction meal. J. Sci. Food Agric. 22, 22-23.

PRATT, H. K. and MENDOZA JR., D. B. 1980. Influence of nut removal on growth and ripening of the cashew apple. J. Am. Sci. Hortic. Sci. 105(4) 540-542.

PRUTHI, J. S., CHAKRABARTHY, R. N., SONDHI, S. P., SASTRY, L. V. L. and SIDDAPPA, G. S. 1963. Studies on concentrating the juice of the cashew apple (*Anacardium occidentale*) Food Technol. 17(11) 95-97.

PRUTHI, J. S., RAO, B. A. S. and SIDDAPPA, G. S. 1963. Determination of the optimal dose of chemical preservatives for the preservation of cashew apple fruit and juice. Sci. Cult. 20(8) 416.

ROCHA GONSALVES, A. M. D'A. and SANTOS COSTA, A. M. B. S. R. C. 1975. Chromatography of cashew nut-shell liquid. J. Chromatogr. 104, 225-227.

SASTRY, L. V. L., LAKSHMINARAYANA SETTY, SATYAVATHI, V. K., PRUTHI, J. S. and SIDDAPPA, G. S. 1962. Polyphenol constituents in cashew apple juice as influenced by region, strain and selection. Indian J. App. Chem. 25, 119-122.

SONDHI, S. P. 1962. Studies on physico-chemical and technological aspects of cashew apple—some observations. Thesis CFTRI, Mysore, India. Cited by Subba Rao. 1984.

STEPHENS, S. E. 1935. Some tropical fruits, II. The cashew nut. Queensland Agric. J. 44, 488-489.

SUBBA RAO, M. S. 1984. Scope for development of alcoholic beverage from cashew apple. *In* Cashew Research and Development. E. V. Bhaskara Rao and H. Hameed Kahan (Editors). Indian Society for Plantation Crops, Kerala, India.

SWAMINATHAN, M. S. 1984. Keynote address. *In* Cashew Research and Development. E. V. Bhaskara Rao and H. Hameed Kahan (Editors). Indian Society for Plantation Crops, Kerala, India.

TYMAN, J. H. P and FRANCE, A. P. 1984. Compositional studies on natural Indian cashew nut shell liquid. *In* Cashew Research and Development. E. V. Bhaskara Rao and H. Hameed Kahan (Editors). Indian Socity for Plantation Crops, Kerala, India.

TYMAN, J. H. P. and MORRIS, L. J. 1967. The composition of cashew nut-shell liquid (CNSL) and the detection of a novel phenolic ingredient. J. Chromatogr. 27, 287-288.

VILASA CHANDRAN, T. and DAMODARAN, V. K. 1984. Physico-chemical qualties of cashew apples of high yielding types. *In* Cashew Research and Development. E. V. Bhaskara Rao and H. Hameed Kahan (Editors). Indian Society for Plantation Crops, Kerala, India.

WAIT, A. J. and JAMIESON, G. I. 1986. The cashew: its botany and cultivation. Queensl. Agric. J. 112(5) 253-257.

XAVIER FILHO, J., HOLLANDA LIMA, I. and VENTRUA, M. M. 1962. Free amino acids in some Brazilian fruits. Phyton (Buenos Aires) 19, 121-125.

3

Marula

Ingrid A. G. Weinert
Pieter J. van Wyk
Lucas C. Holtzhausen

The marula is one of the most commonly utilized wild fruit of southern Africa. Archeological evidence indicates that the fruit of the marula tree was known and consumed by man in Africa 9,000-10,000 years B.C. (Friede and Pienaar 1974). Drought resistance, exceptional yield of fruit per tree, the possibility to utilize both the fruit and the nut contained within the seed, ease of harvesting tall trees and the exotic flavor, and nutritive value of the fruit are some of the characteristics which have led recently to its commercial utilization. Since 1980, considerable attention has been paid to the selection of wild plants suitable for domestication (Holtzhausen and Van der Laarse 1984; Von Teichman 1982) resulting in the recent establishment of the first marula orchards. Although initial problems with seed germination were overcome and propagation techniques were developed, there remains a great deal to be learnt about the cultivation of the tree as well as the chemical composition and processing characteristics of its fruit.

DISTRIBUTION AND BOTANICAL CLASSIFICATION

The marula tree is widely distributed through Africa: its southernmost location is the lowlands of Natal (Republic of South Africa) from where it occurs northwards through tropical Africa into Ethiopia and the Sudan (Palmer and Pitman 1972; Van Wyk 1974). It prefers a warm frost-free climate, but is also found in areas up to 1370 m above sea level, where temperatures may drop below freezing point for a very short period in winter. The marula is a member of the Anacardiaceae. The plant was first described in 1832 and named *Spondias birrea*. In 1844 it was renamed *Sclerocarya birrea* and *Sclerocarya caffra* in 1850 (Shone 1979). It has recently been reclassified as *Sclerocarya birrea* (A. Rich.) Hochst. subspecies *caffra* (Sond.) (Kokwaro and Gillett 1980). Whereas the closely related *Sclerocarya birrea* occurs in the northwestern and northeastern parts of Africa, its subspecies *caffra* is found mostly south of the Zambesi river (Fernandez and Fernandez 1966). The name *Sclerocarya* is derived from the two Greek words "skleros" and "karyon", meaning "hard" and "nut" respectively (Von Teichman 1982) and refers to the hard stone of the fruit.

Owing to its wide distribution and numerous uses, various vernacular names exist for the tree in southern Africa (Shone 1979; Fox and Young 1982; Quin 1959; Palmer and Pitman 1972; Von Teichman 1982).

DESCRIPTION

Sclerocarya birrea subsp. *caffra* is a medium to large tree (usually 9 m tall, but up to 18 m have been recorded), single stemmed with a wide

Courtesy of L. C. Holtzhausen

FIG. 3.1. MARULA TREE (*SCLEROCARYA BIRREA* SUBSP. *CAFFRA*) IN SWANAMIBIA dense crown and deciduous shiny foliage (Fig. 3.1). The bark is grey and usually peels off in flat round disks, exposing the underlying light yellow tissue. Young twigs are thick and digitaliform with spirally arranged composite leaves at their ends (Van Wyk 1974; Palmer and Pitman 1972).

Leaves are 180-250 mm long and 80-150 mm wide. They are composite, containing from two to twenty-three leaflets, averaging eleven. The leaflets are oblong and elliptic with petioles ranging from very short to 20 mm in length (Shone 1979).

Although male and female flowers occasionally occur on the same tree, it is considered to be dioecious. Flowering takes place in early spring. Male flowers are borne in groups of three on racemes below new leaves. They are dark red when young, turning pink or white when open. The female flowerbuds are blood red, but change color from purple to white after opening. They occur below the leaves on long peduncles and consist of four curling petals, numerous infertile stamens and a long shiny ovary (Van Wyk 1974).

Fruit is borne in clusters of up to three at the end of the twigs (Shone 1979) and always on the new growth (Fig. 3.2). It is a round or oval drupe and usually wider than it is long with a diameter of 30-40 mm. The shape and number of nuts per stone determine the final shape of the fruit. Marula fruit have a thick leathery soft exocarp with tiny round spots enclosing a white juicy mucilaginous flesh which adheres tightly to the stone and can only be removed by sucking. The flesh tastes tart, sweet and refreshing, whereas the fruit has a slightly turpentine-like aroma and can give off a very unpleasant smell when decaying (Palmer and Pitman 1972).

Each fruit contains an exceedingly hard seed which is covered by fibrous matter (Shone 1979). It usually is trilocular, but sometimes only bilocular. Each seed locule contains a single light nut filling the entire cavity (Van Wyk 1974). Every locule also has an opening close to one tip, which is sealed by a round hard disk protecting the embryo until germination.

The marula tree is a prolific bearer: between 0.2 and 1.5 metric tons of fruit per tree have been collected in one season in the wild (Quin 1959). Mature fruit drop when still green and ripen to a yellow color on the ground, usually during January and February. Harvesting is done by picking up the fallen fruit.

The proximate contribution of the various parts of the fruit to its total weight is shown in Table 3.1. The wide variation in weight of the fruit and its parts may be due to climatic and geographical influences or even genetic differences.

Courtesy of L. C. Holtzhausen

FIG. 3.2. MARULA TWIG BEARING MATURE UNRIPE FRUIT

TABLE 3.1. PROXIMATE WEIGHT DISTRIBUTION OF RIPE MARULA FRUIT

	Source	
	Quin (1959)	Shone (1979)
	g	
Fruit (average)	17.9	30
Peel	7.4	9
Stone	9.4	—
Stone shell	—	3
Nut	—	1
Edible flesh	1.1	12
Inedible fibrous matter	—	5

TRADITIONAL UTILIZATION

Food Uses

Ripe marula fruit can be consumed by biting or cutting throught the thick leathery skin and sucking the juice or chewing the mucilaginous flesh after removal of the skin.

A popular fermented alcoholic beverage is prepared from ripe fruit. In some cases the skin is removed and the juice fermented together with the flesh still on the seed. In other cases the skin is only cut and the whole fruit allowed to ferment. The yeasts occurring naturally on the fruit are normally utilized for a spontaneous fermentation. This beverage is commonly known as marula beer, but could more appropriately be called a "marula wine" (Shone 1979). It has approximately twice as much ascorbic acid as orange juice and thus has excellent antiscorbutic activity.

Marula nuts found inside the seed of the fruit are regarded as a delicacy in the regions of the tree's natural habitat. The nut is approximately 15 mm long and 5 to 10 mm in diameter and the yield per seed on average only 10% (Shone 1979). Women spend hours cracking the hard shell of the seed between two stones (Fig. 3.3) and picking the nuts from the debris. Sometimes a specially shaped tool is used to aid the extraction of the nut from the cracked seed. Nuts are commonly used to supplement the diet during winter. They are mixed with vegetables or meat or may be pounded and formed into a cake before consumption. In some households the nuts are used in biscuits (Shone 1979).

More recently the fruit has been used to prepare jelly or jam which is sold on a small scale. At present limited amounts of juice are processed which is also utilized for the manufacture of liqueur.

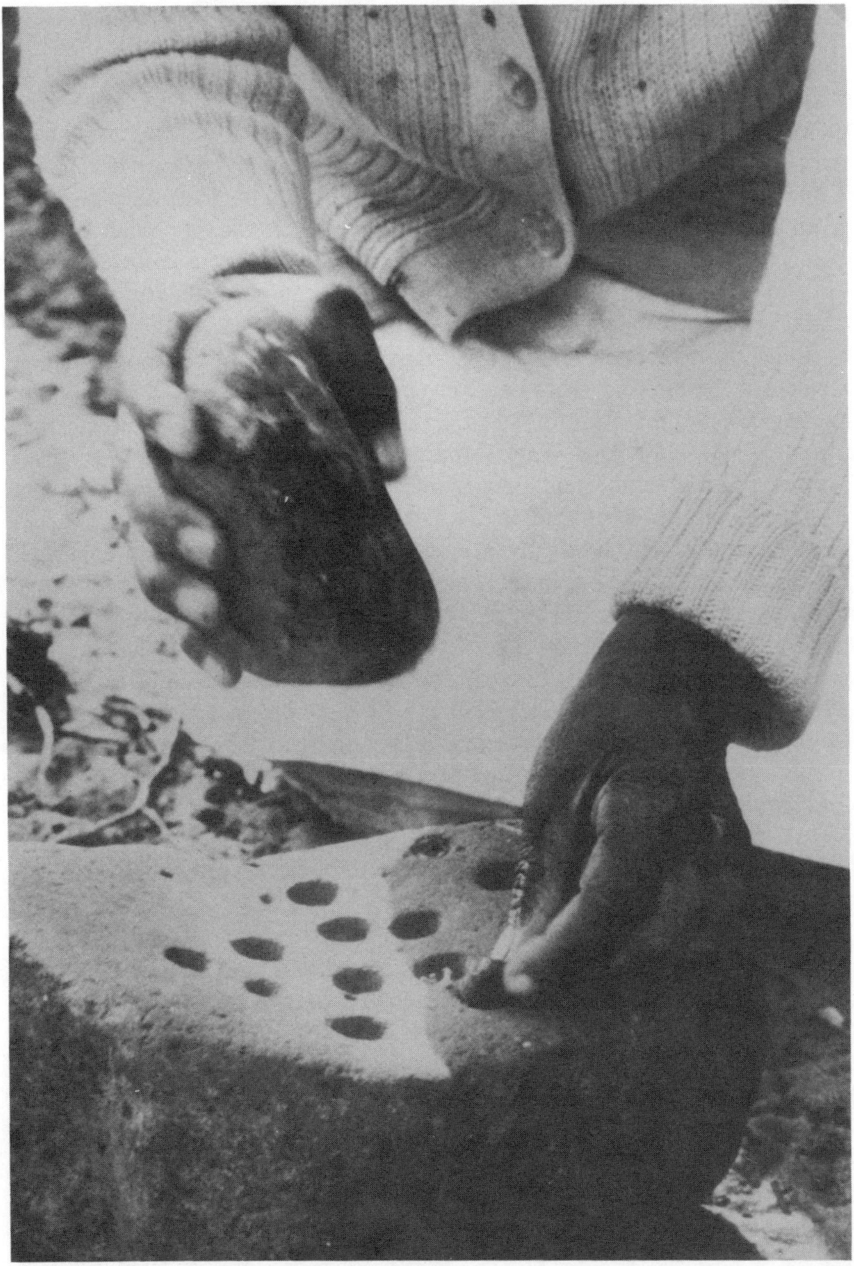

Courtesy of NFRI, CSIR

FIG. 3.3. WOMAN CRACKING MARULA SEEDS
Note cavities formed in stone by seeds through extended usage

Non-food Uses and Characteristics

Medicinal properties against boils, malaria, diarrhoea, etc. are attributed to infusions of the bark, leaves or sap of the marula tree (Shone 1979). It is claimed that blood circulation is aided by a steam bath of marula bark extracts mixed with extracts from other plants and roots (Von Teichman 1983). Claims are also made that the fruit may be used as an insecticide as well as a germicide and that the juice can relieve stomach ailments, whereas the seed oil is sometimes used as a cosmetic and to preserve meat (Shone 1979). Tree gum can be dissolved in water, mixed with soot and used as an ink (Van Wyk 1974).

Since the timber of the tree is easy to work yet fairly tough, it has been used for making crates and other wooden objects. A large timber shortage resulted in the felling of thousands of marula trees during and after the Second World War. Commercial utilization of the wood was, however, halted when the tree was officially declared a protected species throughout South Africa in 1962.

Anderson *et al.* (1986) investigated the composition of the colorless and friable gum of the marula. They found that the gum was acidic and had a low intrinsic viscosity, low molecular weight and high methoxyl content. The main sugar component was galactose (63%), whereas 4-0-methylglucuronic acid (12.5%), relatively little arabinose (14%), glucuronic acid (8%) and galacturonic acid were present. It contained no rhamnose, which sometimes occurs in large amounts in other gums. The marula gum had a relatively low proteinaceous N-content (0.06%). Hydroxyproline and serine accounted for 40% of the amino acid content.

CONSTITUENTS OF MARULA FRUIT

Nutrient and Proximate Composition

The nutrient and proximate composition of destoned fruit and of skins or flesh of marulas collected from various regions in southern Africa are given in Table 3.2. The variation of these results can be explained by differences in ripeness of fruit, geographical origin as well as methods of analyses.

On average, the marula has a moisture content of approximately 85%, with the moisture of the skin slightly less than that of the flesh. The fat and protein contents of the fruit are low and it can contain more than 2.0% crude fiber. Carbohydrates (7-14%) consist of sucrose and smaller quantities of glucose and fructose. Citric acid is the most abundant of the organic acids, excluding ascorbic acid. The presence of malic and tartaric acid has also been reported.

TABLE 3.2. NUTRIENT AND PROXIMATE COMPOSITION OF MARULA FRUIT (OR PARTS THEREOF) FROM DIFFERENT REGIONS IN SOUTHERN AFRICA

	Zimbabwe[1]	Botswana[2]	Sibasa[3]	SWA-Namibia[4]		SWA-Namibia[5]	
	Fruit	Fruit	Fruit	Flesh	Skin	Flesh	Skin
	g/100 g						
Moisture	87	91.7	86.4	85.2	80.7	85.5	78.8
Protein		0.5	0.3	0.5	0.8	0.6	0.6
Fat		0.1	0.1	0.4	0.6	0.4	0.4
Ash		0.2	0.4	0.9	1.0	1.3	1.0
Crude fiber		0.5	0.8	1.2	2.4	2.0	2.6
Carbohydrate		7.0[6]		12.0	14.4		
Fructose	2.3[7]			0.97 (Juice)			
Glucose				0.75 (Juice)			
Sucrose	5.9			5.95 (Juice)			
	mg/100 g						
Calcium		6.2	10.4	20.1	44.7	117.9	126.7
Magnesium		10.5	14.8	25.3	33.5	74.7	41.5
Phosphorus		8.7	9.6	11.5	14.2	7.78	3.94
Iron		0.10	0.24	0.5	0.55	0.30	0.27
Copper		0.04	0.11	0.07	0.08	0.45	0.07
Zinc			0.17	0.10	0.17	0.34	0.12
Sodium		trace	0.64	2.24	1.71	2.53	2.0
Potassium		54.8	163	317.0	345.0	490.2	417.3
Thiamin		0.03		0.03	0.03		
Riboflavin		0.05		0.02	0.02	0.05	0.02
Nicotinic acid		0.25		0.27	0.37	0.48	0.37
Malic acid				0.22 (Fruit)			
Tartaric acid				0.75 (Fruit)			
Citric acid				11.6 (Fruit)			

[1]Carr (1957); [2]Wehmeyer (1967); [3]NFRI (1972); [4]Shone (1979); [5]NFRI (1977).
[6]Calculated by difference.
[7]% invert sugar.

The mineral composition of the fruit seems to vary with its geographical origin. Most notable are the relatively high concentrations of K, Ca and Mg (Grivetti 1982).

Of all the nutrients in the marula, the vitamin C content has attracted the most attention. The vitamin C content of marula fruit, its various parts and products made thereof, is presented in Table 3.3.

Ripe fruit has an average vitamin C content of 168 mg per 100 g which is approximately three times that of oranges and comparable to the amount present in guavas (Wilson 1980) and blackcurrants (Herrmann 1978). Carr (1957) found that the total vitamin C content of the fruit and

TABLE 3.3. VITAMIN C[1] CONTENT OF MARULA FRUIT OR ITS PARTS AND SOME PRODUCTS PREPARED FROM THE FRUIT

	Vitamin C content (mg/100g)						
Reference	Fruit	Skin	Flesh	Juice	Beer	Jelly	Jam
Carr (1957)	179.1				97.1	70.7	105.0
Unripe	127.2						
Ripe	133.7						
Overripe	52.7						
Grivetti (1982)	61				140.0		
Wehmeyer(1967)	67.9						
NFRI (1967)					49.0		
NFRI (1971)	166.0						
NFRI (1976)							
Unripe		150.4	198.2				
Ripe		237.6	194.4				
NFRI (Shone 1979)		230.0	194.0				
NFRI (1979)		227.0	161.0	192.0			
NFRI (1980)				129.0			

[1]Ascorbic acid + dehydroascorbic acid.

the pulp remained constant over a storage period of up to 22 days. This observation has not been studied further or confirmed by other workers.

The value of the vitamin C in the marula was noted in 1937 when an outbreak of scurvy due to a severe drought in the Letaba district (South Africa) was relieved as soon as the affected population could again obtain marula fruit and produce marula beer (Shone 1979).

Relatively small amounts of other vitamins like thiamin, riboflavin and nicotinic acid were found in marulas (Table 3.2).

Volatile Components

The flavor of the marula has been described as pleasant, sour-sweet, guava-like (Quin 1959) and tart (Palmer and Pitman 1972). Shone (1979) mentions its turpentine-like aroma, which increases during ripening and may even be nauseating in overripe fruit.

Two studies have been conducted on the volatile components of marula aroma. Ballschmieter and Torline (1973) found that the aroma was mainly contained in the skin and investigated different methods for its isolation, such as vacuum stripping, carbon absorption of volatiles after vacuum stripping and extraction of the skin with either freon 11 or freon 12. Experiments showed that steam distillation of the skins caused a "cooked aroma", whereas extraction of the skins with methylene chloride resulted in a loss of "freshness". They thus investigated the aroma obtained by vacuum stripping of the skins: 12 major components (all sesquiterpene hydrocarbons) were found by capillary gas chromatographic separation in the methylene chloride extract of the aqueous condensate obtained through vacuum stripping. Three were identified by GC-MS analysis as caryophyllene, α-humulene and copaene.

Over 100 components were identified in the aroma fraction obtained after carbon absorption, but the overall note was sour and unbalanced. The authors concluded that extraction of the skins with freon 12 gave the most authentic marula flavour. Thirty-nine components which were identified in the freon 12 extract are given in Table 3.4.

Pretorius *et al.* (1985) investigated the volatile aroma of marula juice obtained by liquid-liquid extraction or headspace analysis followed by GC and GC-MS identification. The authors separated the aroma extract into 153 compounds of which 64 with retention times lower than pentadecan-2-one were reported (Table 3.4). The compounds were identified either tentatively or completely.

Pretorius and co-workers found that 38 components (α-ylangene to benzenemethanol in Table 3.4) accounted for over 95% of the volatiles of the extracts. Similar to the results obtained from the extracts, sesquiterpene hydrocarbons and benzyl alcohol were the major aroma components in the headspace of the juices.

The identified compounds included alcohols, esters, an organic acid, furans, sesquiterpenes and a few monoterpenes. Sesquiterpene hydrocarbons were the most abundant components of the volatile aroma. Benzyl alcohol was the nonterpenoid compound of highest concentration and benzaldehyde was present only in trace amounts. No qualitative differences existed between the aroma constitution of the fruit from two trees growing in the same region. Quantitatively, however, the one sample contained no α-ylangene and only trace amounts of (E,E)-α-farnesene, which occurred in large amounts in the other.

No characteristic aroma impact compound could be identified, which indicates that an unknown number of compounds contribute to the typical aroma. Marula and grapefruit have certain aroma compounds in common, but the authors state that the similarity in flavor of these two fruits should probably be attributed to nonvolatiles causing a bitter taste. Also, the resemblance which was observed between the aromas of marula and pineapple cannot be due only to the few volatile components (e.g. ethyl acetate, benzaldehyde, linalool) which occur in both fruits. It is clear that further studies will have to be conducted in order to obtain a better understanding of the components which contribute meaningfully to the aroma of the marula.

TABLE 3.4. VOLATILE AROMA COMPOUNDS IDENTIFIED IN MARULA SKINS AND JUICE

Skins[1]	Juice[2]
n-pentane	ethylacetate
n-hexane	ethyl-3-methylbutanoate
benzene	3-methyl-1-butanol
2-octene	pentan-1-ol
1,5-hexadiene	styrene
diethylbenzene	3-hydroxybutan-2-one
	3-methylbutyl-3-methylbutanoate
methanol	hexan-1-ol
n-pentanol	(E + Z)-3-hexen-1-ol
3-methoxy-2-butanol	an ethylester
2-methyl-1-pentanol	trans-linalool oxide furanoid
2-ethoxypropanol	furfural; an alkylbenzene
2-ethyl-3-hexen-1-ol	α-cubebene
	δ-elemene
acetone	α-ylangene
	α-copaene
acetaldehyde	β-bourbonene
glycolaldehyde	an ethyl ester
crotonaldehyde	β-cubebene
n-pentanal	benzaldehyde
3-methylbutanal	linalool
2-hexenal	α-bergamotene
n-hexanal	β-caryophyllene
n-heptanal	a sesquiterpene hydrocarbon
n-octanal	tetrahydro-2-H-pyran-2-one
2-ethylbutanal	aromadendrene
	γ-elemene
ethylformate	(Z)-β-farnesene
ethylpropanoate	3-methylbutanoic acid
ethylisobutyrate	α-humulene
ethylvalerate	a sesquiterpene hydrocarbon
ethylbutanoate	(E)-β-farnesene
ethylisovalerate	γ-amorphene
ethylcaproate	a sesquiterpene hydrocarbon
benzylacetate	γ-muurolene
	three sesquiterpene hydrocarbons
glycolic acid	α-muurolene
oxalic acid	(Z,Z)-α-farnesene
2-methylpropanoic acid	a α-farnesene isomer; two sesquiterpene hydrocarbons
2-methylbutanoic acid	
	(E,E)-α-farnesene
ethyl-2-propenylether	δ-cadinene
ethylisopropenylether	γ-cadinene
methylamine	a sesquiterpene hydrocarbon
acetamide	ethylnicotinoate
	geraniol

TABLE 3.4. VOLATILE AROMA COMPOUNDS IDENTIFIED IN MARULA SKINS AND JUICE (Continued)

Skins[1]	Juice[2]
	a sesquiterpene hydrocarbon
	benzenemethanol
	benzene-ethanol
	calcorene
	2,5-furandialdehyde
	a sesquiterpene hydrocarbon
	dodecan-1-ol
	pentadecan-2-one
	heptadecan-2-one
	nonadecan-2-one
	5-(hydroxymethyl)-2-furan-carboxaldehyde

[1]Ballschmieter and Torline (1973); [2]Pretorius et al. (1985).

CONSTITUENTS OF MARULA NUTS

Proximate and Nutrient Composition

Since marula nuts (Fig. 3.4) are commonly consumed by man the proximate and nutrient composition of nuts obtained from different regions have been analyzed by various researchers (Ferrao and Xabregas 1960; Shone 1979; NFRI 1971; Oliveira 1974). The results are summarized in Table 3.5.

On average the nut contains 56% oil which is comparable to the oil contents of nuts like almonds, walnuts and pistachios. The protein content (28%) of the marula nut is, however, much higher than those of the above-mentioned nuts.

The minerals Ca, Mg, K and P contribute significantly to the average ash content of 3.9%. In general the marula nut's mineral composition is comparable to other tree nuts with the exception of P which is notably higher (Oliveira 1974). Likewise the vitamin composition of the marula nut is similar to that of the most common tree nuts.

Proteins and Amino Acids

Protein contents of 54 to 70% have been reported for defatted marula nuts (Rijke and Joubert 1967; Burger *et al.* 1987). Rijke and Joubert (1967) found that the largest part of these proteins consist of globulins which are soluble in a 10% aqueous NaCl solution.

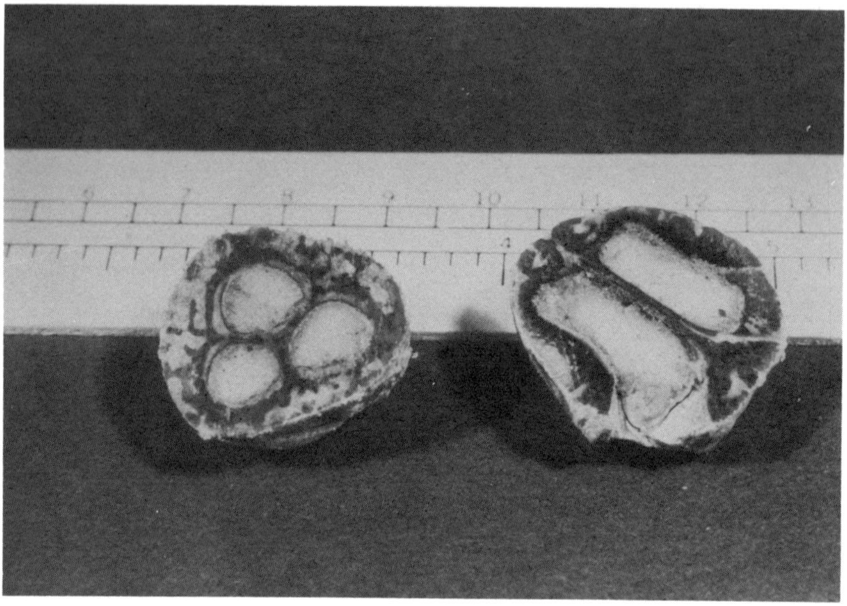

Courtesy of NFRI, CSIR

FIG. 3.4. MARULA SEEDS SHOWING NUTS IN LENGTH- AND CROSSWISE CUT

By means of analytical ultracentrifugation Rijke and Joubert (1967) could discern two major protein components (designated $F_{12.8}$ and $F_{1.6}$) and three minor components in the salt-soluble (globulin) fraction. Both major components which represent approximately 90% of the total protein in the nut, were purified by using a combination of precipitation techniques and chromatographic separation on an agarose column. The purified proteins were further characterized in terms of physical constants, molecular weights (Table 3.6) and amino acid composition (Table 3.7). The latter Table also contains the amino acid composition of marula nuts as reported by Ferrao and Xabregas (1960), Oliveira (1974) and Burger *et al.* (1987).

In comparison to other nuts, the marula nut contains small amounts of lysine and would therefore not be suitable to supplement cereal diets which normally lack lysine (Ferrao and Xabregas 1960). Rijke and Joubert (1967) found the marula nut to be rich in arginine, aspartic acid and glutamic acid. They also compared its essential amino acid content with that of human milk and whole hen's egg. The amino acid patterns were very similar with the exception of lysine which was significantly deficient in the marula.

TABLE 3.5. PROXIMATE AND NUTRIENT COMPOSITION OF MARULA NUTS FROM DIFFERENT REGIONS OF SOUTHERN AFRICA

	Region			
	Angola[1]	South Africa[2]	South Africa[3]	Mocambique[4]
	g/100 g			
Moisture	3.9	4.0	3.9	9.0
Protein	27.6	30.9	25.9	27.6
Oil	56.2	57.0	57.6	54.3
Ash	4.1	4.2	3.5	3.7
Fiber	4.5[5]	2.4	3.5	2.8[5]
Carbohydrate		1.5[6]	5.7[6]	5.3[6]
Caloric value[7]			645.4	
	mg/100 g			
Ca	161.0	106.0	130.4	93.0
Mg		467.0	456.6	329.0
Na		3.4	4.2	81.0
K		677.0	525.3	675.0
Cu		1.99	3.62	8.1
Fe		0.42	9.32	4.4
Zn		4.89	4.89	2.9
P	1907.0	836.0	779.1	774.0
Thiamine		0.4	0.43	
Riboflavin		0.12	0.12	0.20
Nicotinic acid		0.71	0.72	0.74

[1]Ferrao and Xabregas (1960); [2]Shone (1979); [3]NFRI (1971); [4]Oliveria (1974).
[5]Value reported as "cellulose".
[6]Calculated by difference.
[7]kcal/100 g.

TABLE 3.6. PHYSICAL CONSTANTS OF TWO PROTEIN FRACTIONS ISOLATED FROM MARULA NUTS[1]

	Fraction	
	$F_{12.8}$	$F_{1.6}$
S°20 Svedberg units	12.79	1.62
D°20 cm²sec⁻²	3.47×10^{-7}	12.04×10^{-7}
Partial specific volume	0.722	0.682
Mol. wt. from s°20 and D°20	359,000	11,200
Mol. wt. (rapid sed. method)	318,000	14,300

Mol. wt. = molecular weight.
sed. = sedimentation.
[1]Rijke and Joubert (1967).

TABLE 3.7. AMINO ACID COMPOSITION OF MARULA NUTS, DEFATTED MARULA MEAL AND TWO FRACTIONS OF NUT PROTEIN ($F_{12.8}$ and $F_{1.6}$)

	Amino acid content			Amino acid nitrogen (% total N)		
Amino Acid	Nut[1] g/16gN	Nut[2] g/16gN	Defatted Meal[3] % prot.	Defatted meal[4]	$F_{12.8}$	$F_{1.6}$
Lysine	1.81	2.43	1.6	2.25	2.37	0.53
Histidine	2.28		2.5	3.96	4.04	3.88
Ammonia				11.48	10.50	14.06
Arginine	14.15		14.2	27.22	29.10	31.62
Aspartic acid			6.4	4.79	6.79	1.78
Threonine	1.38	2.09	1.8	1.56	1.79	0.61
Serine			3.8	3.49	3.85	1.65
Glutamic acid			23.2	14.58	12.17	14.86
Proline			2.1	2.37	2.51	1.86
Glycine			4.2	5.30	5.20	4.33
Alanine			2.7	2.96	3.21	1.83
Cystine + 2-cysteine		2.30[5]	1.6[6]	2.05	0.00	7.66
Valine	12.42	6.52	3.9	3.40	4.31	1.82
Methionine	2.28	1.89	1.5	0.92	0.75	1.44
Isoleucine	5.47	5.03	3.2	2.53	3.32	1.57
Leucine	4.74	5.58	5.0	3.77	4.36	2.75
Tyrosine		2.30	2.6	1.31	1.48	1.14
Phenylalanine	3.37	4.72	3.5	2.15	2.41	1.70
Glucosamine				0.47	0.00	0.27
Tryptophan	2.64	1.31		0.21	0.60	0.38
% N in Sample				11.71	16.95	17.28

[1]Ferrao and Xabregas (1960); [2]Oliveira (1974); [3]Burger *et al.* (1987); [4]Rijke and Joubert (1967).
[5]only cystine.
[6]half cystine.

Lipids

Marula oil extracted with hexane from the nut, is of an attractive pale yellow color with an acceptable aroma which is faintly reminiscent of the fruit. Pressed oil is darker and has a nutty aroma. Physicochemical characterization of this oil was carried out by the Imperial Institute (1920) and Salama (1973) (Table 3.8). Its specific gravity of 0.9167 is comparable to that of other common oils like cottonseed (0.915-0.921), peanut (0.912-0.920) and olive oil (0.910-0.918) (Considine 1982). The iodine value of marula oil (71.2) is relatively low compared to safflower (140-152) and cottonseed (109-116), but near to that of peanut (88-98) and

TABLE 3.8. PHYSICOCHEMICAL CHARACTERISTICS OF MARULA OIL FROM TWO REGIONS IN AFRICA

	South Africa[1]	Sudan[2]
% oil in kernel	56.2	63.2
Color	pale yellow-brown	very pale yellow
Refractive index	1.460 at 40°C	1.4598 at 30°C
Optical rotation		0.0
Specific gravity (15°C)	0.9167	
Acid value	3.7	1.3
Saponification value	193.5	199.8
Hydroxyl value		2.6
Iodine value	76.6	65.7
Unsaponifiable matter	0.6	0.82
Sterols (mg/100 g)		410.0

[1]Imperial Institute (1920).
[2]Salama (1973).

olive oil (77-88). Marula oil has a saponification value (196.7) which is comparable to those of the common oils (e.g. olive oil: 188-196).

Engelter and Wehmeyer (1970) and Burger *et al.* (1987) analyzed the fatty acid composition of the oil extracted from nuts of *Sclerocarya birrea* subsp. *caffra* by GLC, whereas Salama (1973) reported the fatty acid composition of the oil obtained from the closely related *Sclerocarya birrea* which occurs in the Sudan (Table 3.9). The results of Burger and co-workers represent average values obtained in two consecutive years from trees selected for possible future propagation.

The yield of extracted oil was on average 45.9% (wt/wt of the nut) with a range of 33.9-54.1%. This corresponds to an average oil content of 3.7% per seed (range 3.2-5.8%). A difference in the oil content per nut was observed in seeds from trees growing in different regions. There were no significant differences in the fatty acid composition of the marula cultivars under investigation and the values compare well to those reported by Salama (1973) for marula oil prepared in the Sudan as well as those of Engelter and Wehmeyer (1970) (Table 3.9). The general fatty acid composition of marula oil is very similar to that of olive oil (Burger *et al.* 1987).

Burger *et al.* (1987) tested the stability of marula oils in comparison to other oils in a Rancimat at 120°C and an airflow of 20 liters/hr. Two of the marula oils had induction periods of more than 48 hours (average value 33 hours), whereas olive oil started deteriorating at 4.6, cottonseed oil after 3.1 and sunflower oil after 1.9 hours at the same tempera-

TABLE 3.9. FATTY ACID COMPOSITION OF OIL EXTRACTED FROM MARULA NUTS FROM DIFFERENT REGIONS OF AFRICA

Fatty acid	South Africa[1]	South Africa[2] Phalaborwa[4]	Grey[4]	De Klerk[4]	Sudan[3]
	% (wt/wt)				
C14:0	trace	0.09	0.07	0.07	0.20
C16:0	12.00	12.77	12.97	12.73	17.10
C16:1	trace	0.25	0.07	0.17	
C17:0		0.13	0.12	0.14	
C17:1		0.06	0.06	0.08	
C18:0	9.20	6.76	5.10	6.03	10.90
C18:1	69.90	73.61	72.50	70.47	66.00
C18:2	7.80	4.96	7.90	8.95	4.30
C18:3		0.05	0.04	0.05	—
C20:0	0.60	0.61	0.50	0.60	
C20:1	0.30	0.32	0.35	0.38	
C20:4					0.90
C22:0		0.80	0.10	0.14	
C24:0		0.27	0.14	0.14	
%oil/nut[5]			45.9 (33.9-54.1)		

[1]Engelter and Wehmeyer (1970).
[2]Burger *et al.* (1987).
[3]Salama (1973).
[4]samples from single tree in the wild.
[5]average of nuts from nine trees.

ture. Mixtures of marula oil with less stable oils, e.g. sunflower or maize (corn), increased the stability of the mixture relative to that of the pure unstable oils. This increase was in direct relation to the percentage of marula oil present.

The sterol and tocopherol contents of the same oils are given in Table 3.10 and Table 3.11, respectively. The sterol content of the marula oil from different trees is very similar and typical for oils of plant origin. Differences in the 4- and 4,4-dimethyl sterols may be due to factors such as ripeness of the seeds and the fact that these components are precursors of free sterols. Two as yet unidentified sterols were found (Table 3.10) which, however, did not occur in all the oil samples or both seasons, and were present at low concentrations. The main tocopherol in marula oil is γ-tocopherol (Table 3.11).

The excellent stability of the oil cannot be explained in terms of either the tocopherol or sterol composition. Δ^5-Avenasterol, which has antioxidative properties (Boskou and Morton 1975) and the most effective antioxidative tocopherol isomer (α-tocopherol) are both present at too

TABLE 3.10. STEROL COMPOSITION OF EXTRACTED MARULA OIL[1]

Sterol	Origin[2] Phalaborwa	Grey	De Klerk
	mg/kg oil		
Campesterol	149	162	215
Stigmasterol	52	56	61
β-sitosterol	1317	1416	1666
Δ^5-avenasterol	241	216	223
unknown (RT = 0.89)	51[3]	98[3]	67[3]
Obtusifoliol	3[4]	5[4]	15
Cycloartenol	4	15	15
24-methylene cycloartenol	35	20	22
Citrostadienol	37	28	41
unknown (RT = 1.67)	19	18	14

[1]Burger *et al.* 1987.
[2]Sample from single tree in the wild.
[3]Value of 1985; compound absent in 1986.
[4]Value of 1986; compound absent or trace in 1985.

low concentrations to be of significance. It therefore seems as if the fatty acid constitution of the oil (high oleic acid content) is the main reason for its stability.

According to the authors the stable marula oil could be of value in the food industry where it can be used as coating of dried fruit, as a frying oil or as a substitute for high-oleic safflower oil in baby foods. However, the competition from newly-developed high-oleic sunflower oil, the current difficulty to collect marula seeds from scattered wild plants and the lack of appropriate processing techniques are prohibitive factors for its present commercial utilization.

TABLE 3.11. TOCOPHEROL COMPOSITION OF EXTRACTED MARULA OIL[1]

Tocopherol	Origin[2] Phalaborwa	Grey	De Klerk
	mg/100g oil		
α-tocopherol	0.22	0.30	0.42
γ-tocopherol	21.69	23.74	25.75
β-tocotrienol	0.06[3]	0.04[3]	0.06[3]
γ-tocotrienol	0.03	0.10	0.04
δ-tocopherol	0.46	0.52	0.93

[1]Burger *et al.* (1987).
[2]Sample from single tree in the wild.
[3]Value of 1985; compound absent in 1986.

HARVESTING AND STORAGE

Redelinghuys *et al.* (1976) studied changes during ripening at 23°C in marulas which were picked from two trees growing on the Magalies Berg mountains near Pretoria. They found that the fruit showed a distinct climacterium in its ripening process with concurrent changes in the pH, total acids and total soluble solids. The climacteric peak of the fruit occurred around the seventh day of storage. There was a fast decrease in the total acids before the climacterium (from 4.3 to 1.3% citric acid wt/wt), after which the decrease was slower. A simultaneous rise in pH from 2.7 to 3.85 occurred, while there was a steady increase in the total soluble solids from 10 to 13.5%.

The respiration rates of picked and fallen fruit were also compared. Fallen fruit was already past the climacteric minimum, which had not been reached by picked fruit at that stage (Fig. 3.5). They found that the riper (fallen) fruit had a storage life of 14 days at 12.8°C at which stage 12% of the fruit was already rotten. After 21 days the percentage rotten fruit had increased to 89%. Redelinghuys *et al.* (1976) reported that storage temperatures of 9°C or below cause damage to marulas, but since then the fruit has been stored at 4°C without apparent damage (Holtzhausen 1987). Marulas could be stored at 4°C for 16 days before rot sets in, whereas the period was reduced to 6 days at 25°C.

TECHNOLOGY

The attention which has been focused on the manifold uses of the marula as well as calls for the commercial utilization of both the fruit and the nut (Ballschmieter and Torline 1973; Redelinghuys *et al.* 1976; Shone 1979; Taylor 1985) has recently resulted in limited commercial production of marula juice. It is estimated that approximately 600 metric tons of marula juice was processed in South Africa during 1986. This amount of juice is apparently still below current market demand and production can be expected to increase dramatically if the availability of fruit can be guaranteed. The latter is hampered to a large extent by the fact that harvesting is restricted to collection of fruit from wild trees and that the crop is influenced by uncontrollable growth conditions. As can be expected both yield and quality of juice from fruit obtained from unselected wild trees are variable.

To overcome the above-mentioned problems and to aid in the selection of trees suitable for propagation, research is being carried out on the chemical composition, flavor and processing properties of fruit from marula trees which have been identified as prolific fruit bearers. As part of this program, juice extraction methods and characteristics of juice

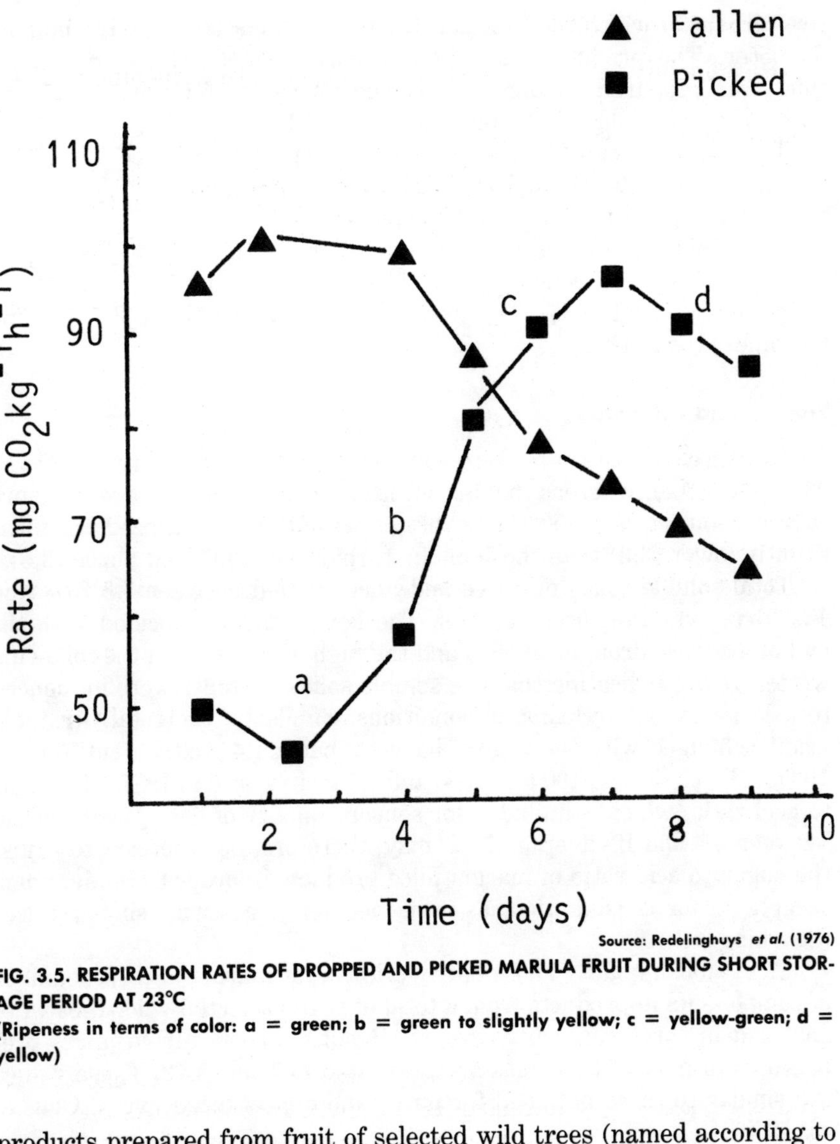

Source: Redelinghuys *et al.* (1976)

FIG. 3.5. RESPIRATION RATES OF DROPPED AND PICKED MARULA FRUIT DURING SHORT STORAGE PERIOD AT 23°C
(Ripeness in terms of color: a = green; b = green to slightly yellow; c = yellow-green; d = yellow)

products prepared from fruit of selected wild trees (named according to their location) were studied during three consecutive years (Gous *et al.* 1988).

A turbid juice was prepared by steam blanching the fruit, cutting through the tough skins with knives and expressing the juice. The juice was then pasteurized (80°C for 15 sec) in bottles or in a tube pasteurizer and cooled. A puree was prepared by passing fruit consecutively through two screens of which the first had 12 mm and the second a 1.14 mm pore diameter. The product was frozen or used immediately. Depectinized juice was prepared by digesting cooled puree at 45°C for 1 hour and pasteurizing the slightly turbid juice at 80°C/15 sec.

The yields of turbid juice, puree and depectinized juice varied between 34 and 54%, 31 and 50%, 21 and 38%, respectively. These low values were partly due to small sample sizes.

In a trial run, samples of (depectinized) puree, turbid as well as depectinized juices were concentrated on a Centritherm CT1B centrifugal evaporator (NFRI 1986). The juices could be concentrated 2.7, 3.4 and 6.6 times, respectively.

Product Characteristics

Moisture contents of the puree and turbid juices were between 85 and 93% (avg. 89%), whereas the depectinized juice had an average moisture of 88% (Gous *et al.* 1988). The average pH of the depectinized juice was slightly lower (3.3) than those of the turbid juice (3.6) and puree (3.5).

Total soluble solids of puree and juices varied between 7.5°Brix and 15.5°Brix over the three seasons. The lower values coincided with the end of a severe drought in 1985 and the higher values with the following wetter years, indicating that the soluble solids contents were influenced to a large extent by climatic conditions. Similarly, the total titratable acidity changed with the year of harvest, being 1.4 g citric acid/100 g in 1985, 1.8 g citric acid/100 g in 1986 and 1.7 g citric acid in 1987. The sugar to acid ratio, which is an index for sensory quality of fruit juices, varied between 3.4 and 12 (average 7). It may, therefore, be necessary to adjust the sugar to acid ratio of marula juice products in order to render them acceptable for consumption, as is needed for juices with similar ratios (Schobinger 1978).

To establish quality and identity criteria for marula juices, a number of components in products from a total of seven selected wild trees were analyzed in three consecutive years. Results for the turbid juices prepared from fruit of two trees are presented in Table 3.12. These values are similar to those obtained for puree and depectinized juices (Gous *et al.* 1988).

Sugars were determined by HPLC on a HPX-87 monosaccharide column (BioRAD). Sucrose was the most abundant sugar present. Glucose and fructose occurred in smaller amounts as well as in similar ratios in all juice products and all cultivars. In some products an as yet unidentified sugar with a retention time similar to that of fructose was present, but either did not occur or could not be resolved from fructose in the

TABLE 3.12. CHEMICAL COMPOSITION OF TURBID JUICE PREPARED FROM THE FRUIT OF TWO MARULA TREES[1]

	Origin[2]			
	Grey		Phalaborwa	
	Average[3]	Range	Average[3]	Range
Sugars (g/100 g)				
Sucrose	3.16	3.00-3.30	4.43	3.80-5.50
Glucose	0.79	0.61-0.97	0.70	0.54-0.86
Fructose	1.06	0.69-1.30	1.13	0.88-1.40
Unknown[4]	0.28	0.20-0.35	0.07	0 -1.13
Acids (g/100 g)				
Citric	1.84	1.52-2.01	1.63	1.44-1.90
Malic	0.28	0.10-0.42	0.28	0.13-0.39
Pectins (g GA/ml)				
Total	3.06	2.25-3.71	2.44	2.14-2.87
Water soluble	2.90	2.25-3.47	2.24	1.88-2.62
Oxalate soluble	0.02	0 -0.08	0.09	0.08-0.10
Alkali soluble	0.04	0 -0.08	0.12	0.09-0.15
Polyphenols (g TA/100 ml)				
Total	0.24	0.05-0.34	0.28	0.08-0.43

[1]Gous *et al.* (1988).
[2]Sample from single tree in the wild.
[3]Average of three years.
[4]Not sucrose, glucose, fructose, stacchyose, raffinose, mannose, sorbose or rhamnose.

remaining samples. By using standards it was established that the sugar is neither sucrose, glucose, fructose, stacchyose, raffinose or mannose.

Organic acids were separated by HPLC on an Aminex HPX-87H column with 0.0025 M HNO_3 as mobile phase. Citric acid and malic acid were present at ratios of between 3 : 1 and 10 : 1. One selection contained practically only citric acid. The presence of tartaric acid, which was previously reported to occur in marulas (Table 3.2) could not be confirmed.

The puree had a total pectin content of about thrice the concentration present in either the turbid or depectinized juices. Accordingly, the apparent viscosity of the puree (e.g. 76 cP at shear rate (D) = $2785 s^{-1}$ at 20°C) was at least three times higher than those of the juices, which makes handling and processing of the former more difficult and results in low concentration factors (see above). The pectins consisted mainly of water-soluble fractions with 10 to 100 times less oxalate and alkali-soluble pectins (Table 3.12). Pressed turbid juice contained between 10 and 50% less water-soluble pectins, between 25 and 100% less oxalate-soluble and between 90 and 100% less alkali-soluble pectins than puree. Enzymatic depectinization of marula puree reduced only the amount of alkali-soluble pectins compared to the turbid juice, whereas the qualitative and quantitative pattern of water and oxalate-soluble pectins was comparable to that of the turbid juice.

A modified Folin-Ciocalteau method (Mcgrath *et al.* 1982) was used to determine the total polyphenol content as g tannic acid/100 ml juice. All three marula juice products contained approximately the same large amount of polyphenols.

The mineral and amino acid compositions of turbid juice prepared from fruit of the Phalaborwa and Grey selections are listed in Tables 3.13 and 3.14, respectively (Gous *et al.* 1988).

The qualitative mineral composition of the marula juices is comparable to that of the fruit (Table 3.2), i.e. Ca, Mg, K and P are the most abundant minerals, whereas Na, Fe, Cu, Zn and Mn are present in smaller amounts. Varying amounts of each mineral pass into the juices during processing. The mineral compositions of the purees and juices did not differ much between samples or between seasons.

Amino acids (Table 3.14) were determined by HPLC on a Lichrosorb RP 18 column with gradient elution (Cilliers *et al.* 1985). The general amino acid pattern was similar in all the marula juice products, but quantitative differences existed between samples and seasons. Asparagine, glutamine and arginine were the amino acids of highest concentration;

TABLE 3.13. MINERAL COMPOSITION OF TURBID JUICE PREPARED FROM THE FRUIT OF TWO MARULA TREES (MG/100 G)[1]

	Origin[2]			
	Grey		Phalaborwa	
Mineral	Average[3]	Range	Average[3]	Range
Calcium	16.90	14.3 - 18.30	16.60	13.40- 19.60
Magnesium	15.80	14.40- 17.90	20.83	20.10- 22.20
Sodium	0.96	0.85- 1.10	1.43	1.20- 1.70
Potassium	261.00	256.00-271.00	291.33	271.00-302.00
Iron	0.22	0.19- 0.26	0.25	0.19- 0.29
Copper	0.09	0.06- 0.14	0.09	0.05- 0.13
Zinc	0.15	0.12- 0.20	0.20	0.11- 0.24
Manganese	0.03	0.02- 0.03	0.04	0.03- 0.05
Phosphorus	22.20	21.10- 22.80	29.87	26.40- 34.10

[1]Gous *et al.* (1988).
[2]Sample from single tree in the wild.
[3]Average of three years.

TABLE 3.14. AMINO ACID COMPOSITION OF TURBID JUICE PREPARED FROM THE FRUIT OF TWO MARULA TREES (PPM)[1]

	Origin[2]			
	Grey		Phalaborwa	
Amino acid	Average[3]	Range	Average[3]	Range
Aspartic acid	88	75- 108	67	47- 95
Glutamic acid	268	155- 372	243	188- 372
Serine	109	85- 142	67	56- 88
Glycine	26	12- 51	12	7- 16
Asparagine	960	660-1517	874	312-1244
Glutamine	1073	890-1346	895	637-1079
Threonine	90[4]	52- 127	52	30- 74
Alanine	168	117- 229	129	116- 151
Histidine	25[4]	29- 47	28	24- 46
Y-ABA	152	118- 213	170	121- 220
Proline	23	14- 38	28	22- 32
Arginine	554	297-1029	537	272- 951
Tyrosine	22	10- 45	21	9- 42
Valine	24	7- 36	14	8- 17
Methionine	12	12- 12	6[4]	5- 6
Isoleucine	13	13- 13	7	6- 9
Leucine	8	7- 10	5	3- 10
Phenylalanine	4	3- 5	7	4- 10
Tryptophan	4	3- 5	4	2- 5
Lysine	23	17- 30	10	8- 12
Formol value (ml 0.1 N NaOH/100 ml)	24	20- 30	24	17- 28

[1]Gous *et al.* (1988).
[2]Sample from single tree in the wild.
[3]Average of three years.
[4]Average of two years.
tr = trace.
γ-ABA = γ-aminobutyric acid.

glutamic acid, alanine and serine occurred in intermediate amounts and the other acids at low concentrations. It seems as if the variation of the asparagine, glutamine and γ-aminobutyric acid contents is related to the fruit selection. Formol values serve as an indicator of nitrogenous compounds in fruit juices (IFFJP 1973) and are suggested as a criterion of identification. Their use is sometimes criticized since they may decrease during ripening, change during storage of fruit products and show large differences within the same fruit. However, the values of three of the marula selections were consistently twice as large as those of the rest.

This pattern was qualitatively similar for the three juice products (Gous *et al.* 1988).

Sensory Characteristics of Products

Turbid juices prepared from five promising marula selections of one season were compared by scaling for intensity of color, taste, aroma and mouthfeel by a trained laboratory panel (Gous *et al.* 1988). Significant differences ($p = 0.05$ simultaneous level) between certain samples were found for all tested stimuli. Of the two selections discussed above, Grey was judged by the panelists as light beige to beige color, with a slightly more than weak sweetness, weak to moderate marula aroma and a near viscous mouthfeel. This sample was the lightest, least sweet, most watery and next to least intense in aroma of all juices tested. The Phalaborwa selection was a beige to dark beige color, moderate in sweetness, slightly less than moderately strong in aroma and with a viscous to thick mouthfeel. Compared to the other samples, Phalaborwa was the darkest and thickest sample, but had an average sweetness and aroma intensity.

Schaefer and McGill (1986) used a sensory panel to determine the flavor characteristics by means of Quantitative Descriptive Analysis of five marula juice samples prepared with a commercial fruit squeezer. The origins of the samples were not clearly identified, but seem to be from the same trees as those investigated by Gous *et al.* (1988). The panel consisted of five selected, trained judges which evaluated the juices according to a total of 14 descriptor terms for odor, flavor, mouthfeel and aftertaste characteristics. All samples differed very little in odor, flavor and aftertaste. Only one sample was judged to have an acceptable sweet to sour balance, whereas the remaining four were extremely sour and lacked sweetness. Another juice sample was found to be much thicker and grainier than the others. The authors concluded that the poor quality and varying degrees of ripeness of the raw material was a major drawback for the sensory quality of the juices and called for more research on determination of optimal ripeness and sensory evaluation.

PRESENT IMPORTANCE AND FUTURE PROSPECTS

Three of the wild trees which had been identified for extensive study on account of their large crops, fruit size and quality are considered to be suitable for registration as cultivars. They will be planted in orchards and tested under horticulturally sound conditions for commercial fruit production. At that stage the technology of juice manufacture should be optimized and alternative processing methods as well as the utilization of fresh fruit will be investigated. The fact that almost the entire marula tree and its fruit can be utilized, offers remarkable opportunities to contribute to the development of agriculturally based industries on the African continent.

REFERENCES

ANDERSON, D. M. W., BELL, P. C., GILL, M. C. L., McDOUGALL, F. J. and McNAB, C. G. A. 1986. The gum exudates from *Chloroxylon swietenia, Sclerocarya caffra, Azadirachta indica* and *Moringa oleifera.* Phytochem. 25 (1) 247-249.

BALLSCHMIETER, H. M. B. and TORLINE, P. 1973. Marula has distinct commercial appeal. Food Ind. South Africa. Sept. 22-23, 25.

BURGER, A. E. C., DE VILLIERS, J. B. M. and DU PLESSIS, L. M. 1987. The marula seed: a potential source of oil and protein. S. Afr. J. Sci. 83, 733-735.

BOSKOU, D. and MORTON, I. D. 1975. Changes in the sterol composition of olive oil on heating. J. Sci. Fd. Agric. 26. 1149-1153.

CARR, W. R. 1957. Notes on some Southern Rhodesian indigenous fruits, with particular reference to their ascorbic acid content. Food Res. 22 (6) 590-596.

CILLIERS, J. J. L., SMIT, S. C. C. and OOSTHUIZEN, M. P. 1985. The determination of free amino acids in fruit juices. Proceedings SAAFot Congress, Pretoria, Rep. South Africa. (ISBN 0 7988 3547 8). Vol. 1. 266-275. (Afrikaans)

CONSIDINE, D. M. (ed.) 1982. *In* Foods and Food Production Encyclopedia. Van Nostrand Reinhold Co., New York.

ENGELTER, C. and WEHMEYER, A. S. 1970. Fatty acid composition of oils of some edible seeds of wild plants. J. Agr. Food Chem. 18 (1) 25-26.

FERNANDEZ, R. and FERNANDEZ, A. 1966. Anacardiaceae. *In* Flora Zambesiaca. Vol. II. A. W. Excell *et al.* (Editors). Crown Agents for Overseas Governments and Administrations, London.

FERRAO, J. and XABREGAS, J. 1960. National agricultural investigations in relation to African nutrition and foods. Agr. Angolana, Luanda 12, 78-86. (Portuguese)

FOX, F. W. and YOUNG, M. E. N. 1982. *In* Food From The Veld. Delta Books, Johannesburg. 79-81.

FRIEDE, H. and PIENAAR, J. N. 1974. Application of the froth flotation process to Kruger Cave deposit. S. Afr. J. Sci. 70(12) 375-376.

GOUS, F., WEINERT, I. A. G. and VAN WYK, P. J. 1988. Selection and processing Marula fruit (*Sclerocarya birrea* subsp. *caffra*). Lebensm.-Wiss. u. - Technol. 21, 259-266.

GRIVETTI, L. E. 1982. Morula (*Sclerocarya caffra*): an underexploited food resource of the eastern Kalahari. Botswana 66th Annual Meeting, Federation of American Societies for Experimental Biology (Abstracts of Papers); New Orleans, Louisiana, April 15-23, 1982.

HERRMANN, K. 1978. Chemical composition of fruit, fruit juices and important vegetable juices as well as their nutritive value. *In* Fruit and Vegetable Juices. U. Schobinger (Editor). Verlag Eugen Ulmer, Stuttgart. 32-79. (German)

HOLTZHAUSEN, L. C. 1987. Personal communication. Dept. Horticultural Science, University of Pretoria, Pretoria, South Africa.

HOLTZHAUSEN, L. C. and VAN DER LAARSE, G. 1984. The ennoblement of the marula. Dendron 6 (3 & 4) 52-53. (Afrikaans)

IFFJP. 1973. Determination of Formol value (Analysis no. 30; 1965). *In* Analyses of International Federation of Fruit Juice Producers. Address: P. Dardonville, 10 rue de Luegé, F-75009 Paris.

IMPERIAL INSTITUTE. 1920. *Cited by* A. K. Shone, 1979. Notes on the marula. Bulletin 58, Department of Forestry, Pretoria, Republic of South Africa.
KOKWARO, J. O. and GILLETT, J. B. 1980. Notes on the Anacardiaceae of eastern Africa. Kew Bull. 34, 745-760.
McGRATH, R. M., KALUZA, W. Z., DAIBER, K. H., VAN DER RIET, W. B. and GLENNIE, C. W. 1982. Polyphenols of sorghum grain: their changes during malting and their inhibitory nature. J. Agr. Food Chem. 30, 450-456.
NFRI. 1967. Internal report: Analyses of South African edible wild fruit V. National Food Research Institute, CSIR, Pretoria, South Africa.
NFRI. 1971. Internal report: Analyses of South African edible wild fruit IX. National Food Research Institute, CSIR, Pretoria, South Africa.
NFRI. 1972. Internal report: Analyses of South African edible wild fruit X. National Food Research Institute, CSIR, Pretoria, South Africa.
NFRI. 1976. Internal report: Analyses of South African edible wild fruit XVI. National Food Research Institute, CSIR, Pretoria, South Africa.
NFRI. 1977. Internal report: Analyses of South African edible wild fruit XVII. National Food Research Institute, CSIR, Pretoria, South Africa.
NFRI. 1979. Internal report: Analyses of South African edible wild fruit XIX. National Food Research Institute, CSIR, Pretoria, South Africa.
NFRI. 1980. Internal report: Analyses of South African edible wild fruit XX. National Food Research Institute, CSIR, Pretoria, South Africa.
NFRI. 1986. Internal report: Processing of subtropical Fruit: Marula. National Food Research Institute, CSIR, Pretoria, South Africa.
OLIVEIRA, J. S. 1974. Nutritional value of some Mocambique wild fruit. Port. Acta Biol. Ser. A 13 (1-4) 47-62. (Portuguese)
PALMER, E. and PITMAN, N. 1972. *Sclerocarya caffra*. *In* Trees of Southern Africa. A. A. Balkema, Cape Town. 1189-1193.
PRETORIUS, V., ROHWER, E., RAPP, A., HOLTZHAUSEN, L. C. and MANDERY, H. 1985. Volatile flavour components of marula juice. Z. Lebensm. Unters.-Forsch. 181, 458-461.
QUIN, P. J. 1959. Foods and feeding habits of the Pedi. Univ. Witwatersrand Press, Johannesburg, Rep. South Africa. 86-90.
REDELINGHUYS, H. J. P., WESSELS, F. J. and HEINEN, E. A. 1976. The technological and economical potential of marula juice production. S. A. Food Rev. 3 (3) 75, 77, 79, 81, 85.
RIJKE, A. M. and JOUBERT, F. J. 1967. A physicochemical study of the proteins from the marula nut (*Sclerocarya caffra* Sond.). J. Sci. Food Agric. 18, 200-203.
SALAMA, R. B. 1973. The seed oil of Sclerocarya birrea. Sudan J. Food Sci. Techn. 5, 4-6.
SCHAEFER, G. and McGILL, A. E. J. 1986. Flavor profiling of juice of the marula (*Sclerocarya birrea* subsp. *caffra*) as an index for cultivar selection. Acta Hort. 194, 215-222.
SCHOBINGER, U. 1978. Chemical composition of fruit and fruit and vegetable juices. *In* Fruit and Vegetable Juices. U. Schobinger (Editor). Verlag Eugen Ulmer, Stuttgart. 40. (German)
SHONE, A. K. 1979. Notes on the marula. Bulletin 58, Department of Forestry, Pretoria, Republic of South Africa.
TAYLOR, F. W. 1985. The potential for the commercial utilization of indigenous plants in Botswana. *In* Plants for arid lands. Proceedings, Kew International Conference on economic plants for arid lands. G. E. Wickens (Editor). Allen + Unwin, London. 231-242.

VAN WYK, P. 1974. *Sclerocarya caffra. In* Trees of the Kruger National Park. Vol. II. Purnell, Cape Town. 323-327.

VON TEICHMAN, I. 1982. Notes on the distribution, morphology, importance and uses of indigenous Anacardiaceae: 1. The distribution and morphology of *Sclerocarya birrea* (the marula). Trees in South Africa 34(3) 35-41.

VON TEICHMAN, I. 1983. Notes on the distribution, morphology, importance and uses of indigenous Anacardiaceae: 2. The importance and uses of *Sclerocarya birrea* (the marula). Trees in South Africa 35(1 & 2), 2-7.

WEHMEYER, A. S. 1967. Edible wild fruits of the Transvaal. Food Ind. S.A. 19 (12) 49-53.

WILSON, C. W. III. 1980. Guava. *In* Tropical and Subtropical Fruits: composition, properties and uses. S. Nagy and P. E. Shaw (Editors). Avi Publ. Co., Westport, Connecticut.

4

Spondias: The Red Mombin and Related Fruits

Jorge Leon
Philip E. Shaw

Four species of *Spondias* are widely used in the tropics for their fruit, although none of them is intensively cultivated. Three are from tropical America: the red (purple) mombin, *S. purpurea*, the yellow mombin or hogplum, *S. lutea*, and the imbu, *S. tuberosa*. The fourth species, which comes from Polynesia, is the ambarella or Otahite apple, *S. cytherea*. All of them are used as fresh fruits, in green and ripe stages, and in the preparation of preserves and refreshments. Due to the softness of the pericarp, they are difficult to transport and store, and their export to foreign markets is negligible (Whiley 1984). The genus *Spondias* belongs to the family Anacardiaceae, which includes some well known fruit, such as the mango (*Mangifera*) and the cashew nut (*Anacardium occidentale*).

RED MOMBIN (*Spondias purpurea*)

The red mombin (purple mombin, Spanish plum, jocote, ciruelo) is the most intensively cultivated of the *Spondias*. Its area of origin is probably Mexico and Central America, where several wild populations are still found; it was taken by the Spaniards to South American and the Philippines. The nahuatl (aztec) name xocotl, is used as "jocote" in a large part of Mexico and Central America. The Spaniards called this species "ciruelo" (plum), and it is known by this name in South America and the Philippines. In Mexico the tree is called "ciruelo" and the fruit is generally called "ciruela".

Characteristics

The red mombin (Fig. 4.1) is usually propagated vegetatively by planting woody cuttings 2-3 m long. In a few years the cutting develops a thick trunk, with several long stout primary branches coming out at the top, divided in secondary and tertiary branches. Often the trees are asymetric, but if they develop a regular shape the crown is hemispheric and may attain up to 10 m in diameter. The leaves are pinnate, with 5-12 pairs of leaflets, each 2-3 cm long, with serrated borders. The trees drop most of the leaves at flowering time, with the abscission changing according to the variety. The flowers are borne often in the primary branches, in lateral inflorescences; they are small, reddish and can be male, female or bisexual. The fruits appear solitary or more commonly in groups of 2-6, and have a thin skin (pericarp), yellow to dark red according to the variety; they have a thin layer of pulp (mesocarp) sweet to acid, 1-4 mm thick, and a large "seed", which is really a hard, wrinkled endocarp containing 1-5 seeds, mostly sterile.

Varieties

In Mexico and Central America, in areas with alternate dry and wet seasons, there are some wild populations, which may be the ancestors to

FIG. 4.1. THE RED MOMBIN

the cultivated red mombins. One of them, "jocote de iguana", has round, small red fruits, very attractive, but very acidic. They are eaten by iguanas, monkeys and other animals. Another primitive variety, the "cismoyo" has ellipsoidal, bright yellow fruit with thin, watery pulp. It is too acidic and is seldom eaten.

Apparently the highest diversity in wild and cultivated populations is found in the coastal area of the Gulf of Mexico, where some twenty cultivars have been described. At least six cultivars are grown in Florida (Popenoe 1979).

The cultivated varieties are clones which may be arranged in two groups: a) *dry season mombins*: These fruit produced at the end of the dry season (February to May) are rather small, ellipsoidal, 2.5-3 cm long, smooth, red purple outside, with soft yellow pulp, sweet to slightly acidic. When green they look in color and shape like olives. Most of the clones in this group grow at low elevations, between 0 to 800 m. To this group may belong Tronador, Crillo, Nica, Morado, Rojo acido.

b) *wet season mombins*: These include the choicest clones, with rather large fruits, 3.5-4.5 cm long, yellow or red, smooth or with protuberances, with the pulp firm, yellow, sweet or slightly acidic. The maturation period covers the last part of the rainy season (September to December). Most of the clones in this group are grown at elevations between 800-2000 m and include Corona, red with warts at the apical end; Petapa, yellow, smooth and shiny; Cabeza de loro, large fruits with thick skin, and others. There is no comprehensive treatment of the genetic diversity of this species, even though varietal names are used in several publications (Rose 1899; Souza Novelo 1949; McBryde 1945; Nava-Kuri and Uscanga 1979; Morton 1987).

Cultural practices

The majority of fruit production in the red mombin comes from trees that are planted as living fence posts or grow without any care in backyards, pastures and maize lands. Only in Oaxaca, Mexico, are there well attended groves. Trees growing without any care may become too large, and are often destroyed to be replaced by crops, houses, roads, etc.

Propagation

The best clones are clonally propagated, since they do not produce seeds, due to the lack of pollen formation (Juliano 1932). Cuttings root easily, since they have a good store of nutrients and water, and often produce the first fruits two years after planting. In Oaxaca, cuttings

1-1.5 m long are planted in double rows, in rows 3 m apart. In Guatemala, up to four cuttings 1 m long are planted close together in a slanted position; one half of the cutting is buried under ground.

Management

In Oaxaca the trees are pruned to keep them at around 2 m tall, in order to facilitate the collecting of the fruit, which is harvested like olives by shaking the branches with poles and gathering the fruit from the ground. The appearance of the red mombin groves in Oaxaca, with the well trimmed rows covering the flanks of the hills, is similar to the pear and apple plots in Europe.

Perhaps the most serious problem in red mombin production is the damage produced by fruit flies, both Mexican (*Anastrepha* spp.) and Mediterranean (*Ceretitis capitata*).

Utilization

The main use of the red mombin is as fresh fruit, both green and ripe. Green fruits are sold in the local markets in large quantities; although they are quite acidic, they are eaten by women and children. Ripe fruits are plentiful during the season in the Indian markets of Mexico and Guatemala. The sellers classify them by size and ripening stage, and place them separately in heaps or large baskets.

Ripe fruit are cooked with brown sugar, stored in glass jars and are sold in this form at the markets and stores. They are also dried in the sun, but do not keep well. In Mexico (Rose 1899; Alcorn 1984) the pulp is prepared in "atole", a watery porridge made with corn flour, sugar and other ingredients. Like most tropical fruits, red mombins are used in the preparation of fermented drinks (chicha), wine and cold drinks.

Dried red or purple mombin slices of good quality were prepared by drying for 5-6 hours in a dehydrator oven, while whole red mombin fruit dried for 24 hours gave a poor product by this drying method (Campbell and Campbell 1983).

Composition

The composition of the edible portion of *Spondias* fruit, including red mombin, is shown in Tables 4.1A and 4.1B. A single value was reported by Leung and Flores (1961) for each constituent even though each value is the result of 6-8 analyses. Morton (1987) reported a range of values for each component listed.

TABLE 4.1A. COMPOSITION OF *SPONDIAS* FRUIT PER 100 GRAMS OF EDIBLE PORTION

Species	Flesh in fruit %	Food energy Cal	Moisture %	Prot. g	Fat g	Total Carbohydrate g	Fiber g	Ash g	Ca mg	P mg	Fe mg	References[1]
Red mombin (*Sponidas purpurea*)	50	83	76.2	0.9	0.1	22.0	0.4	0.8	22	40	0.6	1
		—	65.9–86.6	0.096–0.261[2]	0.03–0.17	—	0.2–0.6	0.47–1.13	6.1–23.9	31.5–55.7	0.09–1.22	2,3,4
Yellow mombin (*Spondias mombin*)		70	82.7	0.8	2.1	13.8	1.0	0.6	26	31	2.2	1
	48	21.8–48.1	72.8–88.5	1.28–1.38	0.1–0.56	8.7–10.0	1.16–1.18	0.65–0.66	31.4	—	2.8	2,3,4
						10.2[3]	1.13		1	40		5
												6
												7
												8
												8
												9
Ambarella (*Spondias dulcis*)	65	157.3	59.65–85.47	0.5–0.8	0.28–1.79	8.05–10.54[4]	0.85–3.60	0.44–0.65				2,3,4
Imbu			92.6	0.54			1.6	0.28	10.7	29.3	0.6	10

[1]References: 1) Leung and Flores (1961); 2) Morton (1987); 3) Sturrock (1959); 4) Winton and Winton (1935); 5) Barbosa *et al.* (1981); 6) Fonseca *et al.* (1969); 7) Keshinro (1985); 8) Schmandke and Guerra (1969); 9) Lopez *et al.* (1956); 10) Benk (1981).
[2]Amino acids per g of nitrogen: lysine 316 mg, methionine 178 mg, threonine 219 mg, tryptophan 57 mg.
[3]Degrees Brix.
[4]Sugar.

TABLE 4.1B. COMPOSITION OF *SPONDIAS* FRUIT PER 100 GRAMS OF EDIBLE PORTION

Species	Carotene (Vit. A) µg	Thia-mine mg	Ribo-flavin mg	Niacin mg	Ascorbic acid (Vit. C) mg	Total acids %	pH	Amino acids mg	Reducing Sugars %	Sucrose %	References[1]
Red mombin (*Spondias*	30	0.07	0.03	1.0	45						1
purpurea)	4-89	0.033–	0.014–	0.540–	26.4–				8.08	5.97	2,3,4
		0.103	0.049	1.770	73.0						
Yellow mombin	70	0.08	0.06	0.5	28						1
(*Spondias mombin*)	71	0.095	0.05	—	46.4				9.41	0.00	2,3,4
					11.06	1.65	2.1	0.026	6.74	1.03	5
					30						6
					166						7
					105[2]						8
					33[3]						8
					37.7						9
Ambarella (*Spondias dulcis*)					42	0.47–0.73			7.09	3.45	2,3,4
Imbu (*Spondias tuberosa*)					13.1						10

[1]References: 1) Leung and Flores (1961); 2) Morton (1987); 3) Sturrock (1959); 4) Winton and Winton 1935); 5) Barbosa *et al.* (1981); 6) Fonseca *et al.* (1969); 7) Keshinro (1985); 8) Schmandke and Guerra (1969); 9) Lopez *et al.* (1956); 10) Benk (1981).

[2]Refrigerated sample containing 53.4 mg ascorbic acid and 51.5 mg dehydroascorbic acid per 100g.

[3]Unrefrigerated sample containing 8.3 mg ascorbic acid and 24.7 mg dehydroascorbic acid per 100g.

YELLOW MOMBIN (*Spondias mombin* or *S. lutea*)

The yellow mombin (hogplum, jobo) grows wild all over tropical America and is seldom cultivated. Most of the trees yield fruit of low quality and with the exception of some islands of the Caribbean (Morton and Morton 1946), seldom appear in the markets. It has been introduced to the tropics of the Old World. The majority of the trees come from seedlings and reach up to 30 m in height. The trunk and main branches are covered with thick, corky plates. The leaves are odd-pinnate, just as in the other *Spondias*, with 6-17 leaflets; they are smooth rather than serrated as in the red mombin, and are 4-20 cm long. The flowers appear in terminal racemes, up to 60 cm long. The fruits, ellipsoidal to obovoidal, are 2.5-3 cm long, bright yellow, and contain a layer 3-4 mm thick of acidic yellow pulp surrounding a large, hard "seed" (endocarp) with longitudinal ridges. Inside the endocarp there are 3-5 viable seeds. Many wild animals are fond of the yellow mombin fruits, and spread them widely. Also, fruits of the red and yellow mombins are carried by the water currents and are found in large quantities on the beaches. Some trees in this species have fruits that contain relatively high contents of sugar, and could be clonaly propagated.

Composition

Composition of the edible portion of yellow mombin is listed in Tables 4.1A and 4.1B. Values by Leung and Flores (1961) are the result of 12-18 analyses. The ascorbic acid content was the most widely variable component reported, ranging from 8.3-166 mg per 100 g of edible portion. The study reported by Schmandke and Guerra (1969) showed increased retention of ascorbic acid and dehydroascorbic acid when the fruit was kept refrigerated compared to unrefrigerated fruit. The ascorbic acid and dehydroascorbic acid contents are combined to determine total Vitamin C content. Keshinro (1985) reported a relatively high ascorbic acid value for this and other unconventional tropical fruit. He noted that 94% of 210 adults surveyed did not like yellow mombin fruit, while only 5% of 275 children interviewed did not like this fruit. Generally, the children surveyed had a much higher acceptance of unconventional tropical fruit than did the adults.

The composition of the edible portion of the fruit (Tables 4.1A and 4.1B) and nectar prepared from the pulp were reported by Barbosa *et al.* (1981). They found the best nectar to contain 18% pulp, at 14° Brix and 0.30% titratable acid. No citric acid was added as long as the total acid in the pulp was 1.65% or greater. The nectar also contained 12.2%

sucrose and 67.7% water with a °Brix/acid ratio of 46.6 and a pH of 2.9. The amount of pulp determined by centrifugation was 12%.

Utilization

The technological possibilities for yellow mombin pulp were considered by Joas (1982). Juice can be diluted to one-fourth to one-sixth single strength, depending on varietal differences, to provide a lightly acidic beverage with a pleasant taste. An acceptable rum punch can be prepared from mombin as well. Sherbet prepared from mombin is much appreciated by Europeans. Other possible products from yellow mombin include brandy and preserves.

Antibacterial Properties

Yellow mombin is widespread in the rain forests of Nigeria and is also cultivated for its plum-like fruit. Its leaves have been used in herbal medicine in Nigeria because of their antibacterial properties. The leaves are ground in cold water to extract juice that is used to cleanse wounds. Ajao *et al.* (1984, 1985) studied the antibacterial properties of both aqueous and alcoholic extracts of the leaves and found them to inhibit the growth of both gram-positive and gram-negative bacteria. Thus, the extracts had a relatively broad spectrum of antibacterial activity.

AMBARELLA (*Spondias dulcis* or *S. cytherea*)

The ambarella (Otahite-apple, Jew plum, jocote judio) is cultivated mainly in Oceania, where spontaneous populations are found in several islands, including Samoa and Santa Cruz. From Polynesia the ambarella was introduced to tropical America, Florida and tropical Africa.

The tree reaches up to 25 m in height, with a main trunk and regular branching, which gives it a stately and symmetric appearance. The pinnate leaves, 15-60 cm long, formed by 11-23 leaflets, grow at the end of the branches. The tree sheds its leaves once a year. The leaflets, 5-10 cm long, have borders remotely serrated, and end in an acuminate apex.

The fruits are the largest of any *Spondias*, 5-9 cm long, obovoid, with five shallow longitudinal depressions. The yellow skin is thick and smooth; the pulp is firm, yellow, and juicy, with a particular smell like turpentine in some varieties. The central part of the fruit or "seed" (endocarp) is a large structure, with thin papery walls and long and slender spines that penetrate into the pulp; inside are five cavities, each with one seed.

As most trees come from seed, there is a wide diversity in ambarella. It may also be propagated from cuttings, or by veneer grafting or shield budding, either on seedlings of the same trees or in a close species, *S. pinnata* (Popenoe 1920).

Ambarella fruits are often sold while still green; when ripe, they are used as fresh fruit or to prepare preserves and jellies.

Compositon

Some information on the composition of ambarella was reported by Morton (1987) as listed in Tables 4.1A and 4.1B. It is reportedly a good source of iron, although the amount present in the edible portion was not reported (Morton 1987). The unripe fruit contains 9.76% pectin. The gum from *Spondias dulcis* was precipitated from water, hydrolyzed, and the hydrolysate shown to contain 19.8% galactose, 48.5% arabinose and 20% galacturonic acid (Basu 1980).

Flavor and Food Uses

Ambarella fruit possesses a pineapple-like fragrance and flavor when ripening and still firm. As the fruit softens, it develops a musky aroma and flavor. If the fruit is utilized while still firm the juice can be used in cold beverages. A product resembling apple sauce has been prepared by stewing the firm flesh with sugar and water, followed by straining the mixture. The sauce can be boiled down to a product resembling apple butter. Jelly, pickles or relish can be prepared from unripe fruit (Morton 1987). Both ambarella slices and puree have been dried in a dehydrator oven to produce products having excellent flavors, colors and textures (Campbell and Campbell 1983).

IMBU (*Spondias tuberosa*)

The imbu (umbu) is the least known of the useful *Spondias* and probably the best in fruit quality. It should be introduced to the dry tropics and the tropical areas with alternate seasons.

It grows naturally in northeastern Brazil, in the dry areas called catingas. The tree has special adaptation to drought in the root storage organs. The natives know how to locate these organs in the ground, dig them up and use the watery pulp as a source of food and water.

In its natural habitat the tree sometimes branches at 1.5-2 m above the ground and forms a broad, flat crown 2-8 m in diameter, but often the branching is irregular and the tree reaches up to 4 m in height. The leaves have 5-9 leaflets, which are oblong, serrated, and 2.5-3 cm long.

The fruit are ellipsoidal, 3.5-4 cm long, greenish yellow, smooth, with thick skin. The pulp is soft, yellowish, and sweet, with a flavor that has been compared to a sweet orange. It is rarely cultivated, but the wild fruit is collected in large quantities and often is found in markets outside the area of production. A special dessert is prepared by mixing the strained and sweetened pulp of slightly ripe fruits with milk. It is called imbuzada, and considered one of the best desserts. The fruit is also made into ice cream or jelly (Morton 1987).

The tree is propagated from seeds or woody cuttings. Attempted introduction in Florida has not been successful.

The small amount of data reported on the chemical composition of imbu is listed in Tables 4.1A and 4.1B. It has a relatively low content of ascorbic acid compared to ascorbic acid levels reported for other *Spondias* species.

OTHER CULTIVATED *SPONDIAS*

Three other species of *Spondias* are cultivated: the amra *S. pinnata (S. mangifera)*, wild and cultivated from India to Moluccas. It is very close in foliage characteristics to *S. dulcis*, but the "seed" (endocarp) is rather smooth, without the long and acute spines of the later. As a fruit it is of inferior quality. In some varieties the thin pulp "smells like rotten apples" (Burkhill 1935). Although the fruits in some cultivars are eaten fresh, the most common use is to prepare them in picklings.

Amra is a medicinally important plant in India, where the leaves and bark are used for dysentery and the acidic and astringent pulp is used in bilious dyspepsia (Hannan and Haq 1979). The polysaccharide from its gum exudate was analyzed by Ghosal and Thakur (1981) who found it to be composed of 47% L-galactose, 23% L-arabinose and 30% D-galacturonic acid.

Two other species are said to be cultivated: *S. borbonica* in Mauritius and Reunion, and *S. lakonensis*, in Vietnam.

REFERENCES

AJAO, A. O., SHONUKAN, O. and FEMI-ONADEKO, B. 1984. Antibacterial effect of aqueous and alcoholic extracts of *Spondias mombin* and *Alchornea cordifolia*. Fitoterapia 55, 337-339.

AJAO, A. O., SHONUKAN, O. and FEMI-ONADEKO, B. 1985. Antibacterial effect of aqueous and alcohol extracts of *Spondias mombin* and *Alchornea cordifolia*—two local antimicrobial remedies. Int. J. Crude Drug Res. 23, 67-72.

ALCORN, J. B. 1984. Huastec Mayan Ethnobotany. University of Texas Press, Austin.

BARBOSA, W. C., DE NAZARE, R. F. R. and HASHIMOTO, K. 1981. Nutritional and technological study of the fruits of *Annona muricata* and *Spondias lutea*. Bol. Pesqui. 32, 1-11. (Spanish)

BASU, S. 1980. Some structural studies on degraded *Spondias dulcis* gum. Carbohydr. Res. 81, 200-201.

BENK, E. 1981. Further contribution on exotic fruits. Reichstoffe, Aromen, Kosmet. 31, 234-235. (German)

BURKHILL, I. H. 1935. A Dictionary of the Economic Products of the Malay Peninsula. Crown Publishers, London.

CAMPBELL, B. A. and CAMPBELL, C. W. 1983. Preservation of tropical fruits by drying. Proc. Fla. State Hortic. Soc. 96, 229-231.

FONSECA, H., NOGUEIRA, J. N. and MARCONDES, A. M. S. 1969. Ascorbic acid and beta-carotene contents of Brazilian fruits and vegetables. Arch. Latinoamericanos Nutr. 19, 9-16. (Spanish)

GHOSAL, P. K. and THAKUR, S. 1981. Structural features of the acidic polysaccharide of *Spondias pinnata* gum exudate. Carbohydr. Res. 98, 75-83.

HANNAN, A. and HAQ, Q. N. 1979. Studies on water soluble polysaccharide from the fruits of *Spondias mangifera* Willd. Bangladesh J. Sci. Ind. Res. 14, 246-250.

JOAS, J. 1982. Mombins: some interesting technological possibilities. Fruits 37, 727-729.

JULIANO, J. B. 1932. The cause of sterility of *Spondias purpurea* Linn. Philipp. Agric. 21, 15-24.

KESHINRO, O. C. 1985. The unconventional sources of ascorbic acid in the tropics. Nutr. Rep. Int. 31, 381-387.

LEUNG, W-T. W., and FLORES, M. 1961. Food Composition Table for use in Latin America. Institute of Nutrition of Central America and Panama, and National Institutes of Health. U.S. Gov't. Printing Office, Washington, D.C.

LOPEZ, H., CIMADEVILLA, M., FERNANDEZ, E., NAVIA, J. H., VALIENTE, A., CLEMENT, I. D. and HARRIS, R. S. 1956. Provisional table of nutritive composition of Cuban foods. Bol. Col. Med La Habana 7(10) 334. (Spanish)

McBRYDE, F. W. 1945. Cultural and Historical Geography of Southwest Guatemala. Smithsonian Institution, Washington, D.C.

MORTON, J. F. 1987. Fruits of Warm Climates. J. L. Morton Miami, FL.

MORTON, K. and MORTON, J. 1946. Fifty Tropical Fruits of Nassau. Text House, (Fla.) Inc. Coral Gables, FL.

NAVA-KURI, G. and USCANGA, M. 1979. Physical and chemical study of twelve types of plum (*Spondias* sp) in the state of Veracruz. Proc. Trop. Reg. Am. Soc. Hortic. Sci. 23, 132-136.

POPENOE, J. 1979. The genus *Spondias* in Florida. Proc. Fla. State Hortic. Soc. 92, 277-279.

POPENOE, W. 1920. Manual of Tropical and Subtropical Fruits. Macmillan, New York.

ROSE, J. N. 1899. Notes on useful plants of Mexico. Contrib. U.S. Nat. Herb. 5:209-259.

SCHMANDKE. H. and GUERRA O. O. 1969. On the carotene, L-ascorbic acid, dehydroascorbic acid and tocopherol content of Cuban vegetable products. Nahrung 13, 523-530.

SOUZA NOVELO, N. 1949. The plums of Yucatan. Bol. Soc. Bot. Mex. 9, 5-12. (Spanish)

STURROCK, D. 1959. Fruits for Southern Florida. Southeastern Printing Co. Stuart, FL.

WHILEY, A. W. 1984. Ambarella, hog plum, imbu, red mombin. *In* Tropical Tree Fruits for Australia. P. E. Page (Editor). Queensland Department of Primary Industries Information Series Q183018, Brisbane, Australia.

WINTON, A. L. and WINTON, K. B. 1935. The Structure and Composition of Foods. Vol. II. Vegetables, Legumes and Fruits. John Wiley and Sons, New York.

5

Biriba

Luis Carlos Donadio
Jose Fernando Durigan

The birbia is a fruit native to the Antilles. It belongs to the genus *Rollinia* of the family Annonaceae, which includes several species of trees and shrubs. The most well-known species is biriba, *Rollinia mucosa* (Jacq.) Baill. (Cavalcante 1972; Leon 1983). Among the *Rollinia*, the biriba is the most common and produces the largest fruit. In the Amazon area, the biriba is cultivated in home orchards where it is a common fruit, adapted to various soil types. The distribution of this species includes São Paulo state in the southeast where it is found in home orchards, however, it is not grown commercially. The only named selection is the Regnard, which was introduced in the Philippines (Morton 1987).

CHARACTERISTICS

The biriba tree is a medium size evergreen plant some 5 to 12 m high. The tree has a spreading canopy, but some branches grow upright. The leaves are alternate, the limb angled on the dorsal side, and the leaf size varies from 10 to 15 cm in width, with a pointed apex. The ventral side is pale green with prominent veins and the dorsal side is green. The hermaphroditic flowers are usually isolated, but occur rarely in pairs, with three sepals and 6 petals. They are light green in color and possess a characteristic smell. The flower peduncle is about 2 cm long; the calyx contains 3 parts 0.5 to 0.7 cm long, with aggregated sepals; the corolla has three glossy petals which are laterally compressed into a small annular tube, alternating with three other internal, rudimentary and triangular shaped tubes. Numerous stamens 1 mm high are present, with extrorse anthers and peltate connective under the anthers; the ovary contains unilocular, uniovular parts that adhere to one another by a viscous liquid; the style is capitate and voluminous. The fruit is an ovoid to globose syncarp, formed by the ovaries jointed at maturation. The skin is yellow, with fleshy and scaly points, and generally black at the apex. The pulp is abundant, juicy, with sweet flavor, but sometimes gelatinous in consistency, with numerous dark gray seeds. The fruit is similar to *Annona muricata* in flavor, but is milder and sweeter (Swift and Prentice 1983). Fruit weight varies greatly from 300 to 1300 g and its consumption is normally fresh. The biriba is propagated normally by seeds and only rarely by budding and grafting, thus affording a large variation in fruit shape, size and quality. In one study, fruit set was approximately 32% of flower production (Falcao *et al.* 1981).

Cultivation of biriba is limited by the lack of selected cultivars and the small capacity of fruit storage. Pests and diseases are of no great importance to the culture of biriba. The biriba was tested as a rootstock for another Annonaceae, but in spite of its potential for that purpose, it is incompatible with some *Annona* as a scion. Since the tree is resistant to nematodes, it has some potential as a rootstock (Swift and Prentice 1983).

MATURATION AND HARVESTING

Fruits ripen in February and March in Brazil, whereas they mature in November and December in Florida. However, it is not uncommon to obtain fruits the year-around. Ripening and maturation of biriba fruit have not been studied, but it is well known that the fruit is quite perishable. The fruit can be picked when it reaches full size and begins to change color from green to yellow. Fruit that is green and hard stores better than fully ripe fruit especially when refrigerated for 2-3 days. For fresh consumption, it is preferable to pick yellow, fully ripe fruit, but at this stage, the fruit becomes soft, burns brown and is easily bruised.

USES

Biriba is commonly consumed as dessert fruit by eating the pulp directly with a spoon, after cutting the fruit in half. The fruit is also used to make refreshing beverages and sherberts (Fouque 1972). Other possible ways to use the fruit would be with sugar, lemon juice, or milk, but those are not common. The fruits and leaves are reported to have medicinal uses, as for example, in blood coagulation.

COMPOSITION

The biriba fruit contains an average of 70.5% edible pulp, 10.8% seeds, and 18.7% peel. The proximate composition of ripe biriba fruit is presented in Table 5.1. The contents of moisture, proteins, lipids, and ash were determined by standard methods (AOAC 1970); total and soluble carbohydrates and starches according to Dubois *et al.* (1956); and reducing sugars according to Nelson (1944).

The composition of biriba pulp is characterized by high levels of water and carbohydrates relative to proteins, lipids and ash. Biriba composition is typical of other Annonaceae species (Bueso 1980; Endef/Fibge 1977). Among the carbohydrates, the amount of soluble sugars may be twice as high as the reducing sugars. The average protein content is similar to that of soursop as cited by Bueso (1980). For lipids, biriba manifests higher amounts when compared to soursop. The ash in the pulp of biriba (3.75 g/100 g pulp) is higher than in soursop (0.41 to 0.86%) (Bueso 1980).

The minerals most commonly found in biriba fruit are potassium, calcium, phosphorus, sulfur and magnesium (Table 5.2). Compared to data presented by Bueso (1980) for soursop (in mg per 100 g of pulp), the potassium, phosphorus, and calcium contents are present at higher levels in biriba. For minor elements, biriba contains small quantities of iron, magnesium, zinc, and copper. The methods used to determine the mineral contents were the nitroperchloric acid digestion, and this followed by spectrophotometric absorption and atomic absorption (Sarruge and

TABLE 5.1. PROXIMATE COMPOSITION OF BIRIBA PULP [1]

Pulp composition	Donadio & Durigan	Morton
Calories	—	80
Moisture	67.05	77.2
Protein (total N x 6.25)	1.01	2.8
Amino Acids (mg per g of nitrogen)		
Lysine	—	316
Methionine	—	178
Threonine	—	219
Tryptothan	—	57
Lipids (ether extractable	2.38	0.2
Ash	3.75	0.7
Fiber	—	1.3
Carbohydrates		
Total	25.81	—
Starch	3.68	—
Soluble sugars	9.38	—
Reducing sugars	3.82	—
Vitamin B_1 (mg/100g)	—	0.04
Vitamin B_2 (mg/100 g)	—	0.04
Niacin (mg/100 g)	—	0.5
Vitamin C (mg/100 g)	—	33.0

[1]g/100 g of pulp (except for caloric value, amino acids, and vitamins).

TABLE 5.2. MINERAL COMPOSITION OF RIPE BIRIBA PULP[1]

Minerals	Donadio & Durigan	Morton
Calcium	250	24
Phosphorus	40	26
Potassium	310	—
Magnesium	20	—
Sulfur	40	—
Copper	0.2	—
Iron	15	1.2
Zinc	1.5	—
Magnesium	5.8	—

[1]mg/100 g of pulp.

Haag 1979; Miranda and Battisti 1981). According to the data obtained, the biriba can be considered as a fruit rich in minerals, mainly because of its ash content of 3.75% and its calcium, phosphorus and iron contents (250, 40 and 15 mg/100 g of pulp, respectively). Morton (1987), however, reported considerably lower levels of calcium, iron and ash contents.

REFERENCES

AOAC. 1970. Official Methods of Analysis of The Association of Official Analytical Chemists. 11 ed. AOAC, Washington, DC.

BUESO, C. E. 1980. Soursop, Tamarind and Chironja. *In* Tropical and Subtropical Fruits. S. Nagy and P. E. Shaw (Editors). AVI Publishing Co., Westport, CT.

CAVALCANTE, P. B. 1972. Edible fruits of the Amazon. I. Publicacões Avulsas no 17. Belém, Pará, Brasil. (Portuguese)

DUBOIS, M., GILLES, K. A., HAMILTON, J. K., REBERS, S. A. and SMITH, F. 1956. Colorimetric methods for determination of sugar and related substances. Anal. Chem. 28, 350-356.

ENDEF/FIBGE. 1977. Tables of food composition. Rio de Janeiro, Fund. Inst. Bras. Geografia e Estatistica. (Portuguese)

FALCAO, M. de A., LLERAS, E., KERR, W. E. and CARREIRA, L. M. M. 1981. Phenological, ecological and productive aspects of *Rollinia mucosa* (Jacq.) Baill. Acta Amazonica 11, 297-306.

FOUQUE, A. 1972. Fruit species of tropical American. Fruits 27(1) 62-72.

LEON, J. 1983. Diversity in native fruits of the Caribbean region. Proc. Am. Soc. Hortic. Sci., Trop. Reg. 27(A) 1-20.

MIRANDA, L. C. G. and BATTISTI, C. R. 1981. Use of bromate chemistry. Vicosa, UFV. Bull. No. 98. (Portuguese)

MORTON, J. F. 1987. Fruits of Warm Climates. Julia Morton, Publisher, Miami, FL.

NELSON, N. 1944. Spectrophotometric adaption of the Somogy method for the determination of glucose. J. Biol. Chem. 153, 375-380.

SARRUGE, J. R. and HAAG, H. P. 1979. Chemical analyses on plants. Piraciaba, ESALQ/USP. (Portuguese)

SWIFT, J. F. and PRENTICE, W. E. 1983. Native fruit species of the Ecuadorian Amazon: production techniques and processing requirements. Proc. Am. Soc. Hortic. Sci., Trop. Reg. 27(A) 95-100.

6

Cherimoya

L. Antonio Lizana
Gabino Reginato

The cherimoya or chirimoya, sometimes referred to as custard apple, is one of the most delicately flavored fruits when properly ripened. The flesh has a pleasant blend of sweetness and mild acidity with a consistency of baked custard. This latter characteristic is the reason why cherimoya fruit and other species are sometimes called custard apples. Cherimoya is primarily consumed as fresh fruit. However, it is often added as pulp to ice cream, sherbet, milk shake, yoghurt, cookies and pastries. Cherimoya makes an excellent drink; it is often added diced to white wine punches. Colombians strain cherimoya juice, add a slice of lemon and dilute with ice water to make a refreshing soft drink (Morton 1987).

The cherimoya (*Annona cherimola* Mill.) and annonas in general are native to tropical America but are now widespread in the tropics and frost-free subtropics of other continents. Cherimoya is a subtropical fruiting plant that grows best in frost-free areas. Although it has been mentioned that cherimoya originated on the Andean slopes of Peru and Ecuador (Popenoe 1934), it has spread to other American countries. Many different types of cherimoya trees and fruits are found in Mexico. From its original growing area, it was spread early to Mexico and Central America; later into Chile, Argentina and the rest of South America (de la Rocha 1967). In Chile, cherimoya favorably adapted to the climate conditions in the Quillota and La Cruz areas of the Aconcagua river valley. Most of the newer plantations are located in La Serena (CORFO 1984). In North America, very good conditions to grow cherimoya are found in the Santa Barbara area of California where commercial orchards have been established. Other places where cherimoya trees are found are the Canary Islands, Sri Lanka, India, Algiers, Egypt, Libya, Eritrea and Somalia (Bailey 1963; Morton 1987).

Cherimoya is definitely a subtropical tree; it grows and fruits very poorly under tropical conditions and does not succeed in lowland tropics but may do well in tropical highlands (Samson 1986). Wild cherimoya trees are grown extensively in the Loja Valley, south of Ecuador and form natural forests (Popenoe 1934). Commercially cultivated cherimoya trees are mainly found in Chile, Peru, Spain, Israel and Italy. Cherimoya trees are mainly propagated from seeds and then budded or grafted; they also can be propagated by cuttings (Duarte *et al.* 1974).

Cherimoya trees are usually grown with some help to support the weight of the limb when fruit are of full size. This necessitates very specific types of trellises for commercial orchards. In some areas, some type of self-support training system has also been developed.

BOTANY

The cherimoya belongs to the family the *Annonaceae*, genus *Annona* and the species *cherimola* Mill. (Bailey 1963). *Annona* derives from the latin "annual harvest" and cherimoya from quechua language "shirimuya" that means cold seeds. Species closely related to cherimoya include: *A. squamosa* (sugar apple, anona blanca), *A. muricata* (guanabana, soursop) and *A. reticulata* (bullock's heart, anona colorada or "red anona"). The fruits of *A. reticulata* are heart-shaped (anona corazon), brown in color when approaching maturity and may turn red on one of its sides. Fruit pulp of bullock's heart is white, sweet but insipid, and is generally rated as mediocre when compared to other annonas.

Two wild annona relatives are *A. glabra* (fruit similar to *A. reticulata* in shape and orange color) and *A. montana*. Both of these wild relatives have been recommended as rootstocks for other annonas. *A. diversifolia* (ilama, anona blanca or papauce) is similar to cherimoya and grows mainly from the southwest coast of Mexico to the Pacific coast of Guatemala and El Salvador (Popenoe 1934; Garcia 1956; de la Rocha 1967; Morton 1987). Some important commercial hybrids have been derived by crossing different annona species, for example, atemoya (*A. squamosa* x *A. cherimolia*). For many years, atemoya was mistakenly called custard apple or cherimoya in Australia (Morton 1987).

A very unusual characteristic of the cherimoya tree is being semideciduous. It loses its leaves in late spring when new buds are about to emerge. The trees tend to look bare in late spring or early summer when other trees may be in their lush spring growth. The tree grows vigorously when young but tends to decline in growth early and never becomes very large (heights of 5-9 m). The fruit is botanically a syncarpium or multiple ovary formed by numerous carpels fused on a fleshy floral tube. If all flowers are pollinated, it takes full form and is usually heart-shaped to round; although a conical shape is also usually found.

FRUIT TYPES

Cherimoya fruits are similar in most general characteristics, but present very interesting morphological differences. Early classification systems to separate cherimoya were established by Safford in Standard Cyclopedia of Horticulture (Bailey 1963). Cherimoya is classed according to degree of surface irregularity into five forms:

Fingerprint depressions (botanically known as forma impressa). In Costa Rica, it is called "anona de dedos pintados". The fruit is conoid or subglobose and has a smooth surface covered with U-shaped aeroles resembling fingerprints in wax. Many seedlings of this type are of good quality and contain few seeds.

Smooth (forma laevis). One of the finest types of cherimoyas. It is called "cherimoya lisa" in South America and "anon" in Mexico. This smooth form is often mistaken for *A. glabra* and *A. reticulata* because of the general fruit appearance and the name "anon" which is also applied to *A. reticulata.*

Tuberculate (forma tuberculata). One of the most common forms. The fruit is heart-shaped and has wart-like tubercles near the apex of each aerole. The Golden Russet variety of California belongs to this group.

Mammillate (forma mamillata). In South America it is called "chirimoya de tetillas". This form is common in the Nilgiri hills of southern India and is one of the best forms grown in Madeira.

Umbonate (forma umbonata). It is call "chirimoya de puas" and "anona picuda" in Latin America. The skin is thick, the pulp is more acid than other forms, and seeds are more numerous. The fruit is oblong-conical, with the base somewhat umbilicate. The surface is studded with protuberances and each corresponds to a component carpel.

HORTICULTURE

The cherimoya is a protergynous plant. Pollination is one of the most important factors in the production of quality fruits with proper size and shape. Cherimoya fruits are comprised of many carpels. When the tree is in the flowering stage, all carpels must be pollinated to produce full-shaped fruits. A problem with cherimoya is that the flowers are not attractive to insects. Additionally, the flowers are unable to self-pollinate (McGregor 1976) because the male and female structures do not mature simultaneously. When the petals open the pistil is receptive; however, the stamens are not mature until the following day when the pistil is no longer functional. Physiological maturation of the pistil indicates that it is at least 72 hr ahead of the liberation of pollen (Schwazenberg 1947). The drying of pistilar secretion occurs 12 hr after pollen has germinated. The dramatic difference due to method of pollination is noted in Table 6.1.

The possibilities for a successful commercial production of cherimoyas will strongly depend upon fruit management as related to pollination. Normally, flowers are hand-pollinated; the results could mean at least a doubling in fruit yield (Schwazenberg 1947; Saavedra 1979; Pavez 1985; Sainte-Marie 1987). Since the flowering period is long (November to April), artificial pollination could impact directly on harvest time. Usually only a few flowers are pollinated at a time and the process repeated after five days. Therefore, harvest time will have similar lag periods and the season of ripening is extended.

Coincidence in pollination time may occur when flowers are coming into bloom and the time when the tree drops its leaves. During that period, artificial pollination may not be successful. The very fast dehydration of the stigma may be considered as the primary problem because the leaves are not present to protect flowers from direct sunlight exposure.

The first fruits are noticed in December. Fruit growth follows a double sigmoid pattern similar to stone fruits, i.e., the first phase of growth is about 100 days after bloom (Saavedra 1979; Pavez 1985). In Chile, maturation to harvest takes about 9 months. Depending on the cultivar,

TABLE 6.1. DIFFERENCES IN FRUIT SET, WEIGHT, AND NUMBER OF FRUITS IN THREE CULTIVARS OF CHERIMOYA WHEN FLOWERS WERE NATURALLY OR ARTIFICALLY POLLINATED

	Bronceada		Concha		Concha Lisa	
Pollination	A	B	A	B	A	B
Fruit Set (%)	6.28	81.15	5.22	78.25	4.22	73.07
Fruit weight (kg)	0.24	0.54	0.18	0.60	0.14	0.71
Perfect fruit (%)	42.22	66.40	39.40	71.60	32.90	72.30

Source: Schwazenberg (1947).
A = Natural; B = Artificial

the harvest period may extend from August to December (Saavedra 1979). This period could be modified if trees are grown in warmer interior valleys. Nevertheless, excess heat may cause dehydration and death to terminal twigs.

The best fruits occur when the final period of growth, prior to harvest, is in a warmer month rather than a colder month. When harvested in a warmer month, fruit tend to accumulate slightly more sugar and develop a stronger typical flavor.

Some efforts have been made to produce seedless fruit of cherimoya and other *Annona* species. Use of gibberellic acid (GA) and indole acetic acid (IAA) have been erratic with cherimoya fruit, but not with atemoya fruit (*A.squamosa* x *A. cherimolia*). The atemoya with fewer seeds than cherimoya attains perfect shape and full size, and contains about 30% more pulp in the spaces that are usually occupied by seeds. Commercial use of growth regulators has not been actively pursued (Saavedra 1979; Campbell 1979).

FRUIT MATURITY

The cherimoya is a typical climacteric fruit. During ripening, changes occur rapidly and are metabolically traceable by monitoring carbon dioxide evolution (Fig. 6.1 and 6.2) (Kosiyachinda and Young 1975; Reginato 1980). Unlike most other climacteric fruits, the respiration of cherimoya shows two peaks of carbon dioxide production (Kosiyachinda and Young 1975; Brown *et al.* 1988). The first rise in carbon dioxide (initiation of ripening) occurs before any ethylene change is detected. Less than 13 ppb was detected 24 hr after the climacteric rise began. In some other fruits, it has been shown that the rise in ethylene concentration triggers the climacteric rise and, consequently, the ripening of the fruit. In cherimoya, however, some other regulator may be involved in the initiation of the climacteric stage because the amount of ethylene detected by Kosiyachinda and Young (1975) and Brown *et al.* (1988) is far below the physiological active level (0.1 - 3 ppm).

Once fruit is detached from the tree, ethylene is present in sufficient concentration (10 ppm or more) to induce the climacteric stage. Ethylene can also be used commercially to regulate ripening after harvest. At the detached stage and in the presence of ethylene, the fruit becomes very susceptible to postharvest handling and storage temperatures. It ripens quickly and may deteriorate. The use of ethylene scrubbers, for example, potassium permanganate (Purafil) or different types of activated carbon compounds (StaFresh), have shown some success in extending the postharvest life of cherimoya.

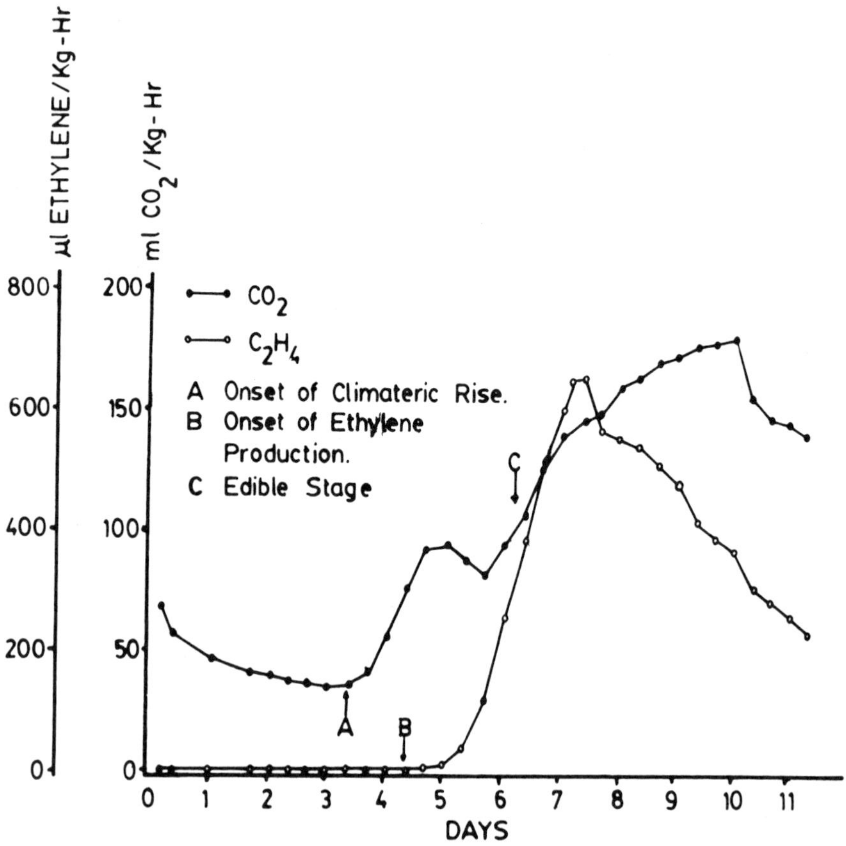

From Kosiyachinda and Young (1975).

FIG. 6.1. RESPIRATION AND EXTERNAL ETHYLENE PRODUCTION OF CHAFFEY CHERIMOYA FRUIT.

An increase in respiratory activity is accompanied by an increase in soluble solids and in the volatiles that impart the pleasant aroma, and a very rapid decrease in pulp firmness (Reginato 1980; Brown *et al.* 1988). Acidity increases up to the 7th day and thereafter decreases (Fig. 6.2). As expected, the pH shows the opposite trend, i.e., it decreases to the 7th day and then increases (Fig. 6.2; Reginato 1980). Taste panels show a very high preference for the fruit after the first climacteric peak; this taste preference coincides with the highest soluble solids content and the highest detectable titratable acidity (Fig. 6.2).

Changes occur in the color and texture of the flesh and skin. The skin changes from yellowish green to light greenish yellow by an apparent partial degradation of chlorophyll. Flesh changes from white to cream in color and the texture from firm to melting (Reginato 1980).

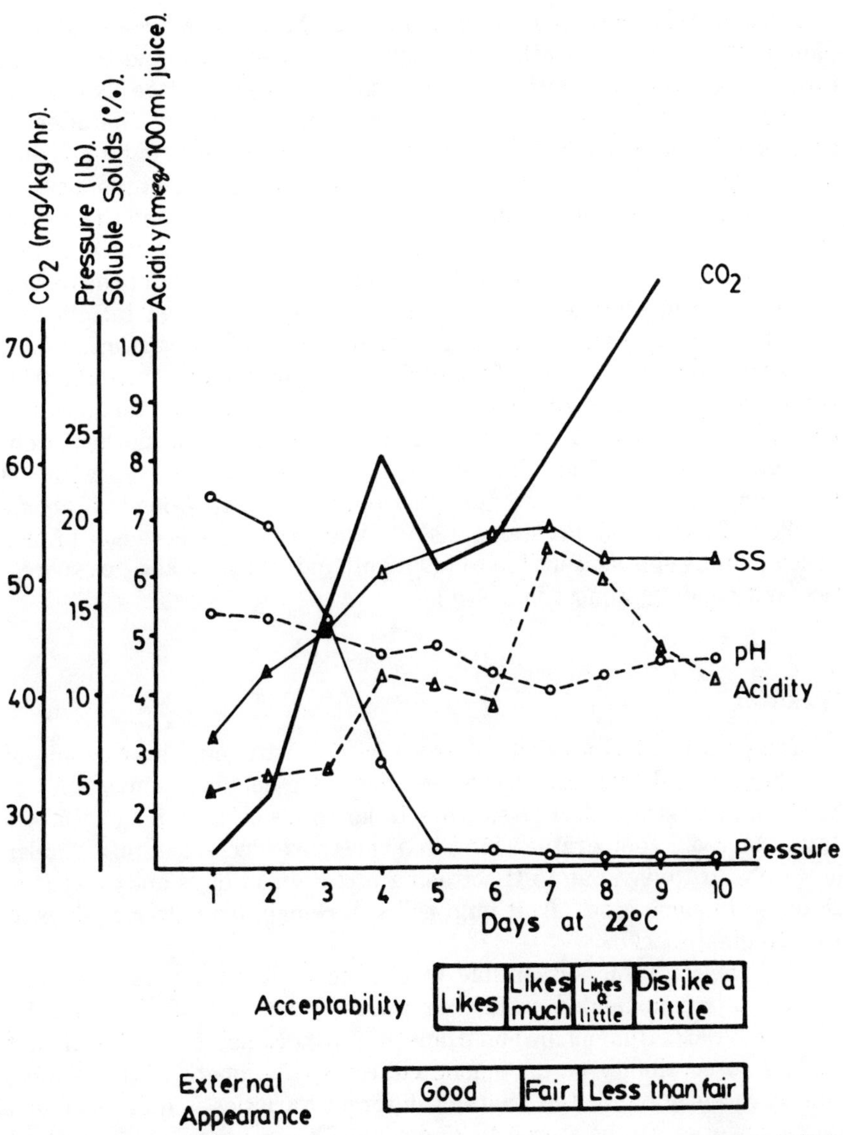

From Reginato (1980).

FIG. 6.2. RESPIRATION AND CHANGES OCCURRING DURING THE RIPENING OF CHERIMOYA FRUITS AT 22°C.

HARVESTING AND HANDLING

Fruits that reach maturity on the tree turn from dark green to light yellow and are readily distinguished. Cherimoyas do not soften on the tree, and at the time of harvest, the fruits are hard (14-18 kg with a 8 mm plunger). The very smooth skin is subject to blemishes due to surface tubercules associated with different cultivars. Any surface damage at harvest time will quickly show as a dark discoloration at the affected area. Additionally, surface damage will allow the entrance of fungus during storage and this is especially true in years of very high humidity. To avoid superficial damage, fruits must be handled carefully and the peduncle clipped as close as possible to prevent damage during handling and transporting (de la Rocha 1967). This will also avoid wounds at the peduncle which during storage could be an entrance point for fungal infection.

If fruit is harvested with a light yellow color, it can be stored under refrigeration up to 35 days. If maintained at room temperature, the fruit will be ready for eating in four or more days depending on temperature and picking maturity (Reginato and Lizana 1980A,B). During the softening period and until the fruit reaches full ripeness (0.7-1.1 kg with a 8 mm plunger), fruits become more susceptible to deep wounds if poorly handled (Fuster and Prestamo 1980). When the skin becomes bronze yellow, the peel is susceptible to falling off and, thus, makes transportation and handling almost impossible.

STORAGE

Temperature has a great influence on the softening and ripening of the cherimoya during its postharvest life. At a contant temperature of 20°C, pulp pressure decreases from 13 kg to less than 0.9 kg in 5 to 6 days. At room temperatue (15°-20°C), pressure decreases to a similar level in 8 to 10 days; at 15°C softening occurs in 14 days and at 11°C in 28 days. In some cases, fruit pulp will still remain firm after 35 days at 7°C (Reginato 1980).

The cherimoya is susceptible to chilling at low temperatures, and storage is recommended at 11°C or above (Everett 1952; Garcia 1956; Chandler 1962; Bunemann and Hansen 1973). It has been shown that chilling susceptibility varies among cultivars, the length of the storage among recently harvested fruits of different varieties is quite striking. Some varieties can be stored to 35 days at 7°C or 7 days at 0°C with no visible detrimental effect (e.g., the concha lisa variety) (Reginato 1980; Peralta 1984). Resistance to low temperature storage among recently harvested fruits of different varieties is quite striking. Some varieties can be stored to 35 days at 7°C or 7 days at 0°C with no visible detrimental effect (e.g., the concha lisa variety) (Reginato 1980; Lizana and Irarrazabal 1984), whereas others will last only 10 days at 7°C or 21 days at

11°C. At the latter temperatures, the most chill-sensitive varieties can be stored without any noticeable disorders (Reginato 1980).

When fruit is exposed to room temperature after storage (the marketing period), the speed of softening will depend on the time and the temperature of storage. The variety, concha lisa, when stored at 7°C for up to 35 days requires at least 6 days to soften at room temperature. When fruit is stored at 11°C, it is more dependent upon length of storage. After 7 days storage, it will need 6 days at room temperature; after 14 days storage, it will need only 3 days to ripen but after 21 days, it will be overripe and with rot problems after 3 days. A similar effect occurs with fruit stored at 15°C for 14 days (Reginato and Lizana 1980A,B).

Fruit stored when fully or partially ripe will tolerate lower storage temperatures than less ripe fruit that has just been harvested. Therefore, less mature fruits that only tolerate storage at 11°C can in a more advanced stage of ripeness be stored at 4°C. Likewise, fruit that normally are stored at 7°C can in a higher stage of ripeness be stored at 0°C (Reginato 1980; Peralta 1984). Fruit that is stored will partially lose its characteristic taste, however, it will never be so severe as to lose its overall acceptability (Reginato 1980).

The USDA (McGregor 1987) recommends that cherimoya fruit be precooled by forced air in a cool room. Storage is recommended at 13°C with a relative humidity of 90-95%. The approximate storage and transit life is 2 - 4 weeks. Fruit should be packaged in a single-layer pack in fiberboard cartons with paper wrapping, foam sleeves or excelsior to reduce bruising.

Dehydration

The cherimoya fruit is highly susceptible to postharvest dehydration which can reduce its weight by up to 18%. Dehydration can be reduced by waxing and/or cold storage (Figure 6.3) (Reginato and Lizana 1980A,B; Lizana and Irarrazabal 1984). The high level of dehydration exhibited by this fruit during ripening and storage affect its appearance; the fruit becomes unattractive as superficial browning occurs. However, flesh characteristics of the ripe fruit are not affected, as well as the texture, flavor and aroma (Reginato and Lizana 1980A,B).

Postharvest Disorders

During cold storage, fruit is susceptible to certain disorders which have been described (Reginato and Lizana 1980A,B). The origins of these disorders are varied; some are associated with storage temperatures, for example, skin mottling, subepidermal browning and flesh mottling. The

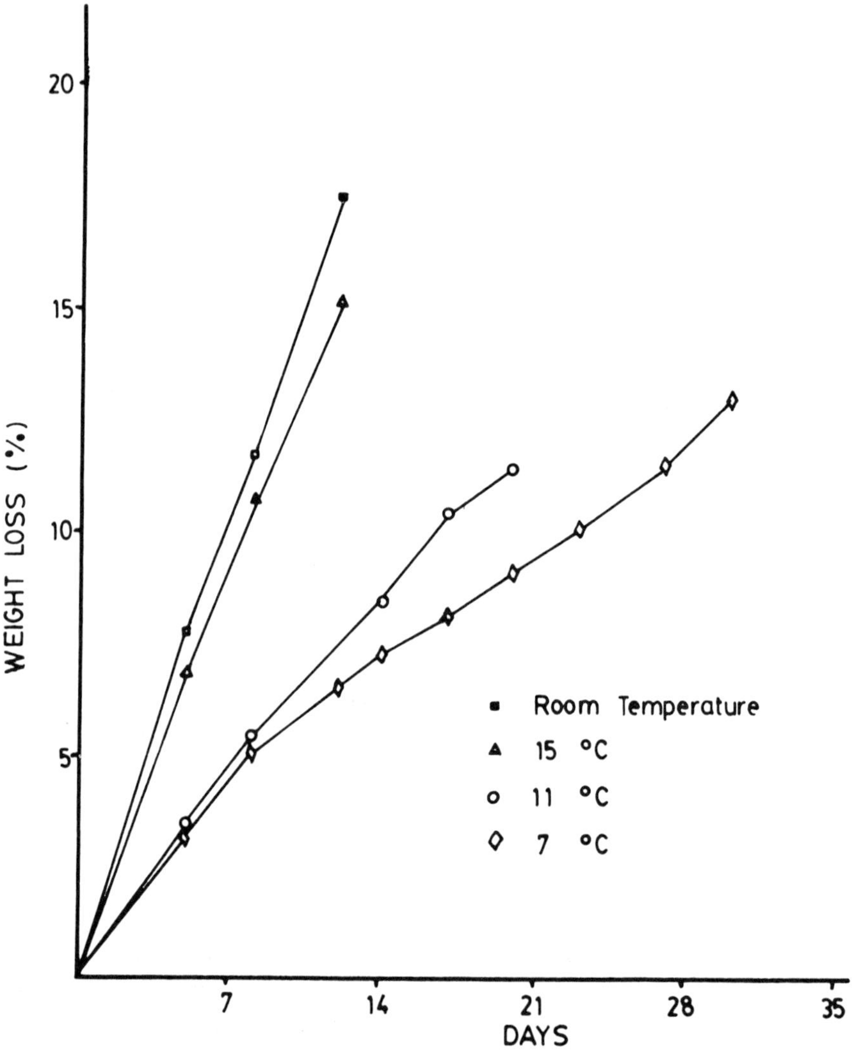

From Reginato (1980).

FIG. 6.3. CHERIMOYA FRUIT DEHYDRATION AT DIFFERENT STORAGE TEMPERATURES.

latter disorder is noted in cultivars susceptible to low storage temperatures and associated with suspension of the ripening process (Reginato 1980). Other disorders, such as epidermal browning, originate from bruises acquired during harvesting and transporting. The damage intensifies during the cold storage period. Histological characterization of some epidermal disorders reveals more brachyesclereids, tannins and starch in damaged tissue than in healthy tissue (Undurraga *et al.* 1988).

Overripening and dehydration of fruit induce the occurrence of other disorders. In overripened and dehydrated fruit, a higher incidence of colored hypodermis, skin russeting, senescent browning and water spot of the flesh are noted. Reginato and Lizana (1980A,B) noted that water spot of the flesh also increases with low storage temperature.

CHEMICAL COMPOSITION

The composition of cherimoya fruit has been extensively studied; however, identification of cultivar type has not always been stated. Additionally, confusion among the many annona species may have inadvertently caused a mixing of data. The chemical composition of the "custard apple" is extensive, but this common name has been applied to *A. squamosa* (sugar apple), *A. reticulata* (bullock's heart), *A. cherimola* (cherimoya) and the hybrid, *A. squamosa* x *A. reticulata* (atemoya).

Table 6.2 lists the chemical composition of authenticated cherimoya fruit from various countries. The edible portion (without skin and seeds) is about 60% (Stahl 1935; Vidal-Valverde *et al.* 1982). Differences in composition may be due to different cultivars and/or to different cultural and climatic conditions.

Benk (1985) recently summarized the composition of the edible part of cherimoya fruit but did not list geographical growing areas. The composition reported by Benk (1985) is similar to values noted in Table 6.2 and includes the following additional data: water (69 - 80%), protein (1.5 - 2.1%), total sugars (13.6%), invert sugars (8.9%), sucrose (4.5%), total acidity (0.39 - 0.77%), crude fiber (1.6 - 2.5%), ash (0.7 - 0.9%), calcium (9 - 15 mg%), phosphate (73 - 122 mg%) and vitamin C (12 - 25 mg%). Benk (1985) reported phosphorus values in phosphate units; thus, the reason for the higher values when contrasted to Table 6.2 for elemental phosphorus.

Carbohydrates

Carbohydrates are the second most abundant components in cherimoya fruit. The main sugars are fructose, glucose and sucrose. Glucose (11.75%) and sucrose (9.4%) were the most prevalent sugars determined by Tamaro (1974) and Kawamata (1977). These values are similar to those reported by Corigliano *et al.* (1968) for sucrose (7.0%) and reducing sugars (11.9%), and by Benk (1985) for sucrose (4.5%) and reducing sugars (8.9%).

Dietary fiber

The dietary fiber content of cherimoya is considered the sum of values obtained for cellulose, hemicellulose, lignin and pectic substances. Vidal-

TABLE 6.2. FOOD COMPONENTS IN 100 G OF THE EDIBLE PART OF CHERIMOYA FRUITS

Component		Peru[1]	Chile[2]	Columbia[3]	Ecuador[4]	Italy[5]	Central America[6]
Calories		81	56	73	—	101	82
Water	(g)	75.7	83.3	77.1	74.6	76.4	76.6
Proteins	(g)	1.0	2.9[7]	1.9	1.4[7]	1.5	1.1
Fat	(g)	0.1	0.5	0.1	0.5	0.4	0.2
Carbohydrates	(g)	22.0	11.7[8]	18.2	—	18.9	21.3
Fiber	(g)	1.8	1.0	2.0	1.5	—	1.9
Ash	(g)	1.0	0.6	0.7	0.6	0.6	0.8
Calcium	(mg)	24.0	24.0	32.0	21.7	—	34
Phosphorus	(mg)	47.0	27.0	37.0	30.2	—	35
Iron	(mg)	0.4	0.6	0.5	0.8	—	0.6
Potassium	(mg)	—	206.0	—	—	—	—
Sodium	(mg)	—	9.0	—	—	—	—
Thiamin	(mg)	0.06	0.09	0.10	0.12	—	0.09
Niacin	(mg)	0.73	0.60	0.90	1.02	—	0.90
Riboflavin	(mg)	-	0.13	0.14	0.11	—	0.13
Vitamin C	(mg)	4.30	5.20	5.00	16.80	9.0	17.0

Sources: [1]de la Rocha (1958).
[2]Schmidt-Hebbel and Pennacchioti (1985).
[3]Cordova (1961).
[4]Morton (1987).
[5]Corigliano *et al.* (1968).
[6]Leung and Flores (1961).
[7]N x 6.25
[8]Non-nitrogenous extract

Valverde *et al.* (1983) showed the contents of dietary fiber and other fibrous components in cherimoya fruit (expressed as g/100 g edible fresh matter) were: dietary fiber (2.40), cellulose (1.70 ± 0.04), hemicellulose (0.12 ± 0.06), lignin (0.15 ± 0.04) and pectic substances (0.43 ± 0.01). In a second study, Vidal-Valverde *et al.* (1982) noted that the pectic substances (expressed as anhydrogalacturonic acid) amounted to about 0.54 ± 0.06% on a fresh weight basis or 1.98 ± 0.22% on a dry weight basis. The degree of ripeness of the fruit did not have a pronounced effect on the contents of pectic substances.

Organic acids

Total acidity, as measured by citric acid, ranges between 0.39 and 0.77% (Stahl 1935; Benk 1985). The principal organic acids are citric and malic. These two acids are also the major acids found in another annona fruit, namely, soursop (Bueso 1980).

Lipids

The lipid content of cherimoya fruit is less than 1% (Table 6.2). Limited information is available on specific lipids of the edible part, but some is available on cherimoya seeds. The oil extracted from seeds was obtained at 26% yield and showed the following properties: d_{20} = 0.9067 - 0.9110; relative viscosity at 20°C = 62.1; refractive index, n_D^{20} × 1.4715 - 1.4717. The oil was comprised of oleic (43%), linoleic (35%), palmitic (12%), stearic (8%), linolenic (1%) and traces of arachidic acid (Jimenez Martin 1971).

Phenolics, Alkaloids, Lactones

Weinges *et al.* (1969) reported that an alcoholic extract of cherimoya fruit contained (+)-catechol, (–)-epicatechol, a mixture of proanthocyanidins (B_1, B_2, B_3, B_4) and polymers of proanthocyanidins. The alkaloids, liriodenine, anonaine, lanuginosine, (+)-reticuline and caffeine, have been isolated from *A. cherimola* seeds (Allen 1943; Villar del Fresno and Rios Canavate 1983). Allen (1943) reported that when the juice of crushed seeds came in contact with the eyes of a human subject, it caused blindness. Morton (1987) reported that the dark-yellow resin isolated from seeds induced dilated pupils, intense photophobia, vomiting, nausea, dryness of the mouth, burning in the throat and other symptoms resembling the effects of atropine. In Mexico, rural people pulverize 1 or 2 seeds and take the powder with water or milk as a potent emetic and cathartic (Morton 1987).

Two new bis-tetrahydrofuranic γ-lactose possessing antimicrobial activity, namely cherimolin and dihydrocherimolin, were recently isolated from cherimoya (Cortes *et al.* 1984).

VOLATILE CONSTITUENTS

One of the most remarkable qualities of cherimoya fruit is its very distinct aroma. The aroma is extremely pleasant and desirable when the fruit is ripe; however, if the fruit becomes overripe the aroma quality may no longer be desirable. This fact suggests that the evolution of volatiles is different and occurs at varying stages during fruit development. During the first stage of growing, the cherimoya fruit do not produce any noticeable aromas. However, during the ripening stage, volatiles intensify with advancing maturity.

The characteristic aroma is derived from many essence volatiles, especially the esters. A very detailed analysis of the volatile composition has been undertaken by Idstein *et al.* (1984, 1985). These workers identified the constituents but did not conduct a parallel sensory evaluation to equate constituents to the characteristic cherimoya aromas. Idstein and co-workers (1984) identified 208 volatiles (Table 6.3) consisting of 23 hydrocarbons, 58 esters, 54 alcohols, 47 carbonyls and 26 volatiles of miscellaneous structures. The major volatiles were the alcohols, 1-butanol, 3-methyl-1-butanol, 1-hexanol and linalool; a series of butanoates and 3-methylbutyl esters were also prevalent. The isomeric pentenyl esters such as butyl and 3-methylbutyl exhibit fruity odors similar to the hexenyl esters. The variety of 3-methylbutyl esters suggests that cherimoya fruit has a forced leucine metabolism and an aroma distribution similar in part to that of the banana (Schreier *et al.* 1985).

Idstein *et al.* (1985) analyzed the volatile acids of cherimoya pulp and identified 47 different acids (Table 6.4). The major acids were hexanoic (3 mg/kg pulp) and octanoic (1 mg/kg pulp).

PROCESSING

The processing of cherimoya fruits was developed to preserve the fruit until it could be added as a natural flavor or as pulp to ice creams,

TABLE 6.3. QUANTITATIVE DISTRIBUTION OF MAIN CONSTITUENTS OF CHIRIMOYA FRUITS VOLATILES

ug/kg pulp[1]				
10-50	50-100	100-500	500-1000	more than 1000
α-thujene	p-cymene	butyl butanoate	1-butanol	3-methyl-1- butanol
comphene	2-methyl-2-butanol	3-methylbutyl butanoate	linalool	1-hexanol
α-terpinene	1-pentanol	hexyl butanoate		
γ-terpinene	1-penten-3-ol			
limonene	(E)-3-penten-1-ol			
3-methylbutyl-3-methylbutanoate	(Z)-2-penten-1-ol			
2-methyl-1-propanol	(Z)-3-hexen-1-ol			
2-methyl-3-buten-2-ol	α-terpineol			
2-pentanol				
(E)-2-hexen-1-ol				
2-ethyl-1-hexanol				
benzyl alcohol				
chrysanthenone				
(Z)-linalool oxide, furanoid				

Source: Idstein *et al.* (1984).
[1]Standard controlled capillary gas chromatographic determinations in fruit pulp, without consideration of calibration factors, i.e., F = 1.00 for all compounds.

TABLE 6.4. VOLATILE ACIDS IN THE CHERIMOYA

ug/kg pulp			
<50	50-100	100-500	>500
2-hydroxyacetic	dodecanoic	2-hydroxypropionic	hexanoic
3-hydroxypropionic	tetradecanoic	butanoic	octanoic
2-methylbutanoic	butan-1,4-dioic	3-methylbutanoic	
E-2-butenoic		pentanoic	
2-methylpentanoic		nonanoic	
4-methylpentanoic		decanoic	
heptanoic		3-hydroxydecanoic	
3-hydroxyoctanoic		5-hydroxy-(Z)-7-decenoic	
9-oxodecanoic		hexadecanoic	
5-hydroxy-(E)-7-decenoic		octadecanoic	
undecanoic		oxalic	
3-hydroxydodecanoic		(E)-cinnamic	
tridecanoic			
pentadecanoic			
heptadecanoic			
(Z)-9-octadecenoic			
2-hydroxybutan-1,4-dioic			
pentan-1,5-dioic			
hexan-1,6-dioic			
heptan-1,7-dioic			
octan-1,8-dioic			
nonan-1,9-dioic			
decan-1,10-dioic			
dodecan-1,12-dioic			
benzoic			
4-methoxybenzoic			
3,4-dimethoxybenzoic			
3-methoxy-4-hydroxybenzoic			
phenylacetic			
3-phenylpropionic			
2-furanoic			

Source: Idstein *et al.* (1985)

yoghurts, syrup and juice. Some processing procedures have proved detrimental because the pulp developed unacceptable color and flavor, e.g., as a result of pasteurization (Abufon 1985).

An industrial possibility is to expand techniques to freeze cherimoya pulp and slices; under frozen conditions the fruit is able to keep for about 120 days. For freezing to become a viable process, the fruit must be picked and left to ripen to some desired level.

The peeling process is conducted either by hand peeling with knives or by chemical treatment with hot caustic soda solutions (20% by weight).

The caustic soda dips are not appropriate for frozen slices because color, texture and appearances are affected (Abufon 1985). Additives, such as ascorbic acid, citric acid or EDTA may be added to the pulped or sliced fruit to prevent oxidation. Freezing of slices can be accomplished by the IQF (Individual Quick Freezing) method and bagged in polyethylene bags. The pulp, with or without sugar added, is bagged prior to freezing (Abufon 1985; Kasahara *et al.* 1986).

REFERENCES

ABUFON, A. J. 1985. Effects of additives, cleaning and type of storage on the quality of pulp and freezing protection of chirimoya (*Annona cherimola* Mill.) cv. concha lisa and bronceada. Agric. Eng. Thesis (Quillota), Catholic University of Valparaiso, Chile. (Spanish)

ALLEN, P. H. 1943. Poisonous and injurious plants of Panama. Supp. to Am. J. Trop. Med. Vol. 23, No. 1. Williams and Wilkins Co., Baltimore, Maryland.

BAILEY, L. H. 1963. Standard Cyclopedia of Horticulture, 3 volumes. Macmillan, New York.

BENK, E. 1985. Preparation of fruit nectars and fruit juice drinks using tropical and subtropical fruits and juices. Fluess. Obst. 52, 358-364. (German)

BUESO, C. 1980. Soursop, tamarind and chironja. *In* Tropical and Subtropical Fruits. S. Nagy and P. E. Shaw (Editors). AVI Publishing Co., Westport, CT.

BUNEMANN, and HANSEN, H. 1973. Fruit and Vegetable Storage. Verlag Eugen Ulmer, Stuttgart, Germany. (German)

BROOKS, M., REID, M. and HESSE, C. O. 1953. The Cherimoya. *In* Western Fruit Gardening. Univ. Calif. Press, Los Angeles.

BROWN, B. I., WONG, L. S., GEORGE, A. P. and NISSEN, R. J. 1988. Comparative studies on the postharvest physiology of fruit from different species of *Annona* (custard apple). J. Hortic. Sci. 63, 521-528.

CAMPBELL, C. W. 1979. Effect of gibberellin treatment and hand pollination on fruit-set of atemoya (Annona hybrid). Proc. Trop. Reg. Am. Soc. Horti. Sci. 23, 122-124.

CHANDLER, W. H. 1962. Evergreen Orchards. UTEHA, Mexico.

CORDOVA, V. J. 1961. The Cherimoya. Agric. Trop. 17, 647-664. (Spanish)

CORFO. 1984. Cultivation of chirimoya, papaya and lucumo in Chile. Manager of Development, Corporation of Development of Production. Santiago, Chile. (Spanish)

CORIGLIANO, F., DI PASQUALE, S., and CALABRO, G. 1968. Composition of pulp and seeds of *Annona cherimolia.* Atti. Soc. Peloritana Sci. Fis. Mat. Natur.14(3) 121-132. (Italian)

CORTES, D, RIOS, J. L., VILLAR, A. VALVERDE, S 1984. Cherimolin and dihydrocherimolin: two new bis- tetrahydrofuranic γ-lactones possessing antimicrobial activity. Tetrahedron Lett. 25, 3199-3202. (French)

DE LA ROCHA, G. 1967. The cultivation of the chirimoya and the results of limited experiments. Regional Center for Technical Assistance. 2a. Minister of Agriculture, Peru. (Spanish)

DUARTE, O., VILLAGARCIA, J. and FRANCIOSI, R. 1974. Effects of some treatments on the propogation of the chirimoya by seeds, stem cuttings and grafts. Proc. Trop. Reg. Am. Soc. Hortic. Sci. 18, 41-48. (Spanish)

EVERETT, P. 1952. The cherimoya: a delicious dessert fruit. N. Z. J. Agric. 84, 168.

FUSTER, C. and PRESTAMO, G. 1980. Variation of cherimoya (*Annona cherimola*) texture during storage as determined with an instrom food testing instrument. J. Food Sci. 45, 142-145.

GARCIA, E. 1956. The chirimoya, *Annona cherimola*. Agric. Exp. Stn. Circ. 71, La Molina, Peru. (Spanish)

HARGOUS, C. 1987. Response of the chirimoya (*Annona cherimolia* Mill.) to treatment with methyl bromide in postharvest. Agric. Eng. Thesis, Catholic University of Chile, Santiago. (Spanish)

IDSTEIN, H., HERRES, W. and SCHREIER, P. 1984. High resolution gas chromatography-mass spectrometry and Fourier transform infrared analyses of chirimoya (*Annona cherimolia* Mill.) volatiles. J. Agric. Food Chem. 32, 383-389.

IDSTEIN, H., BAUER, C. and SCHREIER, P. 1985. Volatile acids from tropical fruits: cherimoya (*Annona cherimolia* Mill.), guava (*Psiodium guajava* L.), mango (*Mangifera indica* L. var. alphonso), papaya (*Carica papaya* L.). Z. Lebensm. Unters Forsch. 180, 394-397. (German)

JIMENEZ MARTIN, J. 1971. Fats of the seeds of the cherimoya tree (*Annona cherimoya*). Ars. Pharm. 12. (Spanish)

KASAHARA, G. I., GERRIDO, F., SIMPSON, P. ALDUNATE, I. and CORNEJO, F. 1986. Kinetics of freezing and thermophysical properties in two species of minor fruit. *In* Topic on Heat Transfer and Thermophysical Properties in Refrigeration and Food Freezing. I. Kashara (Editor). Catholic University of Valparaiso, Chile. (Spanish)

KAWAMATA, S. 1977. Studies to determine the sugar composition of fruits by gas-liquid chromatography. Bull. Tokyo Agric. Exp. Stn. 10, 53-57.

KOSIYACHINDA, S. and YOUNG, R. E. 1975. Relationship of ethylene production to the initiation of the respiratory climacteric in fruits. Plant Cell Physiol. 16, 595-605.

LEUNG, W. T. W. and FLORES, M. 1961. Food Composition Tables for Use in Latin America. Institute of Nutrition of Central America and Panama and National Institutes of Health, Bethesda, Maryland.

LIZANA, L. A. and IRARRAZABAL, M. J. 1984. Postharvest behavior of chirimoya (*Annona cherimola* Mill.) subjected to low temperatures and wax. Am. Soc. Hortic. Sci. Trop. Reg. 28, 63-70. (Spanish)

MCGREGOR, B. M. 1987. Tropical Products Transport Handbook. USDA Agric. Handb. 668.

MCGREGOR, S. E. 1976. Insect pollination of cultivated crop plants. USDA Agric. Handb. 496.

MORTON, J. F. 1987. Fruits of Warm Climates. J. F. Morton Publisher, Miami, FL.

PAVEZ, R. M. 1985. Response of artifical pollinization and determination of physical and chemical changes of chirimoya fruit of different cultivars in the La Cruz area. Agric. Eng. Thesis (Quillota), Catholic University of Valparaiso, Chile. (Spanish)

PERALTA, P. M. 1984. Maturation and storage of chirimoya (*Annona cherimola* Mill.) cvs. concha lisa and bronceada. Agric. Eng. Thesis, University of Chile, Santiago. (Spanish)

POPENOE, W. 1934. Manual of Tropical and Subtropical Fruits. Macmillan and Co., New York.

REGINATO, G. 1980. Behavior of chirimoya to cold (*Annona cherimola* Mill.). Agric. Eng. Thesis, University of Chile, Santiago. (Spanish)

REGINATO, G. and LIZANA, L. A. 1980A. Detectable changes in chirimoya (*Annona cherimola* Mill.) during storage. Inv. Agric. (Chile) 6(3) 97-101. (Spanish)

REGINATO, G. and LIZANA, L. A. 1980B. Behavior of chirimoya (*Annona cherimola* Mill.) cv. concha lisa to storage. Simiente 50(3-4) 138-145. (Spanish)

SAAVEDRA, E. 1977. Influence of pollen grain stage at the time of hand pollination as a factor on fruit set of cherimoya. HortScience 12(2) 117-118.

SAAVEDRA, E. 1979. Set and growth of *Annona cherimola* Mill. fruit obtained by hand pollination and chemical treatments. J. Am. Soc. Horti. Sci. 104, 668-673.

SAINTE-MARIE, H. M.1987. Effect of the cultivar's pollen and of different systems of pollination on fruit set, percentage of fruit harvested, symmetry and form of three cultivars of *Annona cherimola* Mill. and studies of the viability of the pollen and habits of flowering in these same species. Agric. Eng. Thesis (Quillota), Catholic University of Valparaiso, Chile. (Spanish)

SAMSON, J. A. 1986. Tropical Fruits. Longman Inc., New York.

SCHMIDT-HEBBEL, H. and PENNACCHIOTI, M. I. 1985. Table of Chemical Composition of Chilean Foods. Universitaria, Santiago, Chile. (Spanish)

SCHREIER, P., IDSTEIN, H. and HERRES, W. 1985. HRGC-MS and HRGC-FTIR analysis of cherimoya (*Annona cherimolia* Mill.). *In* Semiochemistry Flavors and Pheromones. T. E. Acree and D. M. Soderlund (Editors). Walter de Gruyter, Berlin.

SCHWAZENBERG, C. 1947. Artificial pollination of the chirimoya. Agric. Tecnia (Chile) 6(2) 156-173. (Spanish)

STAHL, A. L. 1943. Poisonous and injurious plants of Panama. Suppl. to Am. J. Trop. Med. Vol. 23, No. 1. Williams and Wilkins Co., Baltimore, Maryland.

STAHL, A. L. 1935. Composition of miscellaneous tropical and subtropical Florida fruits. Univ. Fla. Agric. Exp. Stn. Bull. 283.

TAMARO, D. 1974. Treatise of Fruit Culture. 7a. Gili, Barcelona, Spain. (Spanish)

UNDURRAGA, P., LOYOLA, N. and GAMONAL, A. 1988. Characteristics of some physiological disorders by histological methods in fruits of chirimoya (*Annona cherimola* Mill.) cv. concha lisa. *In* Programs and Abstracts. VII National Meeting of Botany. National Society of Biology of Chile. (Spanish)

VIDAL-VALVERDE, C., BLANCO, I. and ROJAS-HIDALGO, E. 1982. Pectic substances in fresh, dried, desiccated and oleaginous Spanish fruits. J. Agric. Food Chem. 30, 832-835.

VIDAL-VALVERDE, C., HERRANZ, J., BLANCO, I. and ROJAS-HIDALGO, E. 1983. Dietary fiber of Spanish fruits. Fruits 47, 1840-1845.

VILLAR DEL FRESNO, A. and RIOS CANAVATE, J. L. 1983. Alkaloids from *Annona cherimolia* seed. J. Nat. Prod. 46, 438.

WEINGES, K., WILD, R. and KALTENHAEUSER, W. 1969. Structure, development and significance of polyphenols in fruits and fruit juices. XIII. Proanthocyanidins. Z. Lebensm. Unters. Forsch. 140(3) 129-134. (German)

7

Sugar Apple

Freddy Leal

The sugar apple (*Annona squamosa* L.) is a fruit native to the West Indies, Central America and the northern part of South America (Leon 1968). It belongs to the family Annonaceae and to the genus *Annona*, which includes about 120 genera and about 2000 species of chiefly tropical trees and shrubs; some of which are cultivated for their fruits and aromatic oils. The sugar apple is considered a delicate fruit and one of the best of the annonas because it usually produces good quality fruit. It is widely grown throughout the tropics at low and medium altitudes (Purseglove 1968).

The sugar apple is also known by the names anona blanca, sweetsop, anon, anona, rinon, saramuyo, ata, pinha, fruta do conde, ariticum pitaia, atta del Brasil, fruta de condessa, custard apple, cinnamon apple, attier, anone ecailleuse, corossolier ecailleux, pomme-canelle, noi-na, nona seri kaya, srikaya, atis, sitaphal, and banreishi (Naik 1949; Ochse *et al.* 1971; Fouquet 1972; Sastrapradja *et al.* 1980).

CHARACTERISTICS

The sugar apple is a small semi-deciduous tree growing from 4 to 6 m high. The simple leaves are deciduous, oblong-ellipticals, with blunt or taper apex, and are glossy-green. Leaves are 5 to 15 cm in length and 2 to 6 cm in width. The fragrant flowers are borne from young branchlet pendulous, solitary or in bunches of 2 to 4.

The fruit is a syncarpium formed by the fusion of pistils and receptacle into a large fleshy structure. The fruits are yellowish-green, heart shaped, 7 to 10 cm in diameter and covered with rounded, fleshy tubercules, which represent the loosely-cohering carpels, that separate readily when ripe. The fruit surface has a white or bluish bloom (Purseglove 1968). A purple-fruited variety is known (Mowry *et al.* 1958). The pulp is white or creamy-white, soft-granular, custard like, very sweet and has an agreeable flavor. Seeds are numerous, black or dark brown, about 9-12 mm in length and embedded in the pulp.

The sugar apple tree grows well in various types of soils; it shows high tolerance for alkaline soil conditions and resistance to long periods of drought (Sturrock 1959; Salazar 1960). The sugar apple is generally propagated by seed or by budding or grafting onto other species of *Annona* (Sastrapradja *et al.* 1980) or onto its own rootstock (Ochse *et al.* 1971; Pace and Araujo 1979; Sturrock 1959). It flowers from May to July and ripe fruits can be harvested in August to November. The sugar apple bears fruit after 3 years and with minimal horticultural care.

Invariably, the fruits are ready for eating at the time of their respiratory peak (Broughton and Guat 1979). As there is a climacteric rise in respiration, *Annona squamosa* can be classified as a climacteric fruit (Biale 1960). The respiratory climacteric always precedes the ethylene peak, unlike other tropical fruits such as papaya, banana and guava (Broughton and Guat 1979). In a comparative study of sugar apple, cherimoya and atemoya, sugar apple showed only one rise in respiration rate, whereas cherimoya and atemoya showed two successive rises in the respiration rate. Fruit ripened in ethylene-free air had better ripe fruit quality than fruit ripened under propylene (Brown *et al.* 1988).

High temperatures enhance ripening, accelerate the onset of the respiration peak, and correspondingly reduce the shelf life of these fruits. Fruits of sugar apple, ripened under dry conditions, were of good taste and appearance while those under other treatments were prone to fungal attack.

HARVESTING AND HANDLING

Sugar apple should be picked by hand just before maturity to avoid the splitting and bruising that would occur if the fruit fell from the branches. Of the annonas, they are the most difficult to harvest because of their tendency to burst open on the tree if permitted to become overmature. In Malaya, storage trials indicated that the ripening of sugar apples can be delayed by storage at 15°-20° C, a relative humidity of 85-95%, and low room levels of oxygen and ethylene (Morton 1987). Pantastico *et al.* (1975) advocated a relative humidity between 85 and 90% for storage of this fruit. Broughton and Guat (1979) showed that temperatures between 15° and 20° C were ideal for both storage and ripening of this fruit. In India, fruits treated with 50-60 g carbide ripened in 2 days and remained in good condition only 2 days, whereas those packed in straw ripened in 5-6 days and kept well for 4 days (Morton 1987). Immature fruit will not ripen to good flavor after harvest at storage temperatures lower than about 14°-18° C. Observations from Bombay (Hayes 1953) reported that hard, immature sugar apple suffered from chilling injuries at 15° C or below. Chandler (1958) recommended that for fully mature fruit, the best storage temperature region is about 4°-8° C. Although ripe fruits could be held for 6 weeks at 4° C with the pulp in good condition, the skin became brown and unattractive. McGregor (1987) recommends precooling sugar apples with forced air in a cool room. Precooling extends the life of the sugar apple by reducing field heat, the rate of respiration and ripening, the loss of moisture, the production of ethylene and the spread of decay. He recommends a storage temperature of 7° with 85-90% relative humidity. Transit and storage life of the fruit under these conditions is estimated at about 4 weeks. The fruit should be individually wrapped and placed in a single layer in a one-piece fiberboard box (4.5 kg).

USES

The sugar apple can be eaten fresh as a dessert fruit. The flesh is creamy white, sweet and of pleasant flavor. The strained pulp makes an excellent drink when mixed with milk. In using the pulp for making ice cream, it should not be cooked but added to the cooled mixture just before putting it in the freezer (Sturrock 1959). It can also be processed into fermented drinks (wine) and juices (Sastrapradja *et al.* 1980; Perez-Arbelaez 1978).

Among the people of the tropics, the sugar apple is reputed to possess medicinal value. Tea made from the roots is highly purgative, while that made from the leaves is mildly laxative. It is considered to have a tonic effect on the digestive tract (Pittier 1926; Noonan 1954). Toxic and insecticidal characteristics are reported from the seeds (Visneswariah *et al.* 1971). The seeds are acrid and poisonous and contain the alkaloid, anonaine. In India, a paste of seed powder has been used to kill head lice. Oils extracted from seeds of custard apple were found to be highly effective in reducing the survival of the leafhopper, *N. virescens*, and its transmission of the rice tungro virus (Mariappan and Saxena 1983). Additionally, Qadri and Hasan (1978) reported that *A. squamosa* seed extracts showed growth disruption against the rice weevil, *Sitophilus oryzae*.

COMPOSITION

Analysis of sugar apples grown in Florida (Stahl 1935) show fruit weight within the range 210-265 g. The distribution of the major parts of the sugar apple is comprised of seed (31-41%), skin (23-40%) and edible pulp (28-37%). The proximate composition of ripe sugar apple pulp is presented in Table 7.1, and the mineral and vitamin contents in Table 7.2.

Carbohydrates

Following water, the second most abundant components of sugar apple pulp are the carbohydrates which constitute about 18.2 to 26.2% of the pulp (Table 7.1; also Thompson 1920; Sturrock 1959; Aykroyld 1966). The values reported in Table 7.1 represent the combined total of simple (glucose, fructose, sucrose) as well as complex (oligosaccharides, fiber) carbohydrates. Chan and Heu (1975) reported the sugar distribution in sweetsop or sugar apple (g/100 g edible tissue) as fructose (3.49), glucose (5.15) and sucrose (3.11). Of nonsugar carbohydrates, Thompson (1920) reported 1.22% as fiber, whereas Sturrock (1959) reported 2.5%.

Lipids

The fat content of sugar apple pulp varies from 0.1 to 1.1% (Table 7.1). No information is available on the compositional nature of the lipids in pulp as relates to neutral lipids, glycolipids and phospholipids.

Ghanekar and Ayyar (1927), cited by Visneswariah et al. (1971), found oven-dried kernels of seeds contain about 30% oil and possess a specific

TABLE 7.1. COMPOSITION (%) AND FOOD ENERGY (CAL) PER 100 G OF RIPE SUGAR APPLE

Constituent	Thompson cited by Popenoe (1920)	Munsell *et al.* (1950)	Leung & Flores (1961)	Wenkam & Miller (1965)	Aykroyld (1966)	Singh (1969)	Morton (1987)
Water	75.2	71.1	72.8	75.9	68.6	73.5	69.8-75.2
Protein	1.5	1.2[1]	1.6	1.9	1.6	1.6	1.5- 2.4
Lipids	0.5	0.1	0.2	0.6	0.3	0.3	0.3- 1.1
Carbohydrates	18.2	—	23.0[2]	20.8	26.2	23.9	19.2-25.2
Fiber	1.2	1.1	1.6	1.4	2.4	—	1.1- 2.5
Total acidity	0.1	—	—	—	—	—	—
Ash	0.7	0.6	0.8	0.8	—	0.7	0.6- 1.3
Food energy	—	—	96	86	114	—	88-96

[1]Approximate protein content obtained by (.191 g N/100 g) x 6.25 = 1.2 g.
[2]Value reported by Leung and Flores (1961) for total carbohydrates is 24.6%; that value represents both sugars (23.0%) and fiber (1.6%).

TABLE 7.2. MINERAL AND VITAMIN COMPOSITION OF RIPE SUGAR APPLE PULP (mg/100 g)

Constituent	Munsell *et al.* (1950)	Leung & Flores (1961)	Sturrock (1959)	Wenkam & Miller (1965)	Adams (1975)
Minerals					
Ca	19.4	28	44.7	17.0	22
P	23.6	36	55.3	53.6	41.2
Fe	0.3	1.8	1.0	0.3	0.6
Na	—	—	—	—	11.2
K	—	—	—	—	275
Vitamins					
Vitamin A	0.007	tr	0.005	tr	0.004
Thiamine	0.109	0.11	0.10	0.104	0.10
Riboflavin	0.113	0.15	0.167	0.057	0.14
Niacin	0.65	0.90	1.28	0.89	1.00
Ascorbic acid	39.8	35.0	42.2	35.9	34.0

gravity of 0.913 (at 15° C), saponification value of 188, and iodine value of 86. Morton (1987) reported seed kernels to contain 14-49% of a whitish or yellowish, nondrying oil with a saponification value of 186. Recent research by Ansari *et al.* (1985) on *A. squamosa* seeds yielded results shown in Table 7.3. The seeds contained 23% oil and revealed the presence of an unusual hydroxy acid (isoricinoleic acid). This acid was characterized as 9-hydroxy-cis-12-octadecenoic acid.

TABLE 7.3. ANALYTICAL DATA OF *A. SQUAMOSA* SEED OIL

Oil, %	23.0
Moisture, %	5.1
Unsaponifiable, %	1.6
Iodine value	58.8
Saponification value	191.8
Refractive index	1.4826
Fatty acid composition (wt. %)	
Myristic	1.5
Palmitic	25.1
Palmitoleic	3.1
Stearic	9.3
Oleic	37.0
Linoleic	10.9
Arachidic	3.3
Isoricinoleic	9.8

Source: Ansari *et al.* (1985).

Nitrogenous compounds

Stahl reported the organic nitrogen content of sugar apple pulp as 0.180-0.301 g N/100 g wet weight. Munsell *et al.* (1950) reported an average value of 0.191 g N/100 g, whereas Sturrock (1959) reported 0.309 g N/100 g. Morton (1987) reported values for three amino acids, namely, tryptophan (9-10 mg/100 g pulp), methionine (7-8 mg/100 g) and lysine (54-69 mg/100 g).

Phenolics

Traces of (+) catechin and (–) epicatechin, and large quantities of polymerised proanthocyanidins consisting of leucocyanidin and leucodelphinidin have been identified in sugar apple pulp (Venkatasubbaiah and Mathew 1970).

Nutrients

From the data of Table 7.2, it is evident that the sugar apple is a very good source of calcium, phosphorus and iron, and a fair source of vitamin C. Agronomic practices known to affect nutrient content of sugar apple fruits have not been investigated. Broughton and Guat (1979) noted that mature sugar apples contained low contents of ascorbic acid during the pre-climacteric stage. Ascorbic acid increased during fruit ripening and reached a maximum at the climacteric, after which the amount decreased. Broughton and Guat (1979) also noted that the total amount of carbohydrate decreased as the fruit ripened and attributed this effect to its utilization as the main source of energy for respiration and other metabolic processes. The amount of starch decreased while the glucose level increased during ripening. However after the climacteric, the glucose level also decreased.

FLAVOR

The most desirable characteristic of the sugar apple is its strong aroma and exquisite flavor. Limited studies have been found on this topic. Bhonsack (1972) considered the use of tropical and subtropical fruits to have flavors that could be of interest to the candy confectionery manufacturer and stated "Other types of the *Annona* flavor are also available. The *Annona squamosa* (also called Cinnamon Apple) is reminiscent of pear, and has a slight cinnamon note." Oliveros-Belardo (1975) found sugar apple peel to yield 0.1% essential oil and noted the presence of at least 33 essence oil components by gas chromatography. The predominant terpene hydrocarbons were alpha pinene, beta pinene, limonene, beta farnesene and trans ocimene.

PROCESSING AND STORAGE STABILITY OF SUGAR APPLE PULP AND NECTAR

Extraction of the pulp

The fruits are cut by hand, the pulp is removed and the seeds extracted. Holanda *et al.* (1980) studied the storage stability of the pulp and nectar.

Pulp

The pulp was heated for 3 min at 70° C, placed in jars, and doubly boiled for 15 min at 95° C. Cooling was done using running water. Sealed jars were stored at 27° C.

Samples were taken every 30 days, and determinations of pH, Brix, total acidity, total sugars, nonreducing sugars, water-soluble pigments (W.S.P.) and vitamin C were performed.

The mean values (samples) of the determined characteristics are summarized at 30-day intervals (Table 7.4).

During the 150 days storage period, the only significant changes observed were an increase in total acidity, an increase of total and reducing sugars, and a loss of vitamin C content.

Nectar

After washing, sugar apple fruits were divided in two groups: the first group was left at room temperature for 72 hours; the other was placed in a climate chamber at 16° C, 80% R.H. and ethylene applied. Fruit maturity was accomplished in 4 days.

After ripening, pulp was extracted and adjusted with water, sugar and citric acid (pulp 1 kg; water 2.51 kg; sugar 0.37 kg; citric acid l g). The mixture was homogenized, heated for 3 min at 70° C, placed in jars

TABLE 7.4. CHEMICAL COMPOSITION OF SUGAR APPLE PULP

Storage time (days)	0	30	60	90	120	150
pH	3.80	3.70	3.70	3.70	3.70	3.70
Brix	22.00	18.40	23.60	23.80	23.60	24.20
Total acidity (%)	0.60	0.50	0.70	0.80	0.91	0.77
Total sugars (%)	12.80	18.10	17.40	19.30	20.00	20.80
Reducing sugars (%)	12.40	13.60	15.20	14.90	13.60	15.00
W.S.P. (420 nm)	96.00	93.00	93.00	94.00	93.00	93.00
Vitamin C (mg/100 g)	5.80	—[1]	—	—	—	

Source: Holanda *et al.* (1980).
[a]Not detected.

TABLE 7.5. CHEMICAL COMPOSITION OF SUGAR APPLE NECTAR[1]

Storage time (days)	0	30	60	90	120	150
pH	4.00	4.00	4.00	4.00	4.00	4.00
Brix	15.60	16.60	16.00	16.10	15.80	15.50
Total acidity (%)	0.13	0.12	0.13	0.11	0.11	0.11
Total sugars (%)	14.40	16.10	13.70	15.30	15.20	13.40
Reducing sugars (%)	4.50	5.50	4.60	5.80	5.40	6.80
W.S.P. (420 nm)	96.50	96.00	96.00	97.00	97.00	97.00
Vitamin C (mg/100 g)	4.80	—[2]	—	—	—	

Source: Holanda *et al.* (1980).
[1]Nonclimatized fruits.
[2]Not detected.

TABLE 7.6. CHEMICAL COMPOSITION OF SUGAR APPLE NECTAR[1]

Storage time (days)	0	30	60	90	120	150
pH	4.00	4.00	4.00	3.90	3.90	3.80
Brix	15.80	16.40	16.00	15.80	15.40	16.20
Total acidity (%)	0.14	0.13	0.13	0.12	0.11	0.13
Total sugars (%)	15.30	16.10	15.20	15.70	15.70	14.90
Reducing sugars (%)	7.00	7.00	6.70	7.20	7.00	8.20
W.S.P. (420 nm)	99.00	96.00	96.00	97.00	97.00	97.00
Vitamin C (mg/100 g)	—[2]	—	—	—	—	—

Source: Holanda *et al.* (1980).
[1]Climatized fruits.
[2]Not detected.

and doubly boiled for 15 min at 95° C. Cooling was done using running water. Sealed jars were stored at 27° C.

Samples were taken every 30 days and determinations made. The mean values of the determined characteristics from the 5 samples on the first and last day of analysis for nonclimatized and climatized fruits are summarized in Tables 7.5 and 7.6. During the 150-day storage period, the only significant changes observed were an increase of total sugars at the end of one month and the complete loss of vitamin C.

REFERENCES

ADAMS, C. F. 1975. Nutritive Value of Foods in Common Units. U. S. Dept. Agric. Handbk. 456.

ANSARI, M. H., AFAQUE, S. and AHMAD, M. 1985. Isoricinoleic acid in *Annona squamosa* seed oil. J. Am. Oil Chem. Soc. 62, 1514.

AYKROYLD, W. R. 1966. Nutritive Value of Indian Foods. ICMR Publication, New Delhi, India.

BOHNSACK, H. 1972. Exotic flavors add excitement to candy. Candy and Snack Industry 137(4), 30, 58.

BIALE, J. B. 1960 Respiration in fruits. *In* Handbuch der Pflanzenphysiologie 12, 536-592. W. Ruhland (Editor). J. Springer, Heidelberg, Germany.

BROUGHTON, N. J. and GUAT, TAN. 1979. Storage conditions and ripening of the custard apple *Annona squamosa* L. Sci. Hortic. 10(1) 73-82.

BROWN, B. I., WONG, L. S., GEORGE, A. P. and NISSEN, R. J. 1988. Comparative studies on the postharvest physiology of fruit from different species of Annona (custard apple). J. Hortic. Sci. 63, 521-528.

CHAN, H. T. and HEU, R. A. 1975. Identification and determination of sugars in starfruit, sweetsop, green sapote, jackfruit and wi apple. J. Food Sci. 40, 1329-1330.

CHANDLER, W. H. 1958. Evergreen Orchards. Lea and Febiger, Philadelphia.

FOUQUET, A. 1972. Specialty fruits of tropical America. Fruits 27(5), 368-382. (French)

HAYES, N. B. 1953. Fruit Growing in India. 2nd Rev. Edition. Kitabestan, Allahabad, India.

HOLANDA, L. F. F. DE, MAIA, G. A., MARTINS, C. B. and MOURA, J. A. 1980. Study on processing and stability of pulp and nectar of sugar apple (*Annona squamosa* L.) Cien. Agron. 10(1) 137-140. (Portuguese)

LEON, J. 1968. Botanical Fundamentals of Tropical Culture. Instituto Interamericano de Ciencias Agricolas de la OEA. San Jose, Costa Rica. (Spanish)

LEUNG, W. T. W. and FLORES, M. 1961. Food Composition Table for Use in Latin America. National Institutes of Health, U. S. Govt. Printing Office, Washington, D.C.

MARIAPPAN, V. and SAXENA, R. C. 1983. Effect of custard apple oil and neem oil on survival of *Nephotettix virescens* and on rice tungro virus transmission. J. Econ. Ent. 76, 573-576.

MCGREGOR, B. M. 1987. Tropical Products Transport Handbook. USDA Agric. Handbk. 668.

MORTON, J. F. 1987. Fruits of Warm Climates. Julia F. Morton Publisher, Miami, FL.

MOWRY, H., TOY, L. R. and WOLFE, H. S. 1958. Miscellaneous tropical and subtropical fruits. Fla. Agric. Ext. Serv. Bull. 156A.

MUNSELL, H. E., *et al.* 1950. Composition of food plants of Central America. VIII. Guatemala. Food Res. 15, 439-453.

NAIK, K. C. 1949. South Indian Fruits and Their Culture. P. Varadachary & Co., Madras, India.

NOONAN, J. C. 1954. Review of investigations on the annona species. Proc. Fla. State Hortic. Soc. 64, 205-210.

OCHSE, J., SOULE, M. J., DIJKMAN, M. and WEHLBURG, G. 1971. Tropical and Subtropical Agriculture. Editorial Limusa-Wiley S.A., Mexico. (Spanish)

OLIVEROS-BELARDO, L. 1975. The predominant terpenes in the fruit peel oil of *Annona squamosa* L. Lloydia 38, 537.

PACE, C. A. M. and ARAUJO, C. M. 1979. Methods of grafting to the sugar apple (*Annona squamosa*). Anais do V Congreso Brasileiro de Fruticultura 3, 952-956. (Portuguese)

PANTASTICO, ER. B., CHATTOPADHYAY, T. K. and SUBRAMANYAM, H. 1975. Storage and commercial storage operations. *In* Postharvest Physiology, Handling and Utilization of Tropical and Subtropical Fruits and Vegetables. Er. B. Pantastico (Editor). AVI Publishing Co., Westport, Conn.

PEREZ-ARBELAEZ 1978. Useful Plants of Colombia. Litografia Arco, Bogota. (Spanish)

PITTIER, H. 1926. Manual of Common Plants in Venezuela and Its Supplement. Fundacion Eugenio Mendoza, Caracas. (Spanish)

POPENOE, W. 1920. Manual of Tropical and Subtropical Fruits. Macmillan, New York.

PURSEGLOVE, J. W. 1968. Tropical Crops: Dicotyledons, Vol. II. John Wiley & Sons, New York.

QADRI, S. S. H. and HASAN, S. B. 1978. Growth retardant effect of some indigenous plant seeds against rice weevil *Sitophilus oryzae* L. J. Food Sci. Technol. 15, 121-123.

SALAZAR, C. 1960. Some fruits of minor economic importance in Puerto Rico. Rev. Agr. P. R. 47(2) 135-156.

SASTRAPRADJA SETIJATI, USEP JUTISNA GILMOUR PANGGABEAN, JOHANIS PALAR MOGEA, SUKRISTIJOND SUKARDJO and ALOYSIUS TRI SUNARTO 1980. Fruits. EBPGR Secretariet, Rome.

SINGH, R. 1969. Fruits. National Book Trust, New Delhi, India.

STAHL, A. L. 1935. Composition of Miscellaneous Tropical and Subtropical Florida Fruits. Fla. Agric. Exp. Stn. Bull. 283.

STURROCK, D. 1959. Fruits for Southern Florida. South Eastern Printing Co., Stuart, FL.

THOMPSON, A. R. 1920. *Cited by* W. Popenoe. Manual of Tropical and Subtropical Fruits.

VENKATASUBBAIAH, G. and MATHEW, A. G. 1970. Polyphenols of custard apple and their role in pink discoloration. J. Food Sci. Technol. (Mysore) 7(4) 203-204.

VISNESWARIAH, K., JAYARAM, M., KRISHNAPRASAD, N. K. and MAJUMDER, S. K. 1971. Toxicological studies of the seeds of *Annona squamosa*. Indian J. Exp. Biol. 9(4) 519-521.

WENKAM, N. S. and MILLER, C. D. 1965. Composition of Hawaii Fruits. Hawaii Agric. Exp. Stn. Bull. 135.

8

Mangaba

Narendra Narain

The mangaba (*Hancornia speciosa* Gomes) belongs to the family of Apocynaceae. The family contains one hundred and thirty genera, and over one thousand species of herbs, shrubs and trees. Many of its plants exude a milky sap, often poisonous or bitter (Sturrock 1959). However, in addition to mangaba, only a few species of the genus, *Carissa* have been known to be of any commercial value.

For a long time, the mangaba fruit has been designated mango by the Portuguese who found some similarity between the fruit and some types of mango. However, the origin of the name mangaba was derived from the names the native Brazilian Indians had given it as "manga-iba" or "manga-uva". To avoid any misunderstanding, it should be remembered that the Portuguese-language word for mango is "manga".

Mangaba was first cited as early as the year 1587 when Gabriel Soares mentioned that in the neighborhood of the sea in Bahia (state of Brazil) he found some trees in the feeble lands which were of the size of peach trees and locally known as "Mangabeiras" (Texeira 1954). The fruits, being yellow and the size of plums, were eaten without leaving anything. The fruit had a pleasant and strong flavor and was good for digestion, serving well also for people suffering from fever (Texeira 1954).

The plant, "Mangabeira" is native to the northeast region of Brazil where it has proliferated spontaneously in the uncultivated highlands "tabuleiros" (broad, flat sedimentary surfaces between stream valleys) possessing sandy and dry soils (Barros 1970). It is also found to a small extent in the north and eastern regions of Brazil, including the states of Mato Grosso, Minas Gerais and Para. An incision in the epicarp of the shrub results in the formation of latex-containing rubber. It is interesting to observe the first commercial exploitation of mangabeira in the form of rubber manufacture during the second world war when there was a great scarcity of natural rubber resources. A production of about 2000 to 2500 tonnes of rubber has been reported during the war (Bekkedahl and Saffioti 1948).

Although the annual production of mangaba is reported to be more than 1200 MT (Fundação Instituto Brasileiro de Geografia e Estatistica 1986), its actual production is thought to be much higher as its accountability is subject to error due to its being in a natural habitat, and the lack of any systematic plantations. It is the experience of the author who has widely travelled the interior of Paraiba State, the major producer of mangaba in Brazil, that a part of its production goes unaccounted and nonutilized primarily due to remote producing areas, difficulties in harvest, and lack of appropriate storage and transport facilities.

In spite of its traditional and long usage in the relatively larger area of northeast Brazil, mangaba fruit has rarely been cited in publications outside of Brazil. It is, however, hoped that lesser known novel or unconventional fruits will gain importance as the demand for tropical and subtropical fruits increases resulting in the need for new and different flavor-impact aromatic fruits.

BOTANY

The mangaba (*Hancornia speciosa*) is an elegant looking shrub having its branches separated and well shaped. It attains a height of about 5 m and possesses a rough and wrinkled stem. The epicarp is dark and cracked, and the leaves are opposite, speared, small, precise, leathery and glabrous (Barros 1970). The flowers retain a pleasant aroma and are bell-shaped, opening in the form of an eye. The fruit is an ellipsoid drupe up to a longitudinal length of 6 cm having a yellow skin striated with red. The green fruit has an inside milky appearance and must be harvested when swollen since the fruit rapidly ripens within a day (Gomes 1980).

In Brazil, it is customary to classify the fruit into three distinct stages of maturity commonly referred as green, green-mature and mature. The terms green-mature or just mature are used to indicate a fruit that has developed to a point that it will ripen to maximum eating quality, but in which ripening has not yet commenced. Arola (1982) proposed the classification of mangaba based on its visual aspects (Table 8.1). The fruit may contain 2 to 10 seeds which are oval, flat and of the size of small bean (Texeira 1954). However, due to the lack of taxonomical studies, there is no mention in the literature on the existence of other varieties or cultivars of mangaba.

TABLE 8.1. CLASSIFICATION OF MANGABA BASED ON FRUIT MATURITY

	Stage of maturity		
Characteristic	Green	Green-mature	Mature
Skin color	Green	Greenish yellow	Yellow
Stains on skin	Absent	Present	Present
Visual aspect on cutting fruit	White pulp; seeds in formation	White pulp; seeds formed though soft	Yellowish pulp; hard seeds
Fruit texture	Hard	Slightly soft	Soft

Source: Arola (1982).

HORTICULTURE

For several centuries, mangaba has survived in the soils of northeastern South America which clearly distinguishes itself as an arid pocket within the humid sub-continent. The predominant growth area is located south of the equator between 5-15° latitude. This small area stands out as undoubtedly the section of most abrupt geographical diversity due to its great variations in climatic conditions. The mangabeira is mostly found in the area between the coastal region receiving more than 1200 mm of rainfall annually and the "agreste" (a transitional zone between the moist coastal zone and the dry interior) having an annual rainfall between 800-1200 mm. The soils in which it habitates vary from thinner sandier type to ones derived from crystalline rocks. The average minimum and maximum temperature in the area is about 15° and 43°C, respectively.

Barros (1967, 1970) reported that the best time for planting mangaba was during the cold and rainy season. The only medium of propagation is by seeds which should be planted in holes of 0.6 x 0.6 x 0.6 cm size with a distance of 6 m between one plant and another. Pimentel and Santos (1978) studied the effects of the mechanical treatments on the conservation of germinative power of manaba seeds utilizing three treatments: a) cleaning and removal of pulp aggregated with seeds; b) fermentation of seeds for 24 hr; c) cleaning the fermented seeds. They found that fermentation did not promote the germination of seed, and cleaning still is the operation most recommended provided the planting is done soon after the removal of seed from fruit. They observed a decrease of 52% in germination if seeds were stored for three days.

HARVEST AND FRUIT HANDLING

The mangaba is mostly conserved on the tree until it attains the final stages of maturity. When mature, the fruit is unable to bear the load and becomes loose, consequently falling to the ground. Since the height of the shrub is not much (less than 5 m), the losses due to bruising or injury caused by falling fruit do not seem high.

Texeira (1954) stated that it would be a better practice for people living in the proximities of such plants to awaken early, preferably before dawn, and pick up the fresh fruit which fell to the ground. This form of picking has an added advantage as it eliminates the problems associated with the latex handling in manual harvest. However, this practice would be feasible only at places where conditions prevail for its quick commercialization and consumption. Care should also be exerted in not letting the fruits become ripe as the texture turns too soft and it becomes difficult to handle the fruit. Ripe fruit practically dissolves between the fingers.

Since the fruit are quite gummy due to the latex oozing out at the broken end of the fruit, gloves must be used if the harvesting is done by hand, which is still the common practice. This characteristic of the presence of latex is most inherent and peculiar to the family of *Apocynaceae*.

Flowering usually commences in the month of August and extends to October when some of the fruits begin to fall. The first harvest is known as the produce of summer season which lasts from January to March while the second harvest pertaining to winter season extends from April to June (Barros 1967). In the Paraiba state of Brazil, the fruit are transported to distances (within 200 km) in small containers, usually baskets of varying sizes, and brought to the open markets for sale.

COMPOSITION AND NUTRITIONAL VALUE

The mangaba fruit is composed of about 77% pulp, 12% seed and 11% peel (Trevas *et al.* 1971). Very little work has been done on the chemistry and biochemistry of this fruit. Also, there are no reports of any work published on post-harvest ripening behavior or any studies on the effects of storage conditions on the chemical composition of fruit. Furthermore, no reports exist on the enzymatic browning nature of the pulp.

The average nutrient content and proximate composition of mangaba fruit is presented in Table 8.2. The data reported by Trevas *et al.* (1971) is based on the analysis of fruit obtained from three northeastern states of Brazil and, hence, could be considered somewhat representative. The fruit contains about 81 to 86% moisture, 2.8 to 3.6% fat, and 0.4 to 0.96% proteins (Trevas *et al.* 1971). In an earlier work, Tiuba *et al.* (1961)

TABLE 8.2. THE AVERAGE NUTRIENT CONTENT AND PROXIMATE COMPOSITION OF MANGABA FRUIT

Constituent	A	B
pH	2.9	—
Food energy (cal)	54	—
	%	
Water	84.4	83.7
Protein	0.87	0.87
Fat	3.2	1.6
Ash	–	0.5
Reducing sugars	6.7	8.36
Nonreducing sugars	Traces	0.82
Total sugars	—	9.23
Acidity (citric acid)	—	0.85
	mg %	
Calcium	1.6	3.24
Phosphorus	7.5	8.96
Iron	2.4	4.09
Ascorbic acid	128	149.9

Sources: A: Trevas et al. (1971).
B: Granja (1985).

reported about 86% moisture, 3.2% fat, 3-6 mg/100 g iron, and 140 mg/100 g ascorbic acid. The low pH of mangaba pulp, varying from 2.5 to 3.6 (Trevas *et al.* 1971), and a relatively higher fat and acidity could relate to its characteristic sweet acidic flavor. However, no research has been reported on the identification and quantitation of volatile flavor components in mangaba.

Carbohydrates

Trevas *et al.* (1971) reported the presence of nonreducing sugars only in traces, whereas Granja (1985) found them to be 0.8%. Similarly, the average value for reducing sugars reported by Trevas *et al.* (1971) was less (6.7%) when compared to 8.4% reported by Granja (1985).

One of the most important characteristics of mangaba when eaten either as a fresh ripe fruit or in the form of its products, including ice cream, is that it leaves in the mouth a residue which has gummy properties. Arola (1982) undertook a detailed study on the isolation and characterization of this gum from mangaba. He reported a yield of about 3% (wet basis) partially purified gum which showed a nonnewtonian pseudoplastic behavior. The sugars positively identified from the gum hydrolysate were L (+) arabinose, D (+) fructose, D (+) glucose and L (–) sorbose. The effect of various substances commonly utilized in the food industry, such as sugars (sucrose and glucose), salts (sodium chloride, potassium chloride and sodium bicarbonate) and chemical additives (potassium metabisulfite and citric acid) were also studied on the viscosity of mangaba gum. In general, an increase in viscosity was observed. However, any variation in pH from the original pH of mangaba gum as well as an increase in temperature resulted in a decrease in viscosity.

Minerals and Vitamins

From the data presented in Table 8.2, mangaba can be considered as a rich source of iron. Although there is some discrepancy in the values reported for iron in mangaba: 3-6 (Tiuba *et al.* 1961), 2-3 (Trevas *et al.* 1971), 3.8-4.5 mg/100 g pulp (Granja 1985), no study deals with the bioavailability of iron. Still, such a high iron content ranks this fruit as one of the foremost in iron-containing fruit.

In addition, mangaba is also an excellent source for ascorbic acid. The reported values (mg/100 g pulp) range from 102-145 (Trevas *et al.* 1971) and 138-158 (Granja 1985), while Tiuba *et al.* (1961) reported about 140 mg/100 g pulp. It is no wonder that the importance of this fruit in curing certain diseases and, in particular, against fever had been documented four centuries earlier (Texeira 1954).

USES

Mangaba has been industrially utilized in the northeast region of Brazil for the production of juice and ice cream, and on a small scale in the manufacture of cordial, wine, syrup and sweet confections (Barros 1967). Some products such as puree, nectar and fruit in syrup were developed and organoleptically evaluated by Trevas et al. (1971). Since the fruit is small and acidic, it is suitable for jelly making. However, the fruit is particularly suited for the manufacture of ice cream due to the presence of its high gum content which imparts functional properties of binding, flavor retention and inhibition of crystal formation. In Brazil, a small-scale cottage industry for ice cream manufacture is still predominant and which caters to the tastes for many cherishable and distinct flavored fruits. The flavor of mangaba ice cream is very much cherished in the relatively larger area of northeastern Brazil. In addition, the fruit is consumed fresh as well as in the form of cordials.

REFERENCES

AROLA, F. M. 1982. Isolation and Characterization of Mangaba Gum. M.Sc. Thesis. Universidade Federal da Paraíba, João Pessoa, Brazil. (Portuguese)

BARROS, R. da C. 1967. Mangaba tree, Queen of shallow surfaces between stream valleys. Mundo Agricola, São Paulo 16(191) 9-12. (Portuguese)

BARROS, R. da C. 1970. Mangabeira. Folha Florestal, Vicosa. 4(17) 67. (Portuguese)

BEKKEDAHL, N. and SAFFIOTI, W. 1948. Latex and rubber from mangaba tree. Bol. Tec. Inst. Agron. Norte. 13, 1-41. (Portuguese)

FUNDAÇÃO INSTITUTO BRASILEIRO DE GEOGRAFIA E ESTATISTICA. 1986. Statistical Year-book of Brazil. FIBGE, Rio de Janeiro. (Portuguese)

GOMES, R. P. 1980. Brazilian Fruit-culture. Nobel, São Paulo. (Portuguese)

GRANJA, M. L. B. B. 1985. Effect of Preservation Methods and Storage Time on the Quality of Umbu (*Spondias tuberosa* Arr. Cam) and Mangaba (*Hancornia speciosa* Muell). M.Sc. Thesis. Universidade Federal da Paraíba, João Pessoa, Brazil. (Portuguese)

PIMENTEL, M. de L. and SANTOS, E. O. 1978. Preservation of germinating power of mangaba (*Hancornia speciosa* Gomes) seed. Empresa Pernambucana de Pesquisa Agropecuária, Comunicado Técnico 1-6. (Portuguese)

STURROCK, D. 1959. Fruits for Southern Florida. Southeastern Printing Co., Stuart, FL.

TEXEIRA, E. 1954. Fruits from Brazil. Mini stério da Educação e Cultura, Instituto Nacional do Livro, Rio de Janeiro. (Portuguese)

TIUBA, Z. de A., DUCHENE, S. G., BURGER, O. N. and ANDRADE, M. V. 1961. Mangaba. J. Sci. Food Agric. Abstr. 12, i-192.

TREVAS, V. F., FREIRE, G. S., FALCONE, E. G. and TREVAS, I. P. 1971. Report of the Section of Agriculture and Cattle-raising Technology, Research Division and Agriculture and Cattle-raising Experimentation. Convénio do Ministério da Agricultura (Estado da Paraiba)/Suberintenendencia do Desenvolvimento do Nordeste, 105-184. (Portuguese)

9

Mountain Papaya

L. Antonio Lizana

The mountain papaya is a subtropical fruit tree that does not acclimate well under tropic conditions. It is a very interesting fruit because in spite of being climacteric, it does not develop sufficient sugar to be satisfactorily sweet when eaten fresh. Nevertheless, when cooked in syrup it becomes a very desirable fruit, with marked preference by those who know it.

The climate requirements, the tree's growth habit, the fruit type and the general management of the orchards make this plant completely different from its tropical cousin, the common papaya.

BOTANY

There are more than 40 species in the genus *Carica*, but only a few are of economic importance. Most produce fruits that are not edible, like *Carica chilensis* (Planch.) Solms-Laub., also called "fat wood" due to the shape of its trunk. Many other wild species are usually small plants (Munoz-Pizarro 1959). There are four edible papayas: the main one is the tropical papaya followed in importance by the mountain papaya, the babaco (apparently a natural hybrid), and another tropical papaya, *Carica monoica*, found in the amazone forests, where the rainfall is over 3,500 mm and the temperatures are not lower than 10° C. The tropical papaya is classified as *Carica papaya* L., the mountain papaya as *Carica pubescens* Lenne et Koch (A.DC., Solms), *Carica cestriflora* (A.DC., Solms), *Carica chiriquensis* Woodson and *Carica candamarcensis* Hook. (Hooker 1875). This last classification is generally accepted for the mountain papaya. The mountain papaya is also called papaya de alturas because in the tropical areas it is found only between 1,500 and 2,500 m above sea level (Morin 1967).

The mountain papaya originated on the high slopes of Ecuador and Colombia (2,400-3,600 m), whereas *Caria papaya* originated in tropical America, probably Mexico (Popenoe 1927).

The babaco, *Carica pentagona* Heilbr. is apparently a natural hybrid between *Carica candamarcensis* Hook. and *Carica sitpulata* Badillo (Badillo 1967). It produces a very large seedless elongated fruit, sometimes well over 40 cm long, with a whitish pulp, very acidic but pleasant to eat. Apparently, it originated in Ecuador, where it is found growing in temperate valleys of Guallabamba, Tumbaco, Ambato, Patate, Paute, Loja and Ibarra. Babaco fruits have arrived in Los Angeles market within the last years. They are imported from New Zealand, where they were introduced from Ecuador about 1975 (Badillo 1967; Camacho and Rodriquez 1982).

HORTICULTURE

The tree is a big herbaceous succulent plant that can grow to 2.5 to 3 m high with an erect and branched trunk, thick and rounded at the base and thinner to the end. Branches are upright and rigid, and sometimes can produce secondary branches. The surface of the trunk is covered with big symmetric foliar scars.

The plant can live up to 17 years but usually for commercial purposes, after 5 years, the productive value decreases so papaya orchards are replaced at about that age. The plant is very precocious, producing its first fruit about 18 months after planting (Larrain and Letelier 1942; Sanchez 1973).

The first fruits produced by a papaya plant allow for selection since the plants are of the monoic-dioic type. Moreno *et al.* (1979) found that the mountain papaya presents a gradation from absolute male (staminate) to absolute female (pistillate) flower morphology. They classified them into four flower types:

1. Female or pistillate flowers: free petals, no floral tube, no stamens, a very well developed ovary over 10 mm in length and 5 mm diameter; short style and a pentabranched stigma.
2. Carpeloid pistillate flowers: fused petals, short or no floral tube, ovary over 6 mm in length, short style and stigma with variable number of branches (1 to 4); 1 to 5 stamens, 2 to 4 fused to the ovary wall which produces deformity in fruits.
3. Carpeloid staminate flowers: fused corola with 4 to 133 mm tube length and 2 to 3 mm in diameter. Out of 10 stamens, 5 to 9 are free, the rest are fused among themselves or to the ovary wall; the ovary

is 4 to 6 mm in length without sporogenic tissue and the stigma is filiforme and sterile; this type of flower produces only pollen.
4. Male or staminate flowers: fused corola 13 to 14 mm in length, 10 free stamens, pistil and ovary rudiments.

Papaya plants may present these different types of flowers that will produce four types of plants: those that produce only female flowers, those which produce only male flowers and two others that produce a combination of hermaphrodite flowers as influenced by cultural and environmental conditions. It is possible to find some male flowers within female-flower plants and some hermaphrodite flowers within male-flower plants.

For commercial purposes, it is not possible to separate plants by the type of flower they produce before coming to bloom. The common practice is to plant three to four plants per site. When the types can be properly identified, the plants are selected leaving the hermaphrodite and the female plants only. Of the four different types of plants, the hermaphrodite is the least frequently found. Thus, the female-flowered plants account for the main fruit production and quality of the fruit.

The fruit is produced year-round but has seasonal fluctuations. In Chile, the main production occurs from August to March. The shape of the fruit is oblongovoid, about 7 to 10 cm in length and lobulated.

FRUIT

The fruit is a berry with a smooth and tough skin, thin pericarp having numerous seeds in an inner cavity filled with a mucilaginous tissue. The seeds are brownish in color and their surface is rough. The peel color is green during growth and development of the fruit, and turns bright yellow when ripe.

The fruit from hemaphrodite flowers tends to be smaller in size, with more pronounced elongated lobules and thinner pulp than the fruits from female flowers (Lizana *et al.* 1981).

The fruit is rich in papain. This proteolytic enzyme is found in the latex and is very abundant in green fruits. One of the uses of papain is as a meat tenderizer. In the areas of Chile where papaya is grown, fresh green fruit is often used domestically for this purpose. Also, canned papaya in syrup is offered as a dessert.

Papain has very high enzymatic activity, and people that harvest and handle the fruit must wear protective gloves to prevent wounds around the finger nails. The greener the fruit the more active is the papain. Preliminary studies have proven the mountain papaya to be a very important source of this enzyme. The best yields of fresh latex were obtained from 5 razor blade cuts made parallel to the axis of fruit that had finished its growth but still had the green color. An average 1 g of fresh latex per 100 g of green fruit weighing 123 to 157 g (Hermosilla 1979) was obtained. The time of day in which papain is collected, and the amount of light (cloudy or sunny) do not influence the flux of latex. The best ambient temperature is between 18° and 22° C, and the amount increases if the orchard has been irrigated 24 hr prior to the collection of the latex. The best yields obtained by an untrained person was 59 g/hr. The latex collected is very pure, but thorough work for each single fruit is tedious. Yields can be increased using a plastic devise similar to an inverted umbrella where all fruit can be milked simultaneously; up to 140 g of fresh latex per hr (Hermosilla 1979).

The papaya fruit is comprised of about 12 to 18% peel, 32 to 35% seeds and mucilaginous substances that fill the inner cavity, and about 50 to 52% pulp. The green fruit represents a higher portion of utilizable pulp for the canning industry, but the quality of the final product is low, and lacks color, flavor and aroma (Lizana *et al.* 1978). The variability in the amount of peel is due to the softening of the fruit when ripe which makes peeling difficult. Also, the proportion of mucilaginous substances decrease at the end of the ripening period. For these reasons, the processing industry occasionally tends to utilize greener fruit rather than impair end product quality (Kiger 1979).

FRUIT RIPENING

Papaya fruit undergoes maturation while attached to the tree and its progress can be followed by a definitive change of peel color from green to a bright full yellow when ripe. Overripe fruit usually become ochre yellow. Most of the maturity indexes to characterize papaya ripening behave similarly to those described for other fruits, with the exception of soluble solids that increase only slightly while the fruit undergoes ripening. Titratable acidity and pulp pressure decrease. The iodine test does not show any color change that would indicate the presence of starch in any of the five stages of maturity (Lizana 1979; Lizana *et al.* 1981).

Fruit Respiration

When papaya fruit ripens at ambient temperature (18° C), respiratory CO_2 evolution shows a typical climacteric pattern similar to that described for the tropical papaya (Biale 1964). Nevertheless, the intensity of respiraton and the time of the climacteric peak differ slightly in fruits from different type of flowers: fruit from staminate flowers have a lower peak intensity and were earlier than fruits from hermaphrodite flowers that show the highest peak in climacteric respiration. Fruits with morphological intermediate characters show an intermediate climacteric peak intensity but later than the other two (Fig. 9.1). The climacteric at 18° C (ambient temperature) progresses slower than in other fruits similarly treated, the whole process takes about 20 days (Lizana *et al.* 1981).

Peel Color Changes and Fruit Respiration

The ripening process can be divided into five stages following peel color changes: Stage A, CO_2 production increases steadily, the peel is a green color for about 8 days then changes to yellowish-green. Furthermore, the CO_2 evolution increases in stage B for four to five days; the respiratory peak is reached at stage C, 14 to 16 days after ripening has started and yellow becomes the predominant color. Stage D shows a steady decrease in respiratory activity while the fruit manifests a very attractive bright yellow color. In stage E, CO_2 evolution stabilizes and color changes to an ochre-yellow; evidence of an overripe fruit (Lizana *et al.* 1981). Other maturity indexes as pH, range from 4.39 in stage A to

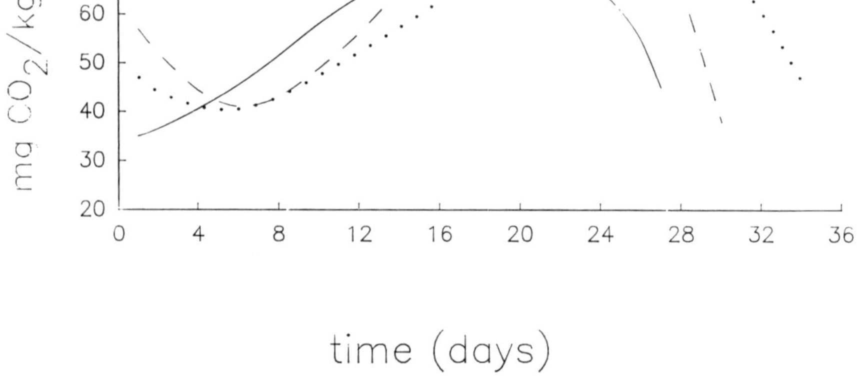

FIG. 9.1. CLIMACTERIC RESPIRATION RATE OF MOUNTAIN PAPAYA FRUIT (*CARICA CANDAMARCENSIS*) FROM DIFFERENT TYPES OF FLOWERS (ADJUSTED TO A 3rd DEGREE EQUATION)

4.51 in stage E. The water content also remains relatively constant (93%).

Organic Acids and Ripening Process

The main organic acids of mountain papayas are malic and citric (Silva *et al.* 1980). They evolve differently while the fruit undergoes the ripening process. Malic acid tends to decrease and citric acid to increase from stage A (green fruit) to stage E (overripe fruit). Moreover, when the fruit was at stage A only malic acid was detected in substantial high amounts. Citric acid starts increasing from stage B when the fruit externally showed color break. Even though citric acid increases continually while malic acid decreases, malic acid always predominates (Fig. 9.2) (Silva *et al.* 1980).

Sugars

Identification of sugars by gas-liquid chromatography shows fructose, glucose and sucrose as the major sugars, with only traces of sorbitol and inositol. During fruit ripening, glucose is the predominant sugar. It initially increases parallel to the increase in the respiratory activity in the first two stages, A (green peel) and B (green to yellow peel), and then decreases continuously towards the end of the process. Fructose, on the other hand, increases stadily to become the major sugar when the fruit is overripe. Sucrose shows an initial small increase while the fruit goes from green to greenish-yellow in color and then remains constant during the rest of the ripening process (Silva *et al.* 1980) (Fig. 9.3).

Sugar production in mountain papaya is unique when compared with other climacteric fruit. The fact that sucrose remains constant during the ripening process and shows very low amounts accounts for the lack of the typical sweet flavor of the ripe fruit. This behaviour might be due to a very fast sucrose hydrolysis. Chan and Kwok (1975) have shown with the tropical papaya (*Carica papaya* L. cv. Solo) that the invertase activity is very high (0.821 invertase units/mg protein), and that 50% of the sucrose could be lost in 160 seconds when preparing fruit tissue for sugar determinations. This was not the case in the analyses of sugars by Silva *et al.* (1980) where they used thermal enzyme inactivation in their procedure. The invertase activity in the mountain papaya fruit (*Carica candamarcensis* Hook.) is unknown, but if similar, an accumulation of glucose and fructose during the ripening process would be expected. Fructose has been shown to accumulate, but glucose increases initially and then decreases steadily. This decrease in glucose could not be attributed to respiratory activity, since glucose decreases simultaneously (Fig. 9.3).

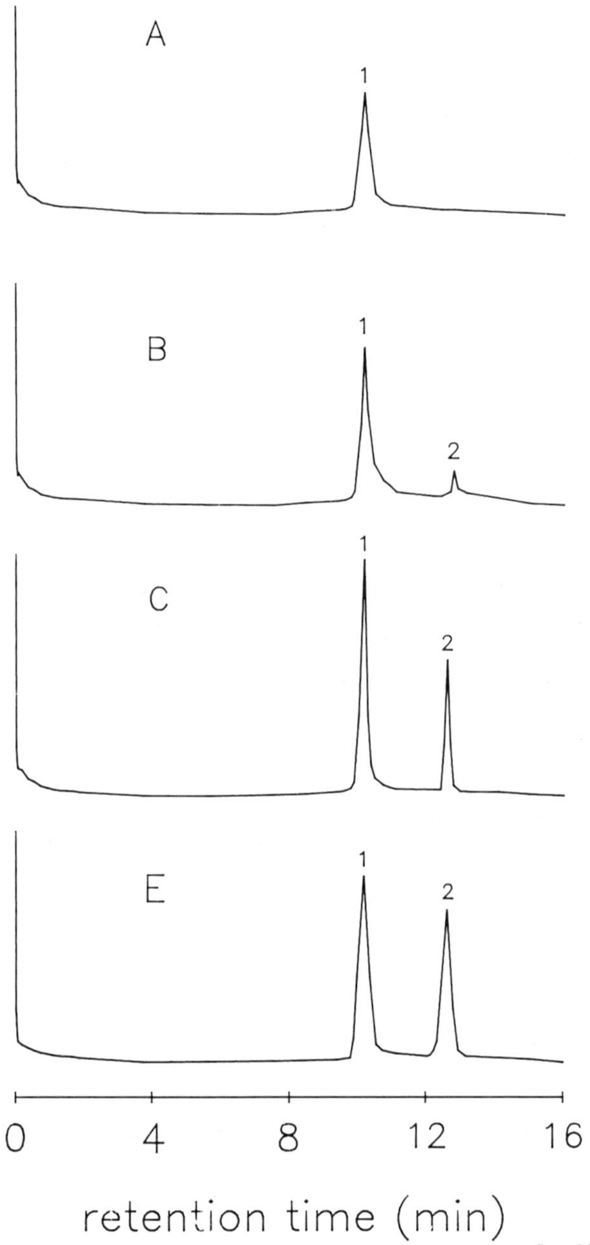

From Silva *et al.* (1980)

FIG. 9.2. GAS-LIQUID CHROMATOGRAMS OF THE EVOLUTION OF MALIC (1) AND CITRIC (2) ACIDS DURING DIFFERENT MATURITY STAGES IN FRUIT OF THE MOUNTAIN PAPAYA. PEEL COLOR: A, GREEN; B, GREEN WITH YELLOW: C, YELLOW WITH GREEN AND E, OCHRE YELLOW

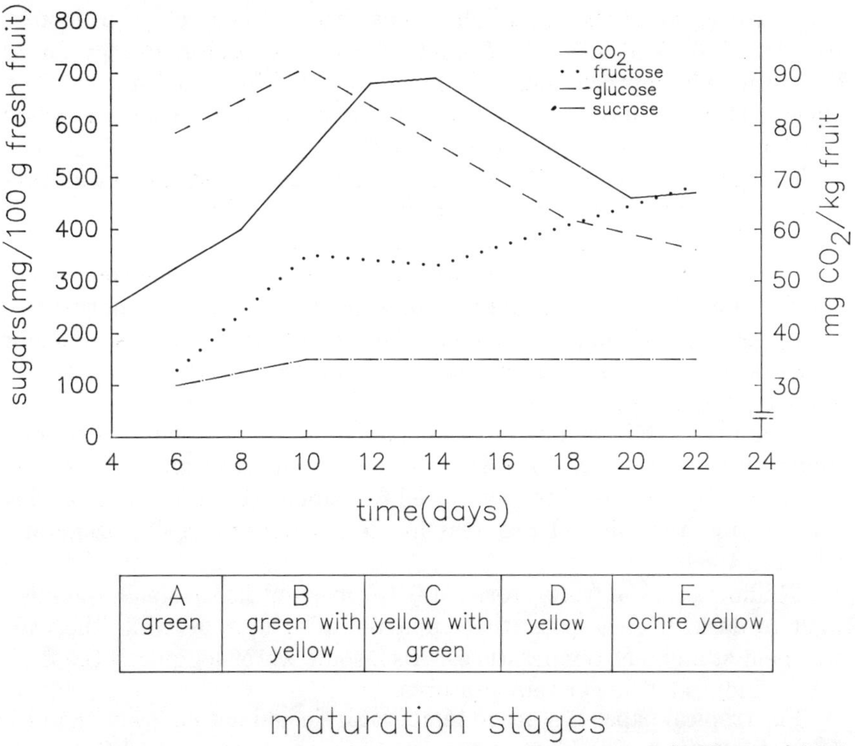

From Lizana *et al.* (1981)

FIG. 9.3. FRUCTOSE, GLUCOSE AND SUCROSE EVOLUTION DURING THE RIPENING OF MOUNTAIN PAPAYA FRUIT (*CARICA CANDAMARCENSIS* HOOK.) AND ITS RELATION WITH 5 STAGES OF MATURATION FOLLOWING PEEL COLOR CHANGES

In addition, mountain papaya fruit has a low starch content similar to plums and peaches.

This is another difference between tropical papaya and mountain papaya where an increase in sugars is constant during the ripening process and the main sugar is sucrose followed by glucose and fructose (Chan and Kwok 1975). Sorbitol and inositol have not been reported either. Total sugar content could accumulate in tropical papaya to 18% or more, whereas in mountain papaya it seldom exceeds 5% (Lizana 1979).

FRUIT STORAGE

The behaviour of mountain papayas held under different temperatures show variation depending on their maturity stage. The storage period during which the fruit kept well was 15 days at 0° C for yellow fruit with green stripes and light-yellow fruit. Green-yellow fruit kept well for 15 days at 4° C. At 7° C all fruit can be held in storage for 30 days. Immediate processing occurs upon removal from cold storage. The riper fruit after 10 days at 7° C or more were affected by severe fungal developments (Lizana and Olivares 1989).

The main pathogen, *Rhizopus*, develops very aggressively during storage at 7° C and ambient temperature; secondarily, *Alternaria*, *Botritis* and *Penicillium* also develop (Olivares 1982).

Fruits at 0° C and 4° C show, after 15 days of storage, some physiological disorders like a sticky and viscous exudate—probably from the rupture of the lacticiferous ducts. When these fruits were placed at ambient temperature afterwards, small scars were present at the site where the exudates were present.

Green fruit did not present this exudation as a response to chilling injury until after 30 days of storage at 0° C; green fruit can then be stored at these temperatures if they are used for papain extraction since at this temperature they do not progress in the ripening process (Lizana and Olivares 1989).

Chilling injury of fruits stored at 0° C consisted also in small, rounded, light to dark, brown spots, that appeared after storage while the fruit was held at ambient temperatures. Also, some water-logging of the flesh can be induced at lower temperatures.

For tropical papaya, storage temperature could not be lower than 13° C or the fruit would show quality impairment (Akamine and Goo 1969; Thompson and Lee 1971; Wardlaw *et al.* 1934). If stored at 7.4° C, *Carica papaya* will show definite chilling injury and does not progress in ripening even if placed at ambient temperature. On the other hand, *Carica candamarcensis* can be successfully stored at 7° C for 30 days without showing chilling injuries. Its ripening process slows during the storage period. It can complete its ripening if placed at ambient temperature for 3 days.

CHEMICAL COMPOSITION

Fruit chemical composition varies depending upon maturity stage, cultural conditions and geographical growing area. Additionally, chemical differences are evident when different types of papayas are compared (Table 9.1). Mountain papayas, when compared to tropical papayas, show lower levels of sugars and ascorbic acid, but contain higher amounts of fiber and vitamin A.

Sugars are the principal carbohydrates in papaya fruits; they are the main source of calories and impart sweetness. The content of sugars

TABLE 9.1. COMPARATIVE FRUIT COMPOSITION OF MOUNTAIN PAPAYA AND TROPICAL PAPAYA (PER 100 G EDIBLE PORTION)

Constituent		Mountain papaya		Tropical papaya		
		A	B	C	D	E
Calories		18	24	—	46	—
Water	(g)	93.2	93.0	88.9	86.8	86.1
Proteins (N x 6.25)	(g)	1.0	0.8	0.6	0.4	0.7
Lipids	(g)	0.3	0.2	0.1	0.1	0.1
Carbohydrates	(g)	3.3	5.4	9.0	12.2	7.7
Fiber	(g)	1.4	1.2	0.9	0.6	—
Ash	(g)	0.8	0.6	0.5	0.6	0.4
Calcium	(mg)	36.0	12.0	30.0	29.9	34.8
Phosphorus	(mg)	28.0	14.0	13.6	11.6	10.1
Iron	(mg)	1.3	0.5	0.3	0.2	0.2
Thiamine	(mg)	0.06	0.02	0.03	0.03	0.04
Riboflavin	(mg)	0.05	0.03	0.03	0.04	0.03
Niacin	(mg)	0.50	1.00	0.39	0.33	0.77
Ascorbic acid	(mg)	26.00	47.00	55.00	84.00	89.61
Vitamin A (retinol)	(mg)	37.00	35.00	—	10.93	2.57

Sources: A = Schmidt-Hebbel and Pennacchioti (1973).
B = Leung and Flores (1961).
C = Arriola *et al.* (1976, 1980).
D = Wenkam and Miller (1965).
E = Beyers *et al.* (1979).

(8-12%) in tropical papaya fruit (Benk 1985; Jagtiani *et al.* 1988) is about twice the amount found in mountain papaya (Leung and Flores 1961). The higher content of sugars in tropical papaya is reflected in a higher caloric value (46 cal) when contrasted to the mountain papaya (18-24 cal). The sugar composition of ripe tropical papayas as determined by Chan and Kwok (1975) shows a sucrose:glucose:fructose distribution pattern of about 2:1:1, whereas Lizana *et al.* (1981) reported a pattern for ripe mountain papaya of about 1:4:4. Sucrose is the dominant sugar in ripe tropical papaya, whereas fructose is dominant in ripe mountain papaya (Fig. 9.3).

The ascorbic acid content of mountain papaya is about 40-50% lower than tropical papaya (Table 9.1), but is similar to yellow passion fruit (Arriola *et al.* 1976). Mountain papaya contains a significantly high content of vitamin A (35-37 mg/100 g) (Jagtiana *et al.* 1988; Arriola *et al.* 1980). The mountain papaya contains the vitamins, thiamine, riboflavin and niacin at levels generally below 1 mg/100 g fruit tissue and, therefore, is not considered a good source for these nutrients.

FRUIT PROCESSING

The largest volume of fruit processed are canned in syrup of various sugar concentrations (22 to 30° Brix). Other products are papaya juice that is obtained from the mucilaginous substance that fills the inner cavity, papaya preserves, papaya honey, papaya candy and papaya pulp that goes into ice creams and pastries.

The fruit can also be dehydrated successfully, but it is not usually done. Lizana *et al.* (1978) studied different fruit maturity and dehydration temperatures; overripe fruit shows the best quality for this process, whereas unripe fruit shows no quality at all. As to drying temperatures, papayas yield more intense aroma and better taste at 45° C, greater acidity at 60° C, and good texture as well as good quality at 75° C.

Barria *et al.* (1983) analyzed four different stages of fruit maturity according to peel color for canning. A better quality final product was obtained when fruit was harvested at full maturity.

REFERENCE

AKAMINE, E. A. and GOO, T. 1969. Effect of controlled atmosphere storage of fresh papaya (*Carica papaya* L. var. "Solo") with special reference to shelf-life extension of fumigated fruit. Hawaii Agr. Exp. Stn. Bull. 144.

AKAMINE, E. A. and GOO, T. 1971. Relationship between surface color development and total soluble solids in papaya. Hortic Sci. 6, 567-568.

ARRIOLA, M. C. de, CALZADA, J. F., MENCHU, J. F., ROLZ, C. and GARCIA, R. 1980. Papaya. *In* Tropical and Subtropical Fruits. S Nagy and P. E. Shaw (Editors) AVI Publishing Co., Westport, CT.

ARRIOLA, M. C. de, MENCHU, J. F. and ROLZ, C. 1976. Characterization, handling and storage of papayas. Cntrl. Amer. Res. Inst. Ind. (ICAITI), Guatemala. (Spanish)

BADILLO, V. M. 1967. Scheme of the Caricaceae. Agron. Trop. (Venezuela) XVII(4) 245-272. (Spanish)

BARRIA, M., LIZANA, L. A., KIGER, F., MORALES, D., and ARAYA, E. 1983. Effect of the stage of initial maturation on the quality of papaya (*Carica candamarcensis* Hook.). Simiente (Chile) 53(3-4) 163-168. (Spanish)

BENK, E. 1985. Preparation of fruit nectars and fruit juice drinks using tropical and subtropical fruits and juices. Fluess. Obst. 52, 358-364. (German)

BEYERS, M., THOMAS, A. C. and VAN TONDER, A. J. 1979. Irradiation of subtropical fruits. 1. Compositional tables of mango, papaya, strawberry and litchi fruits at the edible-ripe stage. J. Agric. Food Chem. 27, 37-42.

BIALE, J. B. 1964. Growth, maturation and senescence in fruits. Science 146, 880-888.

CAMACHO, S. and RODRIQUEZ, V. 1982. Commercial cultivation of babaco (*Carica pentagona* Heilb.) in Ecuador. Proc. Am. Soc. Hortic. Sci., Trop. Reg. 26, 17-20. (Spanish)

CHAN, H. T., JR. and KWOK, S. C. M. 1975. Importance of enzyme inactivation prior to extraction of sugars from papayas. J. Food Sci. 40, 770-771.

HERMOSILLA, J. 1979. Papain, an organic compound of commercial value present in papaya. Symposium on the cultivation of the papaya, 144-146. University of the North, CICA, Coquimbo, Chile. (Spanish)

HOOKER, L. P. 1875. *Carica candamarcensis* Curtis. Bot. Mag. 1, 619.

JAGTIANI, J., CHAN, H. T. and SAKAI, W. S. 1988. Tropical Fruit Processing. Academic Press, New York.

KIGER, F. 1979. New alternative technologies for industrialization of the Chilean papaya. *In* Symposium on the cultivation of the papaya, 132-143. University of the North, CICA, Coquimbo, Chile. (Spanish)

LARRAIN, M. S. and LETELIER, B. L. 1942. Cultivation of the papaya and its industrial exploitation. Faculty of Agronomy Degree Thesis, University of Chile, Santiago. (Spanish)

LEUNG, W. W. T. and FLORES, M. 1961. Food Composition Table for use in Latin America. National Institutes of Health and Institute of Nutrition of Central America and Panama, U.S. Govt. Printing Office, Washington, D.C.

LIZANA, L. A. 1979. Maturity and postharvest of papaya (*Carica candamarcensis* Hook.). Symposium on the cultivation of the papaya, 106-123. University of the North. CICA, Coquimbo, Chile. (Spanish)

LIZANA, L. A., CIENFUEGOS, V., FUENTES, F. and SILVA, C. 1981. Characterization of the maturation of papayas (*Carica candamarcensis* Hook.). Simiente 51(3-4) 174-184. (Spanish)

LIZANA, L. A., KIGER, F. and CRUZ, R. 1978. Influence of the state of maturity and temperatures on the dehydration of papayas (*Carica candamarcensis* Hook). Proc. Am. Soc. Hortic. Sci., Trop. Reg. 22, 48-61. (Spanish)

LIZANA, L. A. and OLIVARES, F. 1989. Influence of the state of maturity on the storage of papayas (*Carica candamarcensis* Hook.) at different temperatures. Inv. Agric. (Chile), in press. (Spanish)

MORENO, R., GOMEZ, G. and PALMA, C. Floral forms and sex determination in the papaya (*Carica pubescens*) (*Carica candamarcensis*). *In* Symposium on the cultivation of the papaya, 20-27. University of the North, CICA, Coquimbo, Chile. (Spanish)

MORIN, C. 1967. The Papaya. *In* Cultivation of Tropical Fruits. 2nd Edition. ABC Libraries. S.A. Pacific Press, Lima, Peru. (Spanish)

MUNOZ-PIZARRO, C. 1959. Synopsis of Chilean Flora. University of Chile, Santiago. (Spanish)

OLIVARES, F. 1982. Storage of papaya (*Carica candamarcensis* Hook.) in the cold. Faculty of Agronomy Degree Thesis, University of Chile, Santiago. (Spanish)

POPENOE, W. 1927. The papaya and its relatives. *In* Manual of Tropical and Subtropical Fruits. Macmillan, New York.

SANCHEZ, C. P. 1973. Cultivation of the papaya. INIA, Program of fruits and wines. Santiago, Chile. (Spanish)

SCHMIDT-HEBBEL, H. and PENNACCHIOTI, M. I. 1973. Table of Chemical Composition of Chilean Foods. Universitaria (Editor). University of Chile, Santiago. (Spanish)

SILVA, C., FUENTES, F., CIENFUEGOS, V. and LIZANA, L. A. 1980. Sugars and constituent acids and their evaluation during maturation of papayas (*Carica candamarcensis* Hook.). Proc. Am. Soc. Hortic. Sci., Trop. Reg. 24, 97-102. (Spanish)

THOMPSON, A. K. and LEE, C. R. 1971. Factors affecting behaviour of papaya fruit. J. Hortic. Sci. 46, 97-102.

WARDLAW, C. W., LEONARD, E. R. and BAKER, R. E. 1934. Observations on the storage of various fruits and vegetables. Trop. Agric. (Trinidad) 11, 9-11.

WENKAM, N. S. and MILLER, C. D. 1965. Composition of Hawaii fruits. Hawaii Agric. Exp. Stn. Bull. 135.

10

Bacuri and Cupuassu

Charles R. Clement
Giorgini A. Venturieri

Eastern Amazonia is not as rich as western Amazonia in terms of species and genetic diversity of cultivated or domesticated fruits but does have a few species with great potential (Clement 1983). Two of these are cupuassu (*Theobroma grandiflorum* (Willd. ex Spreng.) Schum.) and bacuri (*Platonia insignis* Mart.) neither of which have common English names because they have not been taken to other parts of the humid tropics.

Both of these species are popular because of the creamy whitish pulp that envelopes their seeds and which have distinct, strong, acid-sweet, agreeable flavors. Because of flavor strength, neither fruit is commonly consumed in fresh form (except for the parthenocarpic sections of bacuri), but rather they are used in juices, ice creams, sherberts, bakery fillings, jams, and candies. They are the two most popular dicotyl fruit species in eastern Amazonia, especially in the city of Belém, Pará, Brazil, and are currently undergoing industrial processing and commercialization. Unfortunately, limited industrial-oriented, technological research is being done by public institutions of the region and whatever private research is done is generally unavailable to the public. Because of increasing demand, the cupuassu is currently being planted in many areas of Brazilian Amazonia. Frozen pulp is exported to large population centers of southern Brazil and to Europe. Bacuri is also increasing in popularity but demand has not yet exceeded supply; the fruit is almost exclusively from wild or managed plants maintained on farms in the southeast of Pará state and the north of Maranhão state, Brazil.

In the following we shall consider the two species separately. They are at different developmental stages enroute to world agro-industrial markets.

CUPUASSU

Description, Phenology and Yields

The cupuassu is a relative of cocoa (*Theobroma cacao* L.) that Cuatrecasas (1964) classified in the Glossopetalum section of the genus. Its natural distribution is restricted to high and low forests on the south side of the Amazon River, east of the Tapajós River (a major southern tributary). It occurs both on upland and semi-flooded soils of this region. The following description is condensed from Cuatrecasas (1964), Cavalcante (1976) and FAO (1986).

The mature cupuassu tree may attain 20 m in height and 45 cm dbh (diameter at breast height) and is an element of the mid-understory. The cultivated plant (Figure 10.1) grown in more open conditions generally attains only 6 to 8 m in height, but may have a crown 7 m in diameter.

Courtesy C. R. Clement

FIG. 10.1. A FOUR-YEAR-OLD SEEDLING TREE OF CUPUASSU AT THE INPA TROPICAL FRUIT EXPERIMENT STATION, MANAUS, AMAZONAS, BRAZIL.

The principal stem is orthotropic and its psuedo-apical growth occurs in spurts of 1 to 1.5 m, at which point it produces a spiral of three plagiotropic branches, giving it a typically pagoda-like architecture. The short-petioled leathery leaves are juvenally reddish, maturing to dark green, attaining 25 to 35 cm in length by 10 to 15 cm in width.

The axilar or extra-axilar inflorescences have 1 to 5 flowers and are only produced on the plagiotropic branches. The flowers are the largest in the genus, with 5 triangular sepals and 5 reddish petals forming the corola. The 5 stamens each have 6 anthers; the obovoid ovary has 5 locules. The fruit are the largest in the genus (Figure 10.2), varying from 200 to 5000 g. They have a short, thick peduncle that abscises at maturity and are covered with a brown tomentum. The smooth, hard, woody epicarp is about 2-4 mm thick, the "mesoendocarp" (Cuatrecasas' term) is thick (5-7 mm), soft and fleshy at maturity. There are 30-50 seeds, each surrounded by a creamy whitish to yellowish, juicy, fibrous pulp. Each seed is ovoid to ellipsoid-ovoid, 2-3 cm long by 2-2.5 cm wide and 1-1.5 cm thick.

Courtesy C. R. Clement

FIG. 10.2. ONE KG FRUIT OF CUPUASSU. THE OPENED FRUIT SHOWS THE LARGE AMOUNT OF PULP AVAILABLE FOR PROCESSING.

Falcão and Lleras (1983) studied 7-year-old plants near Manaus, Amazonas and found they produced an average of 3,500 flowers per year, most of which are produced during the dry season (July-October in Manaus). Less than 1% of these produce mature fruit, unless heavily fertilized (C. H. Muller, personal communication), in which case about 2% may mature. Yields are therefore low. The main harvest is from January to May.

Calzavara *et al.* (1984) considered average yields, without fertilization, to be between 12 and 20 fruits/plant at five years from seed, each weighing about 1 kg. At 234 plants/ha (7 x 7 m triangular spacing) this is between 2.8 and 4.7 MT/ha/yr. Venturieri *et al.* (1985) suggest that better management and adequate fertilization can raise yields to a level of 20-30 fruits/plant (4.7-7 MT/ha/yr) at five years, increasing to 60-70 (14-15.4 MT/ha/yr) at 7 years. Although this is much lower than most fruit crops, current market prices make it extremely lucrative. A one kg fruit on the Manaus market during harvest is sold for U.S. $1.50-$2.00, making it one of the most expensive local fruits. At least one local farmer profits more from his one hectare of cupuassu than from his 20 ha of cocoa, using the recommended cocoa mangement practices on both species!

A seedless mutant has been found near Cametá, Pará (Calzavara *et al.* 1984; Venturieri 1984). The mutation suppresses the formation of the embryo and endosperm, but not the testa, which results in the pulp being almost ready for use. The empty testa, however, is more difficult to remove from the pulp than a normal seed, so that this mutant is more of a curiosity than a practical cultivar. Additionally, this cultivar has low yields and is exceptionally susceptible to Witch's Broom disease (*Crinipellis perniciosa*), the major disease of both cocoa and cupuassu in Amazonia.

History of Use

Ducke (1946) mentions the cupuassu as a pre-Columbian crop. Although the diversity found in the southeastern Amazonian forest today is no different from that found in cultivation, this would suggest that the Amerindians did not do much in the way of domestication.

Cuatrecasas (1964) mentions the popularity of the pulp in central and eastern Amazonia without going into details. Earlier authors are equally vague while mentioning its popularity. Calzavara *et al.* (1984) mention the growing industrial importance of cupuassu in the Belém market but do not give more than a decade's worth of data. Several industries are actively preparing several cupuassu products and commercializing them in Amazonia and exporting them to southern Brazil and to Europe. This is a new activity, with less than a decade of active processing and commercialization.

Composition

Fruit components vary considerably with size, provenance, harvest period, and separation methods. Table 10.1 lists several examples of fruit component variation (by manual extraction) as influenced by size (Calzavara *et al.* 1984). Manually extracted pulp (which generally means a housewife with a pair of scissors) is preferred by most consumers because this leaves chunks of pulp that impart texture to dishes made from cupuassu. Mechanical separators have been adapted for use with cupuassu, but are generally less efficient than the manual method (Chaar 1980). Gelar S. A., of Belém, obtains only 23% pulp and 19.4% seed from their machinery (Calzavara *et al.* 1984).

Barbosa *et al.* (1979) and Chaar (1980) analysed the pulp for a variety of factors of interest in food technology. These data are presented in Table 10.2. In Table 10.3, Chaar's (1980) and IBGE's (1977) proximate compositions of the cupuassu pulp are presented. Chaar (1980) mentions that Alves studied the volatile components of the cupuassu pulp, which is what makes the fruit so attractive.

Chaar (1980) analysed the composition of the cupuassu seed; Chaar (1980) and Vasconcelos *et al.* (1975) analysed the fatty acid composition of the seed oil, and Vasconcelos *et al.* (1975) analysed the physical and chemical constants of this oil. In Table 10.4, these data are compared with those from the cocoa. While the proximate composition of the two species is reasonably similar, the fatty acid compositions are quite different. The cupuassu fats are less saturated than cocoa and, thus, better meet modern dietary restrictions. Chocolate made from this fat (locally called cupulate) may find a ready market.

TABLE 10.1. VARIATION IN CUPUASSU FRESH FRUIT COMPONENTS, MANUALLY EXTRACTED, AS AFFECTED BY FRUIT WEIGHT

		Fruit weight (g)				
Component		500	1000	1500	2000	mean
Pulp	(%)	30.1	35.0	38.8	36.0	35.0
Seed	(%)	23.2	15.0	16.2	19.0	18.3
Exocarp & waste	(%)	46.7	50.0	45.0	45.0	46.7

Source: Calzavara *et al.* (1984).

TABLE 10.2. PHYSICAL-CHEMICAL ANALYSIS OF CUPUASSU PULP

Component		Source A	Source B
Titratable acidity	(%)	2.15	2.35
Total soluble solids	(°Brix)	10.8	10.5
pH		3.3	3.6
Amino acids	(mg% N)	21.9	—
Vitamin C	(mg%)	23.1	28.3
Pectin	(mg/100g)	390.0	703.0
Pulp	(3,000 rpm/10 min) (%)	80.0	—
Viscosity	(cps 21°C, 100 rpm)	—	6750.0
Ether extract	(%)	0.53	—
Starch	(%)	—	0.96
Total solids	(%)	11.0	—
Volatiles	(%)	89.0	—
Reducing sugars	(%)	3.03	3.0
Nonreducing sugars	(%)	—	5.81
Fixed mineral residue	(%)	0.67	—
Phosphorus	($\%P_2O_5$)	0.31	—
Calcium	(mg/100g)	2.9	3.1
Magnesium	(mg/100 g)	—	9.31
Total iron	(mg/100 g)	—	1.52

Source: A—Barbosa *et al.* (1979).
B—Chaar (1980).

The number of analyses available is exceedingly small, which limits our understanding of the variability available for selecting plants that best meet industrial requirements. The cupuassu, however, is currently receiving considerable research attention and it is expected that this will rapidly provide a variety of new data, including data for the plant breeder to use in selecting new clones.

Processing

Both the pulp and the seed are potentially important for developing industrial products. The pulp has only been studied as *materia prima* for nectar, although it is used in ice creams, jellies, puree and canned pulp.

TABLE 10.3. PROXIMATE COMPOSITION OF CUPUASSU PULP

Reference	Water	Proteins	Fats	Carbohydrates	Fiber	Ash
	% Fresh weight					
Chaar (1980)	86.8	1.9	0.5	9.9	1.8	0.7
IBGE (1977)	81.3	1.7	1.6	14.7	0.5	0.7

Barbosa *et al.* (1979) prepared a nectar of uncertain composition and pasteurized it at 90° C for 30 seconds, sealed it in 250-ml cans with internal varnish, let it settle for 10 min, cooled it under running water, and stored it at room temperature. They analysed Brix, pH, acidity, amino acids, vitamin C, pulp (see Table 10.2 for initial data), odor and flavor after 1 day, 1 week, 1, 2, 3 and 15 months. Brix rose slightly (from 12.64 to 12.88), whereas pH (3.4 to 3.25), amino acids (4.9 to 4.04) and vitamin C (3.4 to 2.5) dropped slightly during the 15 months. All these minor modifications were expected by the authors. Odor and flavor remained good throughout the period.

Chaar (1980) has made the most complete study of nectar to date. He prepared a nectar with 37% (by weight) cupuassu pulp, 17% refined sugar and 46% water; used a control, four chemical (1-750 ppm sodium benzoate/250 ppm potassium sorbate; 2-500/500; 3-250/750; 4-750/0) and two pasteurization (75° C for 15 min; 85° C for 5 min) treatments. He analysed Brix, pH, acidity and vitamin C, as well as microbial contamination (at 4-day intervals for 28 days) and used a taste testing panel (at 8-day intervals for 28 days). The 75° C for 15 min pasteurization treatment was statistically superior to the others, both in terms of consumer preference

TABLE 10.4. (A) PROXIMATE COMPOSITION OF THE CUPUASSSU AND COCOA SEED, (B) FATTY ACID COMPOSITION AND (C) PHYSICAL-CHEMICAL CONSTANTS OF THEIR OIL

	% Fresh weight	% Dry weight				
(A)	Water	Proteins	Fats	Carbo-hydrates	Fiber	Ash
Cupuassu[1]	56.6	19.9	50.7	15.8	9.6	4.0
Cocoa[2]	—	12.0	46.3	34.7	8.6	3.4

(B)	C_{16}	C_{18}	$C_{18:1}$	$C_{18:2}$	C_{20}
Cupuassu[1]	6.7	35.5	45.0	3.3	9.1
Cupuassu[3]	5.8	38.3	42.8	8.3	4.8
Cocoa[3]	32.8	35.5	29.6	1.0	1.1

				Index			
(C)	pH	Fusion point	Specific density	Re-fractive	Acid	Iodine	Saponi-fication
Cupuassu[3]	5.7	32-34	0.907	1.4583	4.2	45.9	174.6
Cocoa[3]	6.3	30-35	0.973	1.4567	1-4	32-42	192-198

Sources: [1]Chaar (1980).
[2]Leung and Flores (1961).
[3]Vasconcelos *et al.* (1975).

and of all the quality control analyses made. The first chemical treatment also gave good results.

Oliveira *et al.* (1984) prepared three nectars (1 kg pulp : 4 liters water : 0.9 kg sugar) and varied the amount of citric acid (1, 1.5 and 2 g/l) used to lower pH. These were heated for 3 min at 70° C, sealed in 350-ml cans, and autoclaved for 15 min at 100° C. They measured Brix, pH, acidity, reducing and nonreducing sugars, and vitamin C at monthly intervals for 6 months. All quality control analyses showed only slight variation during the 6-month period.

Barbosa *et al.* (1979) and Chaar (1980) concluded that cupuassu's low pH did not require further adjustment to maintain quality during the test periods, whereas Oliveira *et al.* (1984) adjusted the pH as the first step in their study. It is probable that Chaar (1980) is correct in stating that cupuassu is acid enough not to need adjustment of pH. He also states that the pectin levels are ideal for making jams and jellies.

Venturieri and Aguiar (1988) have studied the centesimal composition of home-made cupulate (chocolate made from cupuassu seed) and commercially available chocolate. This product has not been studied technologically. Calzavara (1970) states that Nestle's has expressed an interest in using cupuassu for making white chocolate, but apparently nothing has come onto the market to date, probably because of uneven quality and supply of the *materia prima.*

Calzavara *et al.* (1984) present 61 recipes for using cupuassu in drinks, pastries, sweets, desserts, and even in pizza! Those tasted by us have been uniformly excellent.

By-products

The exocarp and residue from pulp extraction have been suggested as compost for returning to the plantation to improve the soil and add some nutrients. Silva and Silva (1986) have shown that it can be an important source of potassium (1.5% dry weight) and a moderate source of nitrogen (0.7% dry wt.). Calzavara *et al.* (1984) have suggested its use in animal ration, but the fiber level is so high that this seems unlikely to interest animal nutritionists.

Problems

Since the cupuassu is still a minor crop, just starting to become widely planted, it is difficult to predict what problems will become limiting in the future. At the moment, the Witch's Broom disease is the major limiting factor, followed by low yields and lack of research results.

Cupuassu, like cocoa, should be planted in light to medium shade during juvenal development. Because of the high humidity and temperature characteristic of the humid tropics, ecological conditions are excellent for the proliferation of Witch's Broom, as well as of the cupuassu. This disease attacks the apical and lateral meristems, causing an excessive production of new laterals that form a broom-like structure (hence the name); after a period of growth these new laterals die back. If uncontrolled, the plant invests excessive energy in this vegetative growth, to the point that yields are drastically reduced. At a later stage, the disease can also attack the flowers and fruit, effectively eliminating fruit production.

Witch's Broom is easily controlled by eliminating all symptoms and burning the refuse. Strangely enough, because the cupuassu is a native species, most farmers do not consider that it should be treated like a horticultural crop. Cupuassu researchers and industrialists are trying to educate farmers to recognize the need for adequate management and phytosanitary controls in their plantations but have had little success to date. Until this educational program is successful, cupuassu will not expand as rapidly as it could.

Because control is manual and should be done every two weeks, large plantations require heavy investments in training and maintaining a labor force for phytosanitary control. For this reason, most recommendations are to limit areas to the available, generally family, labor force. A five-working-member family can easily control 5 ha of plantation.

With currently available germplasm (i.e. unselected), even well maintained and fertilized plantations are not very productive. The National Research Institute for Amazonia (INPA) has started to develop a germplasm collection and the local units of the Brazilian Enterprise for Agricultural Research (EMBRAPA) are developing clonal collections. INPA also has several spacing and fertilization trials underway, which will provide the necessary agronomic data to increase yields through management (Venturieri *et al.* 1985).

As noted by the general lack of information available, there is an urgent need to expand research effort with this species. INPA and the regional EMBRAPA both consider the species to be a priority, but government funding is weak and irregular. Private research is isolated and only focussed to specific problems of each industrial unit, rather than focussing on improving all phases of knowledge about cupuassu.

Future

The cupuassu's attractive exotic flavor is sufficient to guarantee it a world-wide market. The research and improvement projects currently underway will surely gain momentum during the coming decade and resolve most of the problems that are currently limiting the expansion of this species. Once this happens, the authors have no doubt that the cupuassu will rapidly develop a world-wide popularity equivalent to or greater than passion fruit (*Passiflora edulis* Sims.).

BACURI

Description, Phenology and Yields

The bacuri (Fig. 10.3) is generally considered to be the single member of its genus, but there is argument about its latin name, *P. esculenta* being cited by FAO (1986) and *P. insignis* by Cavalcante (1976). Its natural distribution is along the forest-savana transitional zone south of the Amazon forest, from the Atlantic to Paraguay. It occurs at low frequency (0.5 to 1 individual/ha) throughout this region, with occasional

Courtesy C. R. Clement

FIG. 10.3. FRUIT OF BACURI, AVERAGING 500 G. THE OPENED FRUIT SHOWS SEVERAL SECTIONS WITH SEEDS AND A COUPLE OF PARTHENOCARPIC SECTIONS.

stands of 50-100 individuals per hectare, possibly due to Amerindian management. It is restricted to upland soils and appears to do better on poor soils than most other regional fruit species. The following description is condensed from Cavalcante (1976) and FAO (1986).

The mature bacuri tree may attain 20 plus m in height and 50 cm dbh and is an element of the mid- to upper canopy in these transitional forests. The cultivated plant, grown in more open condition, may attain 15-20 m in height and have a crown of 10-15 m in diameter. In the individual forest trees, there is apparent apical dominance; although this is not always apparent in field-grown plants. Branches arise at an angle of 50-60° C from the trunk. The short-petioled, smooth, leathery leaves mature to a dark green, attaining 8-14 cm in length by 4-8 cm in width.

The bisexual, solitary flowers are born terminally on young branches just after leaf fall. They have 5 sepals and 5 pale rose-colored petals; the numerous stamens are fused into 5 bundles; the superior ovary has 5 locules. The fruit (Figure 10.3) are ovoid to subglobose, 7-15 cm long and 5-15 cm in diameter, weighing between 200 and 1000 g. The rind is thick (1-3 cm), pale yellow to brownish yellow in color, tough and somewhat elastic, and exudes a yellow latex when bruised. The pulp is creamy white, somewhat mucilaginous, fibrous, juicy and envelopes 1 to 5, generally 3 seeds. When a seed does not develop, the pulp forms a parthenocarpic segment that is greatly appreciated by all consumers. The large seeds are oblong-ovoid, somewhat angular, 4-6 cm long and 2-3 cm in diameter.

The bacuri starts flowering at the beginning of the dry season (June to September in Belém). Harvest season extends from December to May in Belém, with greatest intensity in February and March. A vigorous, mature seedling tree (15-20 years old) will produce about 500 fruit per season (Calzavara 1970), although some have been observed to produce 1000. At 115 plants/ha the yield of fresh fruit may exceed 20 MT.

History of Use

Like the cupuassu, the bacuri appears not to have been domesticated by the eastern Amazonian Amerindians; although they surely managed the species and may have selected larger fruited types. The bacuri has always been one of the more popular fruits on the Belém market.

During the last decade several small industries in Belém have started to produce canned parthenocarpic sections, fruit puree, jams and ice creams. Some of these are now being shipped to southern Brazil but we are unaware of any being exported on a continuing basis.

Composition

Much less research has been done with bacuri than with cupuassu. Barbosa *et al.* (1979) list component analysis based on small fruit (213 g average) from the Belém market. They found only 12% pulp, 18% seed and the rest exocarp. At 20 MT/ha/yr, this is only 2.4 MT of pulp, which is just as difficult to remove from the seed as is cupuassu pulp. The seed (3.6 MT/ha/yr), however, may be useful as a source of oil and animal feed. Barbosa *et al.* (1979) also studied the physical-chemical characteristics of the pulp (Table 10.5). The food value of the edible portion of bacuri has recently been reported by Morton (1987) (Table 10.6).

Bentes *et al.* (1986/87) studied the fatty acid composition of the bacuri seed oil, which is about 46% of the fresh weight of the seed (about 28% is water). Their results are presented in Table 10.7. They also found about 10% of tripalmitin, a glycerol ester of palmitic acid.

Processing

Only the pulp is important for developing industrial food products. This pulp has only been studied as *materia prima* for nectar, although it is used in ice creams, jellies, puree and canned parthenocarpic sections.

Barbosa *et al.* (1979) prepared a nectar of uncertain composition and pasteurized it at 90° C for 30 sec, sealed it in 250-ml cans with internal varnish, let it settle for 10 min, cooled it under running water and stored it at room temperature. They analysed Brix, pH, acidity, amino acids, vitamin C, pulp (see Table 10.5 for initial data), odor and flavor after 1 day, 1 week, 1, 2, 3 and 13 months. Brix (from 13.6 to 12.64) and pH (3.25 to 3.2) dropped slightly, whereas amino acids (4.86 to 2.48) dropped

TABLE 10.5. PHYSICAL-CHEMICAL ANALYSIS OF BACURI PULP

Titratable acidity	(%)	1.6
Pulp (3,000 rpm/10 min)	(%)	100.0
Total soluble solids	(°Brix)	16.4
Reducing sugars	(%)	3.98
pH		3.5
Ether extract	(%)	0.6
Total solids	(%)	19.3
Fixed mineral residue	(%)	0.4
Pectin	(%)	0.12
Phosphorus	(%P_2O_5)	0.13
Volatiles	(%)	80.8
Calcium	(% CaO)	0.31
Amino acids	(mg% N)	38.8

Source: Barbosa *et al.* (1979).

TABLE 10.6. COMPOSITION OF THE EDIBLE PORTION OF BACURI (PER 100 G)

Calories		105	Calcium	(mg)	20.0
Water	(g)	72.3	Phosphorus	(mg)	36.0
Proteins	(g)	1.9	Iron	(mg)	2.2
Lipids	(g)	2.0	Thiamine	(mg)	0.04
Carbohydrates	(g)	22.8	Riboflavin	(mg)	0.04
Fiber	(g)	7.4	Niacin	(mg)	0.50
Ash	(g)	1.0	Ascorbic acid	(mg)	33.0

Source: Morton (1987).

significantly during the 13 months. All these modifications were expected by the authors. Odor and flavor remained good throughout the period.

By-products

The seed, with its high oil level and high percentages of palmitic and oleic fatty acids, may have some use in industry, especially since a plantation of bacuri can theoretically produce about 3.5 MT/ha/yr of seed or 1.5 MT of oil.

The exocarp is the major by-product in terms of quantity. At the moment, it should be composted and returned to the plantation to improve soil quality and reduce mineral fertilizer inputs. There may be an industrial use for the yellow latex, similar to that found in all Guttiferae.

Problems

So little is known about the agronomy and industry of the bacuri that it is impossible to determine problems that might limit its expansion in Amazonia. Propagation by seed or root cuttings is easy, spacing and the rudiments of fertilization are known (Calzavara 1970). It is probable that phytosanitary problems will develop as cultivation becomes more inten-

TABLE 10.7. FATTY ACID COMPOSITION AND PHYSICAL-CHEMICAL CONSTANTS OF BACURI SEED OIL

Fatty acid	%	Constants	
C_{16}	44.2	Specific density	0.896
$C_{16:1}$	13.2	Refractive index	1.457
C_{18}	2.3	Acidity index	14.1
$C_{18:1}$	37.8	Saponification index	205.1
$C_{18:2}$	2.5	Iodine index	47.0
		Index of nonsaponifiable matter	26.4

Source: Bentes *et al.* (1986/87).

sive, but bacuri has so few pests at this time that it is impossible to guess which might become limiting.

Research is obviously a limiting factor. Germplasm collections are nonexistent and must be a priority. Once germplasm is available, this must be chemically and technologically analysed to guide the plant breeder in selecting those types most suitable for industrialization and commercialization. More complete technological experimentation is necessary to guide the development of new products.

Future

The bacuri's strong attractive flavor guarantees it a place in the world market for exotic fruit flavors. It will not reach the world market, however, until it is agronomically produced. Some research is underway at the local unit of the Brazilian Enterprise for Agricultural Research in Belém and at a state unit in Maranhão. Much more research is required than is currently planned.

FINAL COMMENTS

As with many tropical fruits, both the cupuassu and the bacuri may only become important when an entrepreneur takes a liking to them or they are taken to other areas of the humid tropics. International law restricts the movement of all *Theobroma* germplasm because of the risks of taking an Amazonian pest to other areas. If any entrepreneur wants to take cupuassu to another country, this must be done through the Brazilian National Center for Genetic Resources (CENARGEN), a unit of EMBRAPA, so as to avoid any possibility of distributing *Theobroma* pests to centers of cocoa production. Although the bacuri is not considered to be a problem in this respect, any transferral of germplasm should also be done through CENARGEN.

REFERENCES

BARBOSA, W. C., de NAZARÉ, R. F. R., and NAGATA, I. 1979. Physical and chemical studies of fruits: bacuri (*Platonia insignis*), cupuassu (*Theobroma grandiflorum*) and murici (*Byrsonima crassifolia*). An. Congr. Bras. Fruit. 5, 797-808. (Portuguese)

BENTES, M. H. S., SERRUYA, H., ROCHA, G. N., GODOY, R. L. O., CABRAL, A. S., and MAIA, J. G. S. 1986/87. Chemical study of the seeds of bacuri. Acta Amazonica 16/17, 363-368. (Portuguese)

CALZAVARA, B. B. G. 1970. Fruit trees: Abiu, Mamey, Bacuri, Biriba, Cupuassu. Series: Cultures of Amazônia, Vol. 1, 2, 45-84. IPEAN, Belém. (Portuguese)

CALZAVARA, B. B. G., MULLER, C. H, and KAHWAGE, O. N. C. 1984. Tropical Fruit Culture: Cupuassu tree. Cultivation, benefits and utilization of fruit. EMBRAPA/CPATU, Belém. (Portuguese)

CAVALCANTE, P. B. 1976. Edible fruits of Amazônia, 3rd Ed. INPA, Belém. (Portuguese)

CHAAR, J. M. 1980. Composition of cupuassu (*Theobroma grandiflorum* Schum.) and physical and chemical properties of canned nectar. M.S. Thesis, Rural Federal Universlty of Rio de Janeiro, Rio de Janeiro. (Portuguese)

CLEMENT, C. R. 1983. Underexploited Amazonian fruits. Proc. Am. Soc. Hortic. Sci., Trop. Reg. 27(A) 117-142.

CUATRECASAS, J. 1964. Cacao and its allies, a taxonomic revision of the genus *Theobroma*. Contrib. U.S. Nat. Herb. 35(6) 379-614.

DUCKE, A. 1946. Plants of precolombian culture in Amazônia Brazil. Bol. Tec. Inst. Agron. Norte 8, 1-24. (Portuguese)

FALCÃO, M. A. and LLERAS, E. 1983. Phenological and ecological aspects of cupuassu production (*Theobroma grandiflorum* (Willd. ex Spreng.) Schum.). Acta Amazonica 13, 725-735. (Portuguese)

FAO. 1986. Food and Fruit-bearing Forest Species. 3. Examples from Latin America. FAO Forestry Paper 44/3. FAO, Rome.

IBGE. 1977. Food Composition Tables. Inst. Bras. Geogr. Stat., Rio de Janeiro. (Portuguese)

LEUNG, W. T. W. and FLORES, M. 1961. Food composition tables for use in Latin America. Institute of Nutrition of Central America and Panama and National Institutes of Health, Bethesda, Maryland.

MORTON, J. L. 1987. Fruits of Warm Climates. Julia F. Morton Publisher, Miami, Fl.

OLIVEIRA, M. L. S., DE HOLANDA, L. F. F., MAIA, G. A. and ORIA, H. F. 1984. Study of the stability of cupuassu nectar (*Theobroma grandiflorum* Schum.). Cienc. Agron. 15 (1/2) 75-77. (Portuguese)

SILVA, A. Q. and SILVA, H. 1986. Content of nutrients in cupuassu (*Theobroma grandiflorum*) Tech. Note. An. Congr. Bras. Fruit. 8, 269-272. (Portuguese)

VASCONCELOS, M. N. L., SILVA, M. L., MAIA, J. G. S. and GOTTLIEB, O. R. 1975. Chemical studies of the seeds of cupuassu. Acta Amazonica 5(3) 293-295. (Portuguese)

VENTURIERI, G. A. 1984. Cupuassu tree—typical fruit of Amazônia with the prospect for export. Inf. Soc. Bras. Frutic. 3(4) 7. (Portuguese)

VENTURIERI, G. A. and AGIUAR, J. P. L. 1989. Composition of the chocolate made from cupuassu (*Theobroma grandiflorum*). Acta Amazonica 18(1/2) 3-8. (Portuguese)

VENTURIERI, G. A., ALVES, M. L. B. and NOGUEIRA, M. D. 1985. Cultivation of the cupuassu tree. Inf. Soc. Bras. Frutic. 4(1) 15-17. (Portuguese)

11

Breadfruit and Jackfruit

Pattisapu Narasimham

The breadfruit and jackfruit belong to the genus *Artocarpus* (family Moraceae) and hence, they are botanically closely related. Breadfruit, also called breadnut (seeded form) is indigenous to Polynesia and is widely distributed over the tropics of both hemispheres. The jackfruit is indigenous to India and is widely grown in Bangladesh, Burma, Sri Lanka, Malaysia, Indian Indonesia, Philippines, Brazil, and other countries. The English names for the jackfruit are Jack, Jaka, Jak and Jackfruit, of which the last is the one most widely used.

BREADFRUIT

Breadfruit, a native of Polynesia, is now grown in Western Samoa, Indo-Malay archipelago, Tahiti, Puerto Rico, New Guinea, Colombia, and as far as western Micronesia. It was introduced into the West Indies in the 18th century by Captain Bligh.

BOTANIAL DESCRIPTION

Botanical names of the breadfruit tree are: *Artocarpus communis* Forst, *Artocarpus incisus* Linn (Barrau 1959), but the name most widely accepted today (Stone 1974) is *Artocarpus altilis* (Park) Forsberg.

The tree, with its straight trunk, thick buttresses and branchlets, and dense heavy foilage, reaches a height of 20 to 30 m (Fig. 11.1). The leaves are alternate and crowded at the tip of the branchlets; they are 40-60 cm long and 28 cm wide, thick and leathery, short and stout petioled, ovate in general outline but pinnately lobed with narrow sinuses and acute lobes, the bases narrowed and entire. The inflorescences are

FIG. 11.1. FULL GROWN BREADFRUIT TREE IN THE HOME GARDEN OF A PRIVATE HOUSE IN MYSORE CITY. THE TREE IS ABOUT 15 M IN HEIGHT AND BEARS DURING JUNE

unisexual - monoecious and axillary. The male inflorescences are 15-30 cm long, about 3 cm thick, cleavate, yellow or greenish yellow, drooping or curving downwards. Each flower bears a single stamen, and the stalk is 5-6 cm long. The female inflorescences are globose-ovoid, the flowers being crowded together at the base, free at the tip, and closed to a minute opening.

The fruit is a syncarp, greenish yellow or yellowish brown in color, oblong or nearly globose, spiny, 20-30 cm in diam, 10-30 cm long, and commonly seedless or nearly seedless (Fig. 11.2). The fruit pulp is moist, pale yellow or whitish in color, with a mild flavor. If seeds are present they are about 2.5 cm long, rounded and flattened by compression (Ochse 1931).

HORTICULTURE

The tree is attractive in appearance and yields useful fruits. Hence, breadfruit is a popular home garden plant in south India, as well as in many parts of the world. It can be cultivated as plantations, but is generally cultivated in mixed plantings with other fruit trees because of its lack of popularity as a commercial fruit. Although breadfruit gives the

FIG. 11.2. A CLOSE-UP VIEW OF A BRANCH, WITH LEAVES AND FOUR FRUITS LOCATED IN THE AXILS, OF THE SAME BREADFRUIT TREE SHOWN IN FIG. 11.1.

impression of being a dessert fruit, it is mostly used in the mature, unripe stage for culinary purposes like any other vegetable.

VARIETIES

Although there may be a number of varieties in cultivation, the commercial varieties are not clearly identified. According to Wootton and Tumaalii (1984A) there are over 30 recognized varieties in Samoa. They are distinguished primarily by the presence or absence of seeds in the fruit, the shape and skin texture of the fruit, leaf texture, differences in season of maturity, cooking quality and flavor. From their study, it was established that the chemical composition of the fruit varied from variety to variety. Therefore, clear identification of varieties based on nutritional value and suitability for processing is very much desirable.

SOIL AND CLIMATE

The tree requires a warm, humid maritime climate and thrives best on coastal plains. Young plants grow best under shade, but later need full exposure to sunlight (Munjunath 1948). The plants grow well on rich, deep, well-drained alluvial, open textured or laterite soils.

PROPAGATION

The seeded varieties of breadfruit are generally propagated by seeds. Both seeded and seedless varieties can be propagated by branch and root cuttings. In India and in other places, breadfruit is commonly propagated by root suckers. In Mauritius, successful multiplication of root cuttings set in solar propagation was achieved (Anon 1951). In the Seychelles (Anon 1964), breadfruit has been propagated successfully through root cuttings. Muzik (1948) propagated breadfruit in Liberia by branch cuttings (30-40 cm long, with 3-4 nodes, and about 1 cm diam), after dipping in 1% indole butyric acid solution and planting in sand immediately after treatment. By this method, 80% success was achieved. As with jackfruit, air layering along with growth regulator treatment and proper care after cutting until transplantation, ensured good success. Vegetative propagation is essential to maintain breadfruit composition, nutritive value and processing quality of the fruit.

PLANTING TIME AND METHOD

With a spacing of 10 m x 10 m about 100 plants can be planted per hectare. For planting, 1 m cube pits are dug about 10 days prior to planting and 30 kg of well composted farm yard manure and one-half kg of superphosphate are mixed with the soil to refill the pits. Insecticide (50 g BHC with 10% a.i.) may also be added to protect the young plants from pests. The best time of planting in India is June to August. Planting should coincide with the rainy season for better survival of the plants.

YIELD

Breadfruit trees start to yield in 5 to 6 years after planting. The mature tree bears about 700 fruit per year (Manjunath 1948; Purseglove 1968) weighing 1 to 4 kg per fruit. Bowers (1981) however, made a reasonable estimate that the average yield per tree is about 200 fruit with an average weight of 1 kg per fruit. A similar estimate was given by Van Wissen (1978) for western Samoa, with a total yield of 20 MT per hectare when planting density was 100 trees per hectare. Wootton and Tumaalii (1984B) estimated that this would result in about 6 MT/ha of dry flour; a yield comparable to that of corn or maize.

PESTS AND DISEASES

The literature contains only a few reports on the pests and diseases of breadfruit. However, most of them can be controlled by proper treatment. Johnston (1964) gave a short review of the various pests and diseases of breadfruit. Zaiger and Zentmeyer (1965, 1967) reported a lethal disease of breadfruit trees called 'Pinglap' disease, an epidemic which spread through the western and central Pacific islands. *Phomopsis* spp. was found in the twigs and branches of diseased plants while *Fusarium solani* and *Pythium* spp. were found in the necrotic feeder roots. Those authors postulated the possibility of vascular wilt or root rot as cause for the disease. Saraiva (1965) reported the presence of fruit flies on breadfruit in Cape Verde Islands. Trujillo (1970) reported *Phytophthora* fruit rot of breadfruit.

MATURITY AND HARVEST

Maturation of fruit requires 60-90 days after inflorescence, depending upon the variety, climate, soil and cultural conditions (Purseglove 1968). The fruit is harvested when it is still firm, hard and unripe. Maturity indices are based on exudation of sap (latex), skin color, appearance and color of the pulp. Graham and De Bravo (1981) considered that a fruit is immature if it exhibits little exudation of sap and shows oxidative discoloration when cut. A very mature fruit may have yellowish skin and may exhibit crevices or cracks in the skin with the pulp quite yellow in color, and the fruit may be quite soft when pressed with the fingers. A mature fruit is one which is comparatively larger than an immature one; the skin, depending on the variety, may show some yellowing or less greening. The pulp is white (white heart) or yellow and, when cut, may show little or no discoloration.

The fruit is harvested by hand, often with the aid of simple devices such as poles with small forks at the upper end. Although breadfruit ripens normally with a decrease in starch content accompanied by a tremendous increase in sucrose and total sugars, the fruit is normally harvested in the immature stage for consumption and is never allowed to ripen. Hence, there is a great need for establishing maturity standards for different varieties not only to extend the storage life of fresh fruit, but also to meet processing needs.

FRUIT RIPENING

Biale and Barcus (1970) classified breadfruit as the climacteric type; ripening of the fruit occurs, as in other climacteric fruits, with a significant rise in respiration rate at the peak stage of ripening. However, ripening is not of great importance because the fruit is not allowed to ripen, and is only consumed or processed at an immature or just mature stage. Graham and De Bravo (1981) found that the total sugar content of the different parts of the breadfruit showed an increase accompanied by softening of the fruit (decrease in thumb pressure resistance) as the fruit passes from mature to very mature stage while on the tree. Working on seven Samoan breadfruit varieties, Wootton and Tumaalii (1984B) made a similar observation of increase in sucrose content with a concomitant decrease in starch at a very mature stage of harvest. These studies, therefore, clearly indicate that the fruit will start ripening on the tree if harvest is delayed; hence, there is all the more reason to harvest the fruit at the appropriate stage of maturity to extend post-harvest storage life and maintain processing quality.

STORAGE

Breadfruit, like many other tropical fruits, has a high post-harvest respiration rate with a short post-harvest storage life. Post-harvest storage losses reach as high as 50% (Coursey *et al.* 1976).

Traditional Method of Storage

In the traditional method of storage, the fruit is kept under water (steeped) throughout the period of storage. This method leads to the splitting of the skin of the fruit accompanied by softening, with adverse effects on the skin. However, the pulp is considered acceptable. By adopting this method, the fruit could be kept for about 7 days in Samoa (Thompson *et al.* 1974). It may be possible to overcome these deleterious effects by adjusting the pH of the water and using antimicrobials.

Refrigerated Storage

Due to the high post-harvest perishability of the fruit in India—where it is grown mostly as a home garden plant—breadfruit are picked from the tree as needed, and marketed on the day of harvest. Sometimes, processing is carried out on home/cottage scale to produce dehydrated forms (chips and powder). However, when breadfruit is grown on a large scale it is imperative to have other methods of storage.

Thompson *et al.* (1974) studied the effects of temperature, packing, harvesting method and cultivar on the refrigerated storage life of the fruit. Storage life increased with a decrease in storage temperature but chilling injury appeared at temperatures below 12° C. Hence, the optimum cold storage temperature of breadfruit is 13 ± 1° C with an RH of 90 ± 5%. Storage in polyethylene bags was found beneficial, while storage life was unaffected by the stage of maturity at harvest or harvesting method. Fungal and bacterial infections were not encountered during refrigerated storage, and hence, Thompson *et al.* (1974) considered them as unimportant. Further, they observed that the normal ripening processes take place during storage and storage life of fruit was 14 days at 12° C. These results indicate that they may have used fruit harvested at immature and mature stages and, hence, did not find any significant difference due to stage of maturity at harvest. On the other hand, Mariott *et al.* (1979) determined the factors affecting the refrigerated storage life of fresh breadfruit and reported that the storage period between harvesting and softening of fruit was greater for fruit harvested at a partially mature stage than for fully mature fruits. Fruit which had been harvested by picking or catching had a longer storage life than fruit harvested by traditional methods. In four storage trials and one shipping trial at refrigerated temperature, the majority of the fruit picked at a partially mature stage and packed in polyethylene bags had a storage life of 15 days. Passam *et al.* (1981) successfully stored Trinidad breadfruit cv. Whiteheart, in polyethylene bags at 14° C up to 14 days, but they could not derive any additional benefit by using either the ethylene oxidizing agent (Purafil) or a wax emulsion (Stay Fresh).

Therefore, the stage of maturity of fruit at harvest determines the storage life of the fruit and its post-storage quality. It appears possible to harvest breadfruit at immature stages without any adverse effects on its composition and quality as noted by the composition of fruit harvested at different maturity stages, including a 'very mature' stage (Graham and De Bravo 1981; Wootton and Tumaalii 1984B).

FREEZING

Passam *et al.* (1981) successfully prepared a frozen product from peeled breadfruit cut into segments, blanched for 2 to 5 minutes in boiling water, and then frozen at –15° C. The frozen product was stored for 11 weeks; on cooking, it compared favorably in color, flavor, texture and taste with fresh breadfruit. Improved methods of blanching, quick freezing, and storage at optimum low temperature might impart a longer storage life to the frozen product.

COMPOSITION

Composition of Pulp

Breadfruit could be separated into 3 different parts, namely the skin, stem and heart, and pulp. The fleshy pulp is usually consumed and makes up between 62.8 and 73.9 percent of the fruit weight (Table 11.1) (Graham and De Bravo 1981) irrespective of stage of maturity.

VARIATION OF CHEMICAL COMPOSITION WITH MATURITY

Breadfruit is a rich source of starch, proteins, minerals, and some vitamins. The major constituents of breadfruit, besides moisture, are

TABLE 11.1. PERCENTAGES OF DIFFERENT PARTS OF BREADFRUIT FROM PUERTO RICO

	Part of breadfruit	Percent (W/W) Max	Min	Avg
I.	Seedless (mature)[1]			
	a. Skin	12.5	9.3	11.2
	b. Stem and heart	16.9	14.3	15.6
	c. Pulp[2]	76.9	71.5	73.9
II.	Seeded (breadnut)[3]			
	a. Skin	16.6	14.3	15.5
	b. Core	6.7	5.4	6.0
	c. Pulp	46.7	33.2	40.0
	d. Seeds	46.8	30.1	38.4

[1]Source: Adapted from Graham and De Bravo (1981).
[2]The average pulp content for very immature, immature, mature and over-mature stages ranged from 62.8 to 73.9% with over 70% for immature and mature stages.
[3]Source: Adapted from De Bravo *et al.* (1983).

starches, sugars, proteins and crude fiber. These constituents vary from the very immature stage to the very mature stage of development of fruit on the tree. Graham and De Bravo (1981) studied the composition of fruit from Puerto Rico harvested at very immature, immature, mature and very mature stages, whereas Wootton and Tumaalii (1984B) studied the variations in the composition of seven varieties of breadfruit from Samoa which were also harvested at the same four stages. The chemical compositions of flours made from these fruits are given in Table 11.2.

TABLE 11.2. COMPOSITION (% DRY WEIGHT) OF BREADFRUIT FLOUR FROM PULP OF FRUITS AT DIFFERENT STAGES OF MATURITY

Constituent	Very immature	Immature	Mature	Very mature
Moisture				
Range[1]	11.2	3.4–21.0	2.8–19.0	2.6–14.9
Average[1]	11.2	10.5± 6.7	10.1± 5.7	9.4± 6.7
Average[2]	5.3±0.06	3.5± 0.07	2.5± 0.04	6.0± 0.01
Crude protein				
Range[1]	4.7	3.6– 4.8	2.9– 5.0	4.0– 5.1
Average[1]	4.7	4.3± 0.5	3.8± 0.8	4.4± 0.5
Average[2]	3.9±0.02	3.9± 0.02	4.1± 0.01	4.1± 0.01
Crude fat				
Range[1]	0.8	0.8– 1.3	0.8– 1.9	0.8– 1.6
Average[1]	0.8	1.1± 0.2	1.3± 0.4	1.3± 0.3
Average[2]	1.8±0.07	1.4± 0.09	1.1± 0.07	2.6± 0.06
Starch				
Range[1]	65.4	62.7–75.7	61.5–73.1	53.4–67.9
Average[1]	65.4	69.0± 4.7	68.6± 4.3	57.8± 6.8
Average[2]	60.1±0.45	71.2± 0.03	69.2± 0.41	42.6± 0.70
Crude fiber				
Range[1]	4.7	3.2– 6.9	2.9– 5.3	2.8– 5.2
Average[1]	4.7	4.9± 1.5	4.3± 0.9	4.2± 1.0
Average[2]	8.4±0.20	4.4± 0.14	3.5± 0.03	6.5± 0.01
Ash				
Range[1]	3.1	1.7– 4.1	1.9– 3.2	2.4– 3.9
Average[1]	3.1	3.0± 0.8	2.6± 0.4	3.0± 0.7
Average[2]	3.1±0.05	2.9± 0.03	2.9± 0.03	3.7± 0.04
Total sugars				
Range[1]	19.7	2.2–18.6	11.5–22.7	14.6–31.8
Average[1]	19.7	12.6± 5.8	17.9± 4.2	26.5± 8.1
Average[2]	8.2±0.17	2.8± 0.03	4.1± 0.06	26.8± 0.6

[1]Source: Data adapted from Wootton and Tumaalii (1984B) for 7 Samoan breadfruit varieties; very immature (1 variety); immature and mature (6 varieties) and very mature (4 varieties). Average between varieties with standard deviation presented.

[2]Source: Values adapted from Graham and De Bravo (1981) for one variety of breadfruit from Mayaguez, Puerto Rico.

Proteins

In Samoan breadfruit, protein contents ranged between 2.9 and 5.0%, varying not only with the stage of maturity but also with variety. On the average, irrespective of variety, the protein content showed a slight decrease as the fruit matured from a very immature stage to a mature stage, while the protein content reached a maximum at a very mature stage. In breadfruit from Puerto Rico, the protein contents varied from 3.85 to 4.12% from very immature to very mature stages, with a slight increase at every stage (Table 11.2).

Starch

In both the Samoan and the Puerto Rican breadfruit varieties, the starch content increased from the very immature to the mature stage. At the very mature stage, it showed a significant decrease. At the mature stage, the average starch content in all varieties was about 70% being almost equal to that of the immature stage (Table 11.2).

Total Sugars

In general, total sugars showed a decrease from the very immature to the mature stage; this was followed by a significant increase at the very mature stage. This striking increase in total sugars, accompanied by a significant decrease in starch content at the very mature stage, might be due to the commencement of fruit ripening while on the tree (Table 11.2).

Crude Fiber

In all varieties, the crude fiber content was high at the very immature stage and showed moderate variation during maturation, with a tendency to decrease. The crude fiber content at the mature stage averages around 4% in all varieties (Table 11.2).

Crude Fat

The fat content of breadfruit pulp is very low and in Table 11.2 it ranged between 0.8 and 1.9% on a dry weight basis for the different stages of maturity.

Ash

The ash content of breadfruit in general decreased from the very immature to the mature stage. It ranged between 1.69 and 4.08% with an average value of 3.0% at the mature stage (Table 11.2).

COMPOSITION OF MATURE BREADFRUIT FROM DIFFERENT REGIONS

The variations in the composition of mature breadfruit from Puerto Rico, Samoa and India given in Table 11.3 show that compositional variation of breadfruit exists from region to region. Highest protein content of 7.32% is reported from India. Although the total carbohydrates from Puerto Rico and India are about the same, the mineral content was higher in the Indian than in the Puerto Rican breadfruit. Differences in crude fat, crude fiber and starch were negligible. However, the total sugars were higher in breadfruit from Samoa than from Puerto Rico.

MINERALS AND VITAMINS

The minerals present in breadfruit are calcium, phosphorus, potassium, sodium, magnesium, iron, copper, zinc and manganese; of which calcium, phosphorus, magnesium and iron are present in considerable quantities. Vitamin A, vitamin C, niacin and riboflavin are the vitamins reported to be present in the pulp of breadfruit. Breadfruit appears to be a good source of niacin and riboflavin (Table 11.4). The vitamin and mineral contents of the fruit may vary from region to region. A higher content of minerals was reported in breadfruit from India than from Puerto Rico. Wootton and Tumaalii (1984B) found that Samoan breadfruit had similar amounts of iron, sodium and calcium, about one-half as

TABLE 11.3. COMPOSITION (% DRY WEIGHT) OF BREADFRUIT FROM DIFFERENT REGIONS

	Puerto Rico				
		Seeded[2]			
Constituent	Seedless[1] pulp	Skin and pulp	Seed pulp	Samoa[3] pulp	India[4] pulp
Crude protein	4.1 ± 0.01	6.4	13.2	3.8 ± 0.80	7.3
Crude fat	1.1 ± 0.07	6.4	6.2	1.3 ± 0.42	1.0
Carbohydrates	84.2	76.2	74.6	—	87.3
Starch	69.2 ± 0.41	46.2	53.4	68.6 ± 4.2	—
Crude fiber	3.5 ± 0.03	18.0	2.5	4.3 ± 0.88	—
Total sugars	4.1 ± 0.06	—	—	17.9 ± 4.2	—
Ash	2.9 ± 0.03	1.8	3.7	2.6 ± 0.42	—
Minerals	1.9	*	*	—	4.4

[1]Source: Graham and De Bravo (1981) average values of mature breadfruit.

[2]Source: De Bravo *et al.* (1983); * indicates skin, pulp and seed contained high quantities of Ca, K and Fe and niacin.

[3]Source: Wootton and Tumaalii (1984B) average values for mature fruit of 7 varieties.

[4]Source: Manjunath (1948).

TABLE 11.4. MINERAL AND VITAMIN COMPOSITION (MG PER 100 G DRY WEIGHT) OF BREADFRUIT PULP

Constituent	India[1]	Puerto Rico[2]
Total minerals:	4390	1856
Potassium	—	1630 ± 127
Phosphorus	146	90 ± 0.2
Magnesium	—	80 ± 0.04
Calcium	195	50 ± 0.2
Sodium	—	2.8 ± 0.2
Iron	2.4	1.9 ± 0.2
Copper	—	0.3 ± 0.05
Zinc	—	0.4 ± 0.03
Manganese	—	0.2 ± 0.02
Vitamins:		
Vitamin A (I.U./100 g dry wt)	24.4	—
Niacin	—	2.4
Riboflavin	—	0.2
Ascorbic acid	—	23

[1]Source: Manjunath (1948).
[2]Source: Adapted from Graham and De Bravo (1981).

much potassium, and four times as much phosphorus as the breadfruit from Puerto Rico (Table 11.4) observed by Graham and De Bravo (1981).

PHYSICOCHEMICAL PROPERTIES OF STARCH

Isolation and characterization of starch from pulp of breadfruit was first reported by Correa *et al.* (1970) and later by Loos *et al.* (1981) and Tumaalii (1982). The amylose content in different varieties of breadfruit from Samoa showed a variation from 16.4 to 19.8%. Loos *et al.* (1981) found an amylose content of 18.2% in the starch; they also found that the debranched amylopectin contained chains with degrees of polymerization of 15 and 38.45 in the molar ratio of 4.4:1. High molecular weight (degree of polymerization greater than 600) linear polymers in debranched β-limit dextrin amounted to 15%. The physicochemical properties of the starch show that the starch granules are small in size, ranging from 10 to 20 μm in diameter and that the shape ranges from spherical to oval or elliptical. The swelling and solubility characteristics indicated a two-stage pattern; the bonding between the granules is weak and nonhomogeneous. The iodine affinity test, Brabander Viscosity Curves at low concentrations of starch, and solubility in dimethyl sulfoxide also suggest a low amylose content and a compact or homogeneous structure (Table 11.5). Wootton and Tumaalii (1984A) suggested that further

TABLE 11.5. COMPOSITION AND CHARACTERISTICS OF BREADFRUIT STARCH

	Constituent/property	Source 1	Source 2
I.	Amylose content (%)	18.2	Low
II.	Degree of polymerization of debranched amylopectin	15 and 38.5	—[3]
	Molar ratio	4.4:1	—
	High mol. wt. linear polymers (>600) in debranched β-limit dextrin (%)	15	—
III.	Physicochemical properties:		
	Shape of granule	spherical and segmented (compound)	oval or elliptical
	Size (diameter - μm)	10-20	—
	Swelling and solubility	2 stage pattern	high (indicating non-homogenous bonding within the granules)
	Iodine affinity (%)	—	1.81 (indicates low amylose)
	Brabender viscosity curves at low concentrations	—	indicate similarity to artificially cross linked waxy starches
	Solubility in dimethyl sulphoxide	—	suggests compact or homogeneous structure

[1]Souce: Loos *et al.* (1981).
[2]Source: Correa *et al.* (1970).
[3](—) not determined.

studies on breadfruit starches might reveal properties which might lead to their utilization as a commercial product.

FATTY ACID COMPOSITION

The fat content of breadfruit pulp is low; a study by Achinewhu (1982) on fatty acid composition revealed the presence of palmitic, stearic, oleic and linoleic acid as the predominant fatty acids among the 11 fatty acids detected. Oleic, linoleic and linolenic acids together comprised over 58% of the total fatty acid content (Table 11.6).

COMPOSITION OF SKIN, STEM, AND HEART

The chemical composition of the skin, which forms nearly 11% of the total fruit weight is comparable to that of the pulp in protein, crude fiber, total carbohydrates and minerals (Table 11.7). Graham and De Bravo (1981) suggested that the skin, which becomes available as a by-product

TABLE 11.6. FATTY ACID COMPOSITION OF FAT FROM PULP OF MATURE BREADFRUIT

Fatty acid	Fatty acid content (as % of total oil)
$C_{14:0}$ Myristic	0.22
$C_{16:0}$ Palmitic	19.10
$C_{17:0}$ Margaric	1.20
$C_{18:0}$ Stearic	17.80
$C_{20:0}$ Bebenic	0.93
$C_{18:1}$ Oleic	24.5
$C_{18:2}$ linoleic	30.5
$C_{18:3}$ linolenic	2.4
$C_{20:1}$ —	0.13
$C_{22:1}$ —	0.44
Anything beyond $C_{22:1}$	2.40
Ether extract	11.9

Source: Achinewhu (1982).

when the pulp is converted into flour, can be used as animal feed, while the stem and heart which are similar in composition to the pulp and have a lower polyphenol content than skin, need not be separated from pulp and can be converted along with the pulp into flour for human consumption. This is the normal practice for consumption in other regions.

TABLE 11.7. CHEMICAL COMPOSITION (% DRY WEIGHT) OF NONEDIBLE PARTS OF MATURE BREADFRUITS

Constituent	Skin	Stem and heart	Average
Total carbohydrate	70.7 ± 0.03	80.7 ± 0.91	75.7
Starch	57.4 ± 0.14	63.4 ± 0.20	60.4
Alcohol-soluble carbohydrates	5.3 ± 0.19	5.8 ± 0.4	5.5
Total sugars	3.8 ± 0.21	4.1 ± 0.03	4.0
Reducing sugars	2.7 ± 0.21	2.9 ± 0.14	2.8
Crude protein	5.9 ± 0.12	6.0 ± 0.03	6.0
Crude fat	3.9 ± 0.04	1.6 ± 0.06	2.8
Crude fiber	5.7 ± 0.01	6.6 ± 0.12	6.1
Ash	6.7 ± 0.04	4.0 ± 0.03	5.4
Niacin (mg/100 g)	4.4	3.2	3.8
Riboflavin (mg/100 g)	0.3	0.4	0.4
Ascorbic acid (mg/100 g)	20.0	14.0	17.0

Source: Adapted from Graham and De Bravo (1981). Values represent mean ± standard deviation, determined from dried powders.

YIELD OF SEEDS

In some regions, Colombia, Valle del Canca, Choco and San Andre's islands, the seeded form of breadfruit is grown. In those regions, the pulp of the fruit, though sometimes eaten by humans, is mostly used as animal feed (Quijano and Arango 1979). The seeded form of breadfruit is oblong or globose in appearance, 10-30 cm in diam, with a yellowish-green rind covered with fleshy prickles. The weight of fruit varies between 1 to 4 kg, depending on variety, age of the tree, and the number of fruit per tree. A young, six-year-old tree may yield as low as 33 fruit over a 7-month period (Bennett and Nozzolillo 1987) compared to 700 fruit per tree per year from a mature breadfruit tree (Purseglove 1968). An average yield of about 200 fruit per tree with an average weight of 1 kg/fruit suggested by Bowers (1981) appears to be a quite reasonable yield (Table 11.8).

The seeds fill the entire interior of the fruit. The seeds are brown in color, round or flattened and about 2 cm long. They are chestnut like in size and flavor, and are called Chataigne in French-speaking areas and Castana in Spanish-speaking areas (Bennett and Nozzolillo 1987). The number and weight of seeds varies from region to region. De Bravo *et al.* (1983) reported that the number of seeds for Puerto Rican breadfruit varied between 46 and 94 constituting 300 to 690 g of the fruit weight, with the mean weight of one seed being 7.0 g (average value in Table 11.8). On the other hand, Quijano and Arango (1979) reported that the mean weight of seeds (total) was 809 g as against a mean weight of fruit of 1.52 kg. The average number of seeds was 890 with an average seed weight of 10.1 g (Table 11.8). The yield of seeds was 53% of the fruit weight. Working with a young 6-year-old breadfruit tree from Trinidad, Bennett and Nozzolillo (1987) observed that the mean number of seeds was 59 and the mean total weight of seeds was 350 g per fruit. Each seed, on an average weighed 7.7 g (Table 11.8). In contrast with 70% edible portion (pulp) in unseeded varieties (Purseglove 1968), seeded breadfruit contain about 50% pulp, moist seeds contain between 56 and 65% water (Quijano and Arango 1979; De Bravo *et al.* 1983).

Based on the above information and assuming planting density of 100 trees per hectare, the yield of dry seeds would vary from 1.41 to 3.2 MT with an average yield per hectare of 2.4 MT (Table 11.8). This yield is quite significant especially if one takes into consideration the yield of pulp also; cultivation of seeded breadfruit, if properly utilized, may be even more profitable than cultivation of the seedless form.

TABLE 11.8. BREADFRUIT SEEDS, NUMBER, WEIGHT AND YIELD PER HECTARE FOR BREADFRUIT FROM DIFFERENT REGIONS

Source of fruit	Seeds per fruit	Weight of seed (g)	Total weight of seeds per fruit (g)	Yield of seeds[4]				Yield of material other than seeds (MT/ha)	
				Fresh wt per tree (kg)	MT/ha	Percent dry material	Dry seeds MT/ha	Fresh	Dry[5]
Trinidad[1]	59	7.7	350	70.0	7.0	40.9	2.86	4.7	1.4
Puerto Rico[2]	71	7.0	497	99.4	9.94	35.0	3.48	6.6	2.0
Colombia[3]	80	10.1	809	161.8	16.18	46.9	7.57	10.8	3.2
Average	70	8.3	552	110.4	11.04	40.9	4.64	7.4	2.2

[1]Source: Bennett and Nozzolillo (1987) from six-year-old tree.
[2]Source: De Bravo *et al.* (1983).
[3]Source: Quijano and Arango (1979).
[4]Source: Bowers (1981)—average yield of 200 fruit per tree with 100 plants per hectare.
[5]Source: Wootton and Tumaalii (1984A) on the basis of 40% dry wt of pulp.

COMPOSITION OF SEED

Chemical Constituents

Quijano and Arango (1979) reported (Table 11.9) that moist seeds contain moisture 56.3%, proteins 8.73%, fat 5.59%, fiber 1.69%, minerals 1.50% and other carbohydrates 26.23%. On a dry weight basis, the seed contains 20% protein, 12.8% fat, fiber 3.9%, minerals 3.4% and other carbohydrates 60%. On a dry weight basis the chemical composition of breadfruit seed with respect to proteins and caloric value is, from a nutritional point of view, superior to that of the cereals rice, wheat and maize, and inferior to that of soybean, coconut and groundnut.

Physicochemical Properties of Seed Oil

Quijano and Arango (1979) worked out the physicochemical properties of seed oil. The seed oil is a light yellow colored, viscous liquid at room temperature, having a characteristic penetrating odor similar to that of peanut oil. The physicochemical properties of the oil listed in Table 11.10 indicate properties similar to most oils found in tropical plants. The saponification value and iodine value indicate that the oil is superior to palm oil and coconut oil and is probably equal to olive oil. The other vegetable oils are definitely superior because of higher percentage of unsaturated fatty acids as indicated by higher iodine values.

Proteins

Of the 20% proteins reported by Quijano and Arango (1979), nonglobular proteins constitute 6.4% and globular proteins 13.5%. Further analysis of the globular proteins indicated the presence of albumins 1.8%, globulins 3.7%, prolamines 3.3% and glutelins 4.7%. For use in baking, the protein composition of seeds together with the presence of fats and carbohydrates appears to be very important. The baking quality depends on flour proteins, fats and sugars, and by the partial substitution of wheat flour (Table 11.11).

Amino Acid Composition of Proteins

Breadfruit seed proteins furnish at least 14 amino acids with at least 5 of the 8 essential amino acids in quantities comparable to cereals and pulses (Table 11.12) (Quijano and Arango 1979).

NUTRITIVE VALUE OF FLOURS

The flour from seed is superior to the flour from breadfruit pulp since it supplies 5 times the quantity of protein, along with more calories (Table

TABLE 11.9. COMPARATIVE CHEMICAL COMPOSITION AND CALORIC VALUE OF BREADFRUIT SEED,[1] CEREALS AND PULSES[2] (G PER 100 G MOISTURE FREE MATERIAL)

	Breadfruit Seeds							
Constituent	Fresh[2]	Dry	Rice	Wheat	Coconut	Groundnut	Soybean	Maize
Proteins	8.7	20.0	8.6	13.8	7.1	32.8	47.0	13.0
Fat	5.6	12.8	1.2	1.9	65.1	41.5	21.2	4.2
Fiber	1.7	3.9	0.7	2.2	6.9	3.2	4.0	3.2
Other carbohydrates	26.2	60.0	88.5	79.0	19.2	20.1	22.7	77.8
Minerals	1.5	3.4	1.1	3.1	1.7	2.4	5.0	1.8
Calories	189	435	399	388	690	585	470	401

[1]Source: Adapted from Quijano and Arango (1979). Moisture of fresh seeds is 56.3%.
[2]Source: Adapted from Aykroyd *et al.* (1966).

TABLE 11.10. SOME PHYSICOCHEMICAL PROPERTIES OF BREADFRUIT SEED OIL VERSUS OTHER VEGETABLE OILS

Physicochemical constant	Breadfruit seed oil[1]	Palm oil[2]	Coconut oil[2]	Olive oil[2]	Soyabean oil[2]	Corn oil[2]	Peanut oil[2]	Sunflower oil[2]
Specific gravity	0.914	0.924	0.926	0.916	0.925	0.924	0.921	0.925
Solidification point (°C)	17-16	35-42	14-22	+2 (turbid)	–10 to –16	–10 to –20	3	–17
Saponification value	232.7	200-205	257-262	185-196	189-193	187-193	186-194	188-199
Iodine value	69.6	49-59	6-10	79-88	122-134	111-128	88-98	129-136

[1]Source: Adapted from Quijano and Arango (1979), the other physicochemical properties determined by them are: refractive index: 1.473; melting point: 21-24°C (both detd. at 25°C); soluble fatty acids: 6.22%; and insoluble fatty acids.

[2]Source: Adapted from Lange (1944).

TABLE 11.11. COMPARATIVE CHEMICAL COMPOSITION OF SEED PROTEINS OF BREADFRUIT AND SUNFLOWER, AND PROTEINS OF WHEAT FLOUR (G/100 G MOISTURE FREE MATERIAL)

Constituent	Breadfruit Seeds[1]	Sunflower seeds[2]	Wheat flour[3]
Total proteins	20.0	29.5	10.7[4]
Nonglobular proteins	6.4	0.9	0.3
Globular proteins	13.5	28.5	10.4
Albumins	1.8	6.5	1.2
Globulins	3.7	16.7	0.4
Prolamines	3.3	0.6	4.4[4]
Glutelins	4.7	4.6	4.4

[1]Source: Adapted from Quijano and Arango (1979).
[2]Source: Shamanthaka Sastry (1976).
[3]Source: Venkateswara Rao *et al.* (1986) for commercially grown Indian wheat variety WG 357.
[4]Total recovery of wheat flour is 75% of the total weight of wheat taken for milling. Gliadins are the proteins instead of prolamines.

TABLE 11.12. ESSENTIAL AMINO ACID CONTENTS OF THE PROTEINS OF BREADFRUIT SEED AND SOME IMPORTANT OIL SEEDS AND WHEAT (G/100 G PROTEIN)

Amino acid	Bread-fruit seeds[1]	Peanut kernel[2]	Soy bean	Sun flower seed	Cotton seed	Wheat flour WG-357[3]
Arginine	3.33	10.70	7.15	8.94	11.2	2.9
Histidine	4.56	2.43	2.38	2.16	2.7	1.5
Isoleucine	12.10	4.11	5.38	4.70	3.8	2.7
Leucine	13.04	6.08	7.71	6.40	5.9	5.8
Lysine	[4]	3.57	6.32	3.20	4.3	1.7
Methionine	15.90	0.88	1.34	1.64	1.4	—
Cystine	3.12[5]	1.50	1.78	1.71	1.6	—
Phenylalanine	5.28	5.06	4.94	4.50	5.2	3.6
Tyrosine	7.24	3.58	3.18	2.38	—	2.5
Threonine	3.91	2.69	3.94	3.36	3.5	3.1
Tryptophan	[4]	1.10	1.38	1.26	1.2	—
Valine	[4]	4.98	5.25	4.99	4.9	3.4
Proline	3.62	—	—	—	—	9.1
Alanine	7.68	—	—	—	—	2.6
Glumatic acid	4.93	—	—	—	—	18.9
Serine	10.43	—	—	—	—	3.7
Glycine	4.78	—	—	—	—	2.8

[1]Source: Quijano and Arango (1979).
[2]Source: Swaminathan (1971).
[3]Source: Venkateswara Rao (1983).
[4]Paper chromatography with ninhydrin for color development and quantities determined by densitometer; lysine, tryptophan and valine could not be detected.
[5]Includes cysteine.

11.13). The flour from pulp is superior to cassava flour in having higher protein content and equals sweet potato flour. When compared to corn flour and potato flour, breadfruit pulp flour is inferior as it furnishes only half the quantity of protein and only about one third of the wheat protein. In caloric value all flours are more or less equal. Quijano and Arango (1979) reported that 20% of the dry matter of the seed is protein with a good balance of essential amino acids. De Bravo *et al.* (1983) found them to be a rich source of minerals. Both in protein content and caloric value, the flour from breadfruit seed appears to be superior to even wheat flour (Table 11.13). However, further in-depth studies on the quality of the protein, on the presence or absence of inhibitors of seed proteins, and also on the physicochemical properties of significance for utilization, may enhance the commercial value of breadfruit flours.

PROCESSING OF BREADFRUIT

Large scale processing of breadfruit has not been undertaken anywhere in the world, mainly because of the absence of organized cultivation of breadfruit on a commercial scale. Secondly, the fruit is known only to traditional consumers where it is grown, and it is not known beyond these growing regions. In recent years, due to urbanization in many parts of the world, there is a need for processing of breadfruit; at least to cater to the taste of the traditional consumers in the urbanized areas. Further, in view of the nutritional superiority and the profitability of growing breadfruit on an organized scale, there is a growing awareness of the importance of processing this fruit. There are a number of products prepared traditionally on a home or cottage scale. In addition, researchers have shown the possibility of processing the fruit.

TABLE 11.13. COMPARATIVE NUTRITIVE VALUE OF BREADFRUIT AND OTHER PRINCIPLE FOODS (G PER 100 G EDIBLE PORTION)[1]

Constituent	Breadfruit flour		Cassava flour	Sweet potato flour	Potato flour	Corn flour	Wheat flour
	Pulp	Seed[2]					
Water	2.5	0	8.7	2.8	7.6	12.0	12.0
Crude protein	4.1	20.0	1.3	4.2	8.0	7.8	11.8
Total carbohydrates	84.2	60.0	86.6	90.0	79.9	76.8	74.7
Crude fat	1.1	12.8	0.5	0.6	0.8	2.6	1.1
Crude fiber	3.5	3.9	1.8	3.2	1.6	0.7	0.3
Ash	2.9	1.5	2.9	2.4	3.7	0.8	0.4
Calories	362	435	356	282	358	362	356

[1]Source: Data adapted from Graham and De Bravo (1981) unless otherwise indicated.
[2]Source: Adapted from Quijano and Arango (1979).

TRADITIONAL PRODUCTS

Fermented Products

The oldest process for the long term preservation of breadfruit is by fermentation (Connen and Barrau 1961; MacKenzie *et al.* 1964; Goodman 1972; Cox 1980). Although the process differs in minor details from place to place, it involves essentially an anaerobic fermentation of the breadfruit in ground pits under leaf coverings. The resulting product is a highly acidic paste which can be preserved from months to years. In western Samoa the fermented paste is mixed with coconut cream or condensed milk, and is spread on slices of bread. The paste is also formed into patties wrapped in leaves and is baked to give a baked product which looks like bread (Cox 1980) with a consistency of cheese (Goodman 1972). No detailed investigations were made on the microbiological aspects of the process, or the compositional aspects of the product; information which is essential. It may also be necessary to develop products from the paste that will receive wider acceptance.

Dehydrated Products

In India, and probably other regions, breadfruit, in its immature and mature stages, is cut into 1 cm cubes and sun dried. These dehydrated cubes are used in the off season in the preparation of spiced curries, or are fried and eaten. Further, the dried slices are powdered, and the resulting flour is sieved and used in the preparation of papads. Papad is a traditional preparation normally made from blackgram flour. The flour is mixed with adequate water, and a dough is prepared and leavened. Small balls of dough are then flattened to give round (about 10 to 15 cm diam) then (1 to 2 mm) sheets which are sundried. Before the dough is rolled into papad, the desired quantities of spices, chili or pepper powder, Assofoetida, garlic and salt are added, depending on the local tastes. Papads prepared from breadfruit flour may be fried and eaten as an adjunct to a meal.

Breadfruit flour is prepared in several regions and the process is similar to that described by Wootton and Tumaalii (1984B), and Graham and De Bravo (1981). The process is schematically presented in the form of a flow-chart:

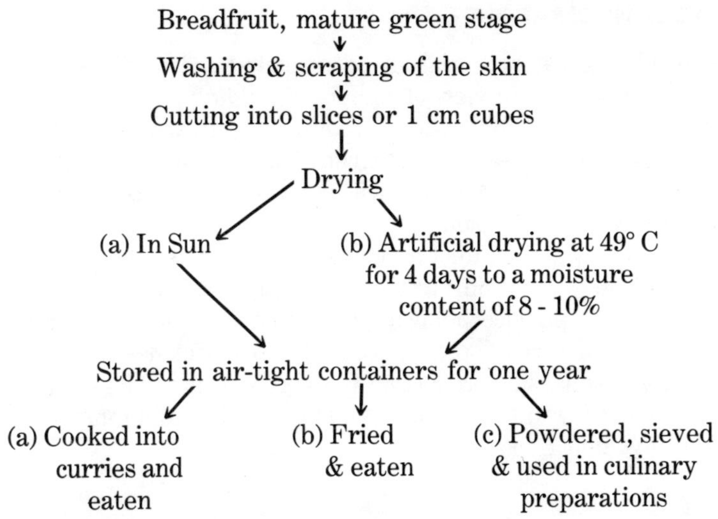

Pickles

Blanched breadfruit slices are used to prepare pickles with spices, salt, chili powder, vinegar or lime juice, and vegetable oil.

Canning

In Jamaica, breadfruit slices are canned in brine for both local and overseas markets (especially England) by a process similar to that used for potatoes.

Breadfruit Seed Utilization

The seeds are removed from the pulp and rind, boiled in salt water or roasted, peeled (the seed coat removed) and eaten (Bennett and Nozzolillo 1987). Canning in brine of peeled seeds can be done in a way similar to that for slices, allowing a somewhat longer processing time.

Preservation of Seeds

Seeds, after removal from pulp, are ground, dried at 36 ± 2° C for 48 hr and reground into a fine flour (sieving if necessary). Seeds can be stored in air-tight containers by a procedure similar to that adopted by Quijano and Arango (1979) for laboratory analysis.

FUTURE CONSIDERATIONS

There is a great potential for both seeded and seedless breadfruit as a major food source, in view of compositional and nutritional values, well recognized in the regions where breadfruit is grown. Available information indicates that breadfruit cultivation is economical, as the yield is as good as that of corn or maize without the enormous annual problems involved in corn cultivation.

There is also a need for further research and development to require an in-depth understanding of the chemical composition of the fruit and seed, and of the physicochemical properties of breadfruit starches which may have great technological potential. Processes need to be standardized for commercial sale production from breadfruit and its seeds of various products—both traditional and sophisticated, including instant or ready-to-use convenience foods.

JACKFRUIT

Jackfruit, although indigenous to India, is now widely grown in most tropical countries—Sri Lanka, Burma, Bangladesh, Malaysia, Indonesia, Philippines and Brazil (Manjunath 1948). In India, over 26,000 ha are cultivated principally in Assam (8,500 ha), Bihar (4,500 ha), Tripura (6,200 ha), Kerala (2,200 ha), Uttar Pradesh (4,600 ha) and in other states like Karnataka, Tamilnadu and Andhra Pradesh (Butani 1978). Considerable quantities of jackfruit are grown in the south and southeast Asian countries but published international reports on production are not available.

Jackfruit is considered the poorman's food in India owing to the numerous culinary uses for the unripe, tender fruit (immature fruit). The value of its versatility is enhanced by its availability during the monsoon period when the supply of other vegetables is small (Singh *et al.* 1963). Unlike breadfruit, jackfruit is consumed both as a vegetable in the unripe stage, and also as a fruit when ripe. There are no seedless varieties. The seeds of jackfruit are consumed in a way similar to breadfruit seeds. The popularity of jackfruit as a commercial fruit has, heretofore, been restricted to the growing regions. The situation is gradually changing in India and other countries and jackfruit is gaining importance commercially, both as a fresh fruit for consumption at places distant from the sites of production, and for processing. In India, it is now transported distances of 1000 to 1500 km from the sites of production to marketing in bigger cities (Anon 1986). Owing to its extremely large size—a hinderance for packaging and transport—and also due to its susceptibility to pressure damage during transit, jackfruit cultivation by and large has been restricted to backyard kitchen gardens, or amidst other fruit tree plantations. Besides the usefulness of its fruit, the jackfruit tree yields good quality timber coming next to teak in quality (Manjunath 1948; Thomas 1980).

BOTANICAL DESCRIPTION

Botanically, jackfruit belongs to the genus *Artocarpus* (family Moraceae). There are three botanical names—*Artocarpus integra* Merr. *Artocarpus integrifolia* Linn. F and *Artocarpus heterophyllus* Lam, of which the last one is now widely accepted (Manjunath 1948). Jackfruit is a small or medium sized tree, 10-25 m high, with a straight, cylindrical, unbuttressed, low-branched trunk, 30-100 cm in diam, a thick, rough gray bark, and a dense irregular or globose crown. The sap is viscid and milky. The twigs are cylindrical, with scattered, deflexed curved hairs when young, becoming glabrous later. The leaves are alternate, petioled, ovate—oblong—elliptic or obovate, the base cunate or obtuse, the apex rounded or obtusely acuminate, entire, stiff, dark green and shiny above, pale green, at first hispidous and later becoming glabrous beneath. They are 10 to 20 cm long and 5 to 10 cm wide.

The inflorescences are peduneled, solitary, in the axils of short, thick branchlets on the trunk and main branches—unlike breadfruit where they are only in the axils. They are usually unisexual, monoecious; the female often have a fleshy ring at the base and are 4 - 15 m long. The female inflorescences are solitary or germinate in the lowest axils, and are larger with thicker peducles than the males. The male inflorescences are much more numerous, occupying the upper axils with longer peducles. The flowers are two lobed and light green having a single stamen with a broad filament and short anther. The tips free with an oblong or compressed ovary, the style apical and obliquely inserted, the stigma clavate.

The fruit, a syncarp, is large, oblong, glabrous, armed with short triangular-hexagonal, pyramidal, acuminate spines, 30-90 cm long and 25-50 cm in diam. The achenes are large and oblong or reniform; the pericarp is golden yellow, coliaceous and juicy (Fig. 11.3 and 11.4) (Ochse 1931).

HORTICULTURE

Although it is generally grown as a home garden plant or in mixed tree fruit plantations, well organized plantations of jackfruit are now appearing due to its increasing commercial importance.

FIG. 11.3. A VIEW OF A JACKFRUIT TREE WITH FRUITS HANGING FROM THE TRUNK AND MAIN BRANCHES. THE HEIGHT OF THE TREE LOCATED IN THE COMPOUND OF A PRIVATE HOME IN MYSORE CITY IS ABOUT 10 M IN HEIGHT.

FIG. 11.4. A CLOSE-UP VIEW OF THE TRUNK AND FRUITS OF ANOTHER JACKFRUIT TREE LOCATED IN THE COMPOUND OF A PRIVATE HOME. TRUNK PARTLY SEEN WITH SOME FRUITS HANGING FROM TRUNK WITH LONG FOOTSTALKS (90-182 CM) AND WITH SOME FRUITS APPROACHING MATURITY

Varieties

Jackfruit, being cross-pollinated and mostly seed propagated, exhibits a wide range of variation in fruit quality. Different cultivated varieties which vary widely in size, shape, bearing, and sensory qualities of fruit have come into commercial cultivation with different local names for the cultivars. However, the cultivated types are broadly classified into two groups: namely (1) soft-flesh varieties—when fully ripe the fruit yields to the thrust of a finger on the rind. The pulp of this variety is very juicy and soft. In the northeastern parts of India (Assam and Tripura etc.) this type is very popular. (2) firm-flesh varieties—in these, the rind does not yield easily to the thrust of a finger, even when fully ripe. The pulp is firm and crispy. This type is very good for freezing and drying (Samaddar 1980).

Naik (1952) reported that even after propagation by seeds, there are two distinct varieties: (1) Rudrakshi—the fruit is of the size of a pomelo. The rind is less spiny than the common jackfruit; and (2) Singapore or Ceylon Jack—this variety was introduced into India from Ceylon. It produces fruits approximately within 2.5 years of planting the seedlings. The size of the fruit is between 5 to 20 kg.

According to Manjunath (1948), *Artocarpus heterophyllus* also has two, more or less well defined varieties Barka, a smaller and inferior variety and Kapa, a large, superior variety. The Barka is locally (in Uttar Pradesh) called Katahali, and yields small fruits which are somewhat sour in taste and weigh 2-8 kg each. The Kapa variety which is called Kathal in Uttar Pradesh and neighboring regions, yields large fruits weighing nearly up to 40 kg and at ripening has fleshy sweet perianth (bulbs). Some fruits may be borne at ground level or below. In this variety 10,000 to 12,000 flowers form the composite fruit and the number of seeds produced vary from 50 to 200 per fruit (Sharma 1964).

Srinivasan (1970) described a local variety from Tamilnadu named Muttam Varikha which produces fruits of an average weight of 7 kg, 46 cm long and 23 cm wide. Another special variety of jackfruit called Kodipalakkai (meaning jackfruit grown at the bottom of the tree) has been recently reported from a village Panruti and is in cultivation in several villages around Panruti (near the town of Cuddalore, Dist. South Arcot, Tamilnadu) since early 1950s (Anon 1986). The characteristic feature of this variety is that the fruits are very large, some weighing up to 55 kg. The fruits are very delicious and, hence, merchants as far away as Bombay (1500 km) purchase and transport these fruits.

SOIL AND CLIMATE

The tree grows from sea level to 1500 m elevation, in regions stretching from South India to the foot hills of Himalayas (Hayes 1957). It flourishes well under humid and warm climates. Soil drainage is important; hence, well drained alluvial soils, or a very open-textured laterite, or loamy soils with adequate moisture are suitable for its cultivation (Amar Singh 1980).

PROPAGATION

In India, and also in some of the Asian countries, jackfruit is normally propagated by seeds, which has resulted in considerable variation between once well defined varieties for important fruit characteristics. As a result, there is a growing interest in vegetative propagation of jackfruit. Several methods have been tried with varying degrees of success. In view of the importance of fruit quality, both for its use as fresh fruit and also for processing, the significant advances made in the propagation of jackfruit are briefly reviewed here.

Seed Viability and Seed Propagation

The viability of jackfruit seeds is hardly 5 weeks (Panggabean 1979) both at low as well as at ambient temperature conditions. In fact the seeds start germinating in the ripe fruit, that is, they are viviparous in nature (Habib 1973) and this may be the reason for the short viability period of the seeds. One good reason for vegetative propagation is that the seeds lose their viability very fast.

VEGETATIVE PROPAGATION

Air-Layering

Ringing, application of a growth regulator (IBA), and air-layering gave 100% rooting; subsequent transplanting resulted in a high survival rate. Success, however, depends on the season (i.e., time of ringing and transplanting), age of the tree and the branch, and the quantity of growth regulator applied. Ringing between May and June with 5000 ppm IBA was best for rooting, with a high percentage of survival after cutting and planting (Sen and Bose 1959). Two-year-old shoots gave good rooting (Singh 1951). The air-layered trees came into bearing at least 5 to 6 years earlier than seedling trees (Singh 1961).

Grafting

Inarching: Naik (1952) reported 60-70% success with *Artocarpus hirsuta* rootstocks. However, in a subsequent trial Kannan and Nair (1960) established that inarching on rootstocks of the same species gave better success, while on *A. hirsuta* the growth was poor. Raman (1957) selected seven superior varieties of jackfruit and grafted seedlings of six of these varieties onto rootstocks of the seventh; he reported a success rate of inarching in the range of 0 - 92%. Raman concluded that individual tree selection would encourage the growing of superior forms and increase the chances of success inarching.

Epicotyl-grafting: Epicotyl grafting with mature, plump terminal scion shoot on germinating jackfruit seedling of about 8-10 days, by the wedge method, during April-May gave 50 to 90% success and the successful epicotyl grafts attained salable size within a year (Gunjate *et al.* 1982).

Bud grafting: Several methods of budding have been tried with different degrees of success. Naik (1949) reported that the modified Forkert budding was very successful in Java. Teaotia *et al.* (1963) achieved 100% success by patch budding. While patch budding gave only 41% success (Samaddar and Yadav 1970), Moti and Chaturvedi (1976) confirmed from a 2-year trial that both patch budding and shield budding in June on 1-year-old stalks was highly successful.

Cuttings

Mukherjee and Chatterjee (1978, 1979) and Chatterjee and Mukherjee (1980A,B, 1982) made extensive studies on invigoration/forcing/stooling, followed by etiolation and then treating with indole butyric acid (1000, 5000 ppm) and reported success of over 75% for rooting and over 70% for survival of cuttings after 1 year. In a later study, Mukherjee *et al.* (1983) combined the above treatments with manure treatment at the time of planting and concluded that farm yard manure gave better growth of plants than manure plus NPK; the latter treatment, however, produced more fruits.

In similar studies, Dhua *et al.* (1983) and Dhua and Sen (1984) established that ringing, etiolation for 15 to 30 days, cutting and treating with growth regulators (3000 ppm IBA plus 2000 ppm ferulic acid) gave 90% success of rooting with 100% survival. They also established the relative roles of etiolation, auxinic and nonauxinic chemicals. Their study showed that etiolation and treatment with IBA alone increased survival. In a similar study, Hamilton *et al.* (1982) reported from Hawaii over 95% success with two cultivars of breadfruit by treating woody, leafless cuttings with 2500 ppm IBA and 5% captan, when planted in pots and propagated under mist in a greenhouse.

Thus, of the various methods tried experimentally, vegetative propagation through cuttings, grafting and air-layering methods are the best. From the point of view of convenience, propagation through cuttings or air-layering can be considered as more practical.

PLANTING

The best time for planting grafts or seedlings in India is June to August (rainy season), since prolonged dry weather after planting affects the survival of plants. Firm pressing of the soil after planting is essential to avoid water-logging in rainy season. A spacing of 12 m x 12 m is adopted in fertile soils which will accommodate 70 plants per hectare. For planting, 1 cubic meter pits are dug about 10 days prior to planting. The pits are refilled with about 30 kg of well composted farm yard manure and 50 g of BHC (10% a i., for protection against insect attack) mixed with soil. The grafts are transplanted into these well prepared, refilled pits and are regularly irrigated initially, until they are established. Irrigation is continued during the dry periods for up to 2 years. Since the jackfruit tree requires a relatively long time (7 to 8 years) to fully develop its canopy, and fully use the land provided, it is common practice in India to use the land for intercropping with either vegetable crops or short duration crops like pulses (Samaddar 1985; Amar Singh 1980).

PESTS AND DISEASES

Over 10 pests and 5 or 6 fungal diseases are known to attack jackfruit trees. However, none of them become very serious if suitable control measures are enforced. The data on pests and diseases, giving the causal organism, symptoms and control measures have been tabulated by Bhutani (1978).

SEX DISTRIBUTION, SEX RATIO AND FRUIT-SET

The distribution of sex on the jackfruit tree, since the flowers are unisexual and monoecious, has attracted a lot of attention related to floral biology and sex-ratio (Sahadevan *et al.* 1950; Sambamurthy and Ramalingam 1954; Shankar and Singh 1965; Sinha 1969; Teaotia and Chauhan 1969; Sinha and Singh 1971; Sinha 1972, 1975; Samaddar and Yadav 1982). The size of the fruit becomes smaller and its shape irregular if flowers on all sides of the inflorescence are not properly pollinated. Pollination is mostly anemophyllus.

The number of fruits per tree is dependent on the number of female spikes (inflorescences) and the ratio of male to female spikes (sex-ratio). Shankar and Singh (1965) compared the flowering behavior and sex-ratios of 10- and 30-year-old trees. The male to female spikes ratio was 4:1 in young trees and 2:1 in old trees. The sex-ratios of 8- and 12-year old trees indicated that the ratio of male to female flower spikes was greater in young trees, and is probably the main reason for the older trees bearing more fruits (Sinha 1972). Teaotia and Chauhan (1969) observed that the sex-ratios varied considerably between varieties.

The size of the fruit is dependent on the number of female flowers in the female spike (inflorescences). Sharma (1964) reported 10,000 to 12,000 female flowers for the Kathal variety of jackfruit. In varieties where the size of the fruit is relatively small, the number of female flowers may be around 2000. However, the final size attained and its shape depends on the number of flowers pollinated and fertilized, and the pattern of pollination of the flowers in the different parts of the inflorescence. The unfertilized flowers also grow but lead to the formation of depression with closely set spikes, giving an irregular shape to the fruit (Sharma 1964).

Improvement in fruit-set, by the application of growth regulators, has been attempted. By spraying with 25 ppm α-napthaleneacetic acid onto newly emerged female spikes, parthenocarpy was achieved. However, this led to reduction in fruit size, seed number and weight, but increased the edible portion (Srivastava 1961). Ethephon (2-chloroethyl phosphonic acid) at a 500 ppm level, when sprayed twice at 15-day intervals onto the foliage of the tree about one month before the normal flowering period, gave the highest number of foot-stalks, female spikes, and pistillate flowers among some of the staminate spikes (male spikes). This clearly indicates its influence on sex reversion and gynoecium (Das *et al.* 1981), as well as the possibilities of increasing the number of fruits per tree.

FRUIT YIELD

The tree generally starts bearing from the seventh or eighth year. The air-layered trees start yielding fruits about one or two years ealier than seedling trees. The Singapore variety of jackfruit, introduced into India through Ceylon, starts bearing within 3 years but gives fruit of average weight (3-15 kg). After about 15 years from the time of planting, when the canopy is well formed and with the sex-ratio altered with the age of the tree, the tree reaches its peak bearing and continues to yield for over 30 years. The yield of fruit per tree, therefore, differs from variety to variety, as well as by tree age and agro-climatic conditions. The number of fruits per tree vary between 100 to 300 and the weight ranges between 5 to 55 kg. The average yield of fruits per tree can be taken as 200, and the average weight of one fruit as 20 kg. The per hectare yield of fruits taking 70 trees per hectare with a conservative estimate of 50 fruits per tree and average weight of 20 kg comes to around 70 MT per hectare. This would afford an edible portion (bulbs) calculated as 30% of the weight, of 21 MT per hectare. The yield of seeds at 10% of fruit weight amounts to 7 MT. Further, there is great interest in utilization of the rind, core, etc., which were heretofore considered nonedible. Therefore, it appears that the cultivation of jackfruit may become commercially very profitable when compared to many other fruit crops.

FRUIT GROWTH AND MATURATION

The composite fruit, or syncarp of *Artocarpus* is highly specialized (Jarret 1959). What is normally called jackfruit is a composite fruit that has developed from the whole inflorescence. It consists primarily of 3 regions: (1) the fruit axis, (2) the persistent perianth, and (3) the true fruit. The fruit axis is the modified and mature stage of the inflorescence axis. It is elongated and dome shaped, consisting of elongated broad parenchymatous cells that are devoid of starch. It has numerous laticiferous cells with milky white latex. The axis region, also called the core of the fruit, is unpalatable and inedible. The perianth is the most important part and forms the bulk of the fruit. It has three regions: (1) the lower fleshy edible region (commonly called bulb), (2) the middle fused region (forms rind of the syncarp), and (3) the upper free and horny nonedible region (spikes). The true fruit, a single seeded achene, is developed from the carpel (ovary) and is surrounded by the fleshy perianth (bulb) (Sharma 1964).

As fruit growth progresses, the relative sizes of the different parts of the fruit increase, attaining maximum size (Table 11.14). The changes in the sizes of the fruit and its different parts clearly indicate that the growth rate is more or less uniform from stage to stage, with a slight indication of slowness between stages 5 and 6, for the overall increase in the length of the fruit. Further, both the whole fruit (syncarp) and the edible region (bulb) seem to increase in size at a faster rate between stages 9 and 10 as the fruit approaches maturity. The perianth of the nonfertilized flowers grows along with the fertilized flowers; although it attains the same length, it does not increase much in breadth and thickness while the perianth of the fertilized flowers attains a size of 3.0 to

TABLE 11.14. MEASUREMENTS OF JACKFRUIT AND ITS DIFFERENT PARTS DURING VARIOUS STAGES OF DEVELOPMENT (IN CENTIMETERS)

Stages of development from fruit-set	Length of fruit	Diameter of fruit	Diameter of fruit axis	Length of perianth			
				Total length	Length of edible region (bulb)	Length of fused region (rind)	Length of free tip region (spikes)
1	4.5	2.5	1.8	0.35	0.08	0.07	0.2
2	7.8	3.6	2.2	0.7	0.18	0.15	0.37
3	12.5	7.5	2.7	2.4	1.0	0.8	0.6
4	16.0	10.0	3.8	3.1	1.6	0.9	0.6
5	26.0	14.0	6.0	4.0	2.5	0.9	0.6
6	28.0	16.0	7.0	4.5	2.9	1.0	0.6
7	35.0	18.5	8.0	5.25	3.5	1.0	0.75
8	40.0	22.5	9.0	6.5	4.55	1.2	0.75
9	44.0	25.0	10.0	7.5	5.6	1.2	0.7
10	50.0	34.0	10.0	12.0	10.0	1.3	0.7

Source: Sharma (1964).

5.0 cm in breadth and 0.5 to 1.0 cm in thickness. Hence, varieties with larger numbers of seeds are considered superior to varieties with few seeds.

HARVEST MATURITY AND HARVESTING

Tree flowering starts in December and may continue up to March, depending upon the age of the tree and agro-climatic conditions. As a consequence, the harvesting season lasts about 4 months in south India, with mature fruits coming into the market, normally in April-May, and continuing up to August. Fruit growth and maturation normally continue up to 5 months, although harvesting may be done after the fourth month. In the north Indian plains and at higher altitudes, maturity may take a longer time (Teaotia and Chauhan 1969).

Tender jackfruits are harvested for use as vegetables during the first 2 to 3 months after fruit-set and until the seeds harden. Mature fruits are harvested for dessert purposes, mostly depending on their size after 4 months from the date of emergence of inflorescence. The normal indices of harvest maturity are the color of the spikes, which changes from dark green to greenish yellow, and the widening area between the spikes. Some can judge the degree of maturation of the fruit by the difference in the sound produced between immature and mature fruits when tapped with the fingers on the rind of the central portion of the fruit. However, since the ripening proceeds normally—as in banana—even if the fruit is harvested a little earlier than at the optimum mature stage, the judgement of optimum maturity is not strictly followed. There is, however, some difference in the taste from fruit to fruit, probably due to variation in maturity. Larger fruits give tastier bulbs than smaller ones.

For the processing of jackfruit, a uniform maturity of harvest is needed to obtain a quality processed product. The data presented in Table 11.15 show that the yield of bulbs and their quality differ with maturity of the fruit. Teaotia and Awasthi (1968) from the data presented in Table 11.15 show that the yield of bulbs and their quality differ with maturity of the fruit. Teaotia and Awasthi (1968) concluded that in north India, where their work was done, the optimum maturity of jackfruit is around 180 days from spike emergence, and these fruit give good finished products. Harvesting is done by cutting the fruit stalks, leaving a portion of the stalk attached to the fruit for handling the fruit after harvest.

HANDLING, TRANSPORT AND MARKETING

The harvested fruits are carried individually by holding the stalk, and are generally loaded into bullock-carts or push-carts and transported to

TABLE 11.15. YIELD OF BULBS AND NONEDIBLE PORTIONS AND PHYSICAL VARIATIONS IN BULBS OF JACKFRUIT HARVESTED AT SIX STAGES OF MATURITY

Harvest date & age of fruit					Variations in bulbs			
	days[1]	Color of rind	Yield of bulbs	Nonedible portion	Texture	Color	Flavor	Discoloration of bulbs
May 22	135	Dark green	45.8	54.2	Hard	White	Poor	High
June 7	150	Dark green	49.5	50.5	Hard to firm	White good	Fairly good	High
June 22	165	Green	55.0	45.0	Firm White	Yellowish	Good	Medium
July 6	180	Green	58.0	42.0	Firm White	Yellowish	Good	Medium
July 21	195	Light green	60.4	39.6	Firm to soft	Whitish yellow	Good	Medium
Aug. 1 (Ripe fruit)	210	Light straw (slight yellow)	60.0	40.0	Soft	Yellow	Ripe	Slight

Source: Adapted from Teaotia and Awasthi (1968).

[1]Spikes appeared on Jan. 8th and the age of fruit was counted from that date. The first harvest was done at 135 days when the seed could be separated from the bulbs.

wholesale merchants in a nearby town where they are sold to consumers or to visiting tradesmen from bigger cities. When distances between the producing centers and city markets are large, the fruits are transported in trucks in the unpacked condition, without any particular order of stacking, as the fruits are large and heavy.

The fruits are either sold wholesale or retail, depending on the size and the reputation of each variety. The retail price of each fruit to a large extent depends on size alone, as the consumer will not be aware of the varieties and their differences.

According to a recent report (Anon 1986), fruit sizes vary from 5 to 55 kg for the varieties of jackfruit grown in Panruti and the neighboring villages in South Arcot District of Tamilnadu (south India). The fruits have gained a good reputation both for their size and quality all over south India, including Bombay. The fruit is transported to Bombay over a distance of 1000 km and sold retail there. The cost of one fruit at the site of production varies from (US $0.3 to 6) depending on its size.

FRUIT RIPENING

Biale and Barcus (1970) classified the respiratory pattern of jackfruit as indeterminate, as there was no clear-cut climacteric rise in respiration. The fruit is judged ripe for cutting when the skin color, at least in part, changes to yellow, when a strong ripe fruit flavor emanates from the fruit, and when there is a marked change in the pressure required to press the rind of the fruit with a finger. The fruit ripens normally at tropical ambient temperatures (20-35° C) in 3 to 10 days depending on the stage of maturity at harvest and with no ripening problems. During ripening, the starch which is the principal storage material in the bulbs of the unripe fruit, is converted to sugars, while the color of the bulbs changes from pale white or light yellow to an attractive golden yellow color and is accompanied by the characteristic, pleasant aroma of the ripe fruit. The intense aroma of the fruit is not restricted to bulbs alone; the rind and the unfertilized floral parts all give out a strong aroma at the ripe stage.

STORAGE

Depending on the maturity, ambient temperature and relative humidity (RH) conditions, jackfruit have a storage life of 3 to 10 days. Refrigerated cold storage of jackfruit has been investigated by Mathur *et al.* (1952) employing seven different temperatures ranging from 0° to 28° C and at various relative humidity conditions ranging from 55 to 90%. From their study, it is evident that the optimum cold storage temperature and

RH conditions are 11° to 13° C and 85-90% RH. Chilling injury occurred at temperatures below 11° C. The cold storage life of jackfruit at optimum cold storage conditions (11-13° C and RH 85-90%) is around 6 weeks. A cumulative weight loss of 17.6% was recorded after 6 weeks of cold storage at optimum conditions (Table 11.16). The authors noted that during storage at temperatures above 11° C the fruit ripened normally. During storage, sucrose content decreased from an initial value of 9.5% to around 5% and was accompanied by a rise in reducing sugars from 2 to 6%. The ascorbic acid content decreased from 8.2 to 3 mg per 100 g fresh material, while the titratable acidity increased from 0.15 to around 0.3% (Table 11.17). Kripal Singh and Mathur (1954) in a later publication confirmed the above findings.

FREEZING

Kripal Singh and Mathur (1954) investigated the freezing of jackfruit bulbs. The edible bulbs from ripe jackfruit (excluding the seeds) were sliced and packed (1) with dry sugar, and (2) in 50% sugar syrup with 0.5% citric acid (on the basis of syrup), into jam cans, leaving a head space of 1.25 cm. The product was frozen at –29° C and subsequently stored at –18° C. Slices packed in 50% sugar syrup containing 0.5% citric acid gave excellent results. The color, taste and flavor of the product were preserved for one year. Moreover, the individuality of the slices was maintained for one year. The product prepared with dry sugar was of lower quality.

COMPOSITION AND NUTRITIVE VALUE

Percent Composition of Different Parts of Fruit

Raw, tender jackfruit, normally cooked as a vegetable when the seeds are not well formed, can be separated into edible and nonedible portions. The nonedible portion comprises the core and spike-containing rind. The nonedible portion weight ranges from 41 to 77% with an average of 53% depending on degree of tenderness of the fruit. The remaining parts of the fruit including the tender seeds form the edible parts of the fruit, which range from 23 to 59% with an average value of 47% (Table 11.18).

Ripe jackfruit can be easily separated into three parts: (1) bulbs, (2) seeds and (3) rind, core and sheath around the seeds (pericarp) and the unfertilized flowers. The third part is considered nonedible. The bulbs form 15 to 50% of the weight of the fruit depending on its size, with an average value of 29%. The weight of seeds ranges between 4 and 16% with an average of 11.5% of fruit weight. Nearly 60% of the fruit weight

TABLE 11.16. PHYSIOLOGICAL LOSS IN WEIGHT, PHYSICAL APPEARANCE OF SKIN AND RIPENING OF JACKFRUIT DURING STORAGE AT DIFFERENT TEMPERATURES AND RELATIVE HUMIDITIES

Storage conditions		Initial wt of fruit[1] (kg)	Physiological loss in weight (cumulative %) after days[1]			Physical appearance of skin of fruits
Temp. °C	RH (%)		14	28	42	
0- 1.6	85-90	7.09 ± 0.02	5.74 ± 1.59	11.52 ± 1.58	—[2]	Brown patches and chilling injury noticeable, accompanied by mold development at later stages.
1.6- 3.3	85-90	8.32	7.22	14.70	21.84	Same as above
5.5- 7.3	85-90	7.92 ± 3.43	5.74 ± 0.28	11.53 ± 1.17	23.70	Brown patches noticeable with extent of chilling injury less compared to lower temperatures.
8.4-10.0	85-90	7.87 ± 1.51	5.40 ± 0.87	11.75 ± 1.18	19.75	Few brown patches noticeable
11.1-12.8	85-90	7.86 ± 1.54	5.18 ± 0.82	10.35 ± 0.32	17.62 ± 2.79	Fruit fresh and getting ripe with progress of time. Fruits fully ripened in 42 days.
15.5-16.6	70-75	5.34 ± 0.77	14.04 ± 0.20	19.83	—	Fruit ripening; normal shriveling noticeable after 14 days; ripening completed in 21 days.
23.3-27.2 (room temperature)	55-70	6.53	17.93	—	—	Fruits fully ripened after 10 days of storage.

Source: Mathur *et al.* (1952).

[1] X ± S.D.

[2] (–) Observations not recorded as either the fruits spoiled/or fully ripened.

TABLE 11.17. INITIAL AND FINAL COMPOSITION OF MOISTURE, SUCROSE, REDUCING SUGAR, ASCORBIC ACID AND ACIDITY OF JACKFRUIT STORED AT DIFFERENT CONDITIONS (VALUES INDICATE G PER 100 G FRESH)

Storage Conditions		Period of storage			Reducing	Ascorbic	Acidity
°C	RH (%)	(days)	Moisture	Sucrose (9.5)[1]	sugar (2.0)[1]	acid (mg) (8.2)[1]	(as % citric) (0.15)
0- 1.6	85-90	35	81.6	—[2]	—	1.10	0.22
1.6- 3.3	85-90	35	76.3	—	—	0.92	0.13
5.5- 7.3	85-90	35	—	2.66	5.20	1.40	0.16
8.4-10.0	85-90	35	79.3	4.48	5.84	3.60	0.16
11.1-12.8	85-90	42	84.0	5.00	5.98	3.50	0.22
15.5-16.6	70-75	21	—	5.45	6.25	3.10	0.33
23.3-27.2 (room storage)	55-70	10	—	6.55	4.63	4.50	0.37

Source: Mathur *et al.* (1952).
[1]Initial values given in parenthensis.
[2](–)indicates observations not recorded.

TABLE 11.18. PERCENTAGES (W/W) OF DIFFERENT COMPONENTS OF RAW AND RIPE JACKFRUIT

	Raw tender fruit[1]			Ripe jackfruit[2]	
Component	Range (%)	Average (%)	Component	Range (%)	Average (%)
1. Edible portion	22.7-59.4	46.9	Bulbs	15.2-49.8	28.7
2. Nonedible portion	40.6-77.3	53.1	Seeds	4.1-15.9	11.5
			Rind, core and sheath around seeds	39.6-75.5	59.1

[1]Bhatia (1953). Tender raw jackfruits of weight 1 to 2 kg are usually used as vegetables. In more tender fruit (i.e. lower in weight) the nonedible portion, peel and core, will be more because of difficulties in separation from edible portions. Fruits weighing 5, 9 and 29 kg were used for ripe jackfruits.

[2]Bhatia *et al.* (1955A). The average composition of different parts of ripe jackfruit is based on analysis of 126 fruits.

is nonedible and is normally used for feeding cattle (Bhatia *et al.* 1955A) (Table 11.18). The data collected by the same authors on the variations in the different parts of the jackfruit collected from a single tree are presented in Table 11.19 and shows an average yield of 40% bulbs and 10% seeds of the fruit weight. From the data, the average number of seeds per fruit are 148 and the weight of a single seed is 7.1 g.

Working with a popular variety from Uttar Pradesh, Teaotia and Awasthi (1968) observed that the yield of bulbs increased from 135 days of maturity of fruit to 195 days. However, at 210 days, the yield decreased, which the authors considered to be due to conversion of bulbs into pulp. The fruits harvested on the 180th day gave the highest yield of bulbs (probably including seeds) of 60.4% of fruit weight. The bulbs had pleasant flavor, fine color and firm texture. Discoloration of the bulbs, which is an important factor for processing, was observed to be high up to the 150th day; later, it decreased with the advance in maturity (see Table 11.15).

VARIATION IN CHEMICAL COMPOSITION WITH MATURITY

The chemical composition of the bulbs showed marked changes as the fruit matured (Table 11.20). Moisture, total soluble solids, acidity, total sugars, starch/sugar, sugar/acid ratios and ascorbic acid increased progressively, while the starch content progressively decreased. The biochemical changes appear to be at a faster rate from the 180th day; at this stage the yield of bulbs was also optimum, and so this stage can probably be taken as the optimum maturity stage. Fruits at the 165 day

TABLE 11.19. PERCENTAGE COMPOSITION OF DIFFERENT PARTS OF RIPE JACKFRUIT FROM A SINGLE TREE[1]

Bulbs[2]	Rind	Core	Seeds	Sheath covering seeds	No. of seeds per fruit	No. of seeds per 45.5 kg of fruit	Weight of 100 seeds (g)
40.1 ± 1.2	37.8 ± 0.45	5.1 ± 0.26	10.1 ± 0.61	6.7 ± 0.57	148 ± 9.5	703 ± 45	710 ± 31.2

[1]Souce: Bhatia *et al.* (1955A). All the fruits from the same tree were used.
[2]The bulb size varied between 2.5 to 6.3 cm in length, 0.6 cm to 3.8 cm in width, with the average weight of 36.9 ± 4.9g.

TABLE 11.20. CHANGES IN CHEMICAL COMPOSITION OF JACKFRUIT BULBS HARVESTED AT SIX MATURITY STAGES

Harvest perid in days from the appearance of spikes		Moisture	Total soluble solids at 20° C ° Brix	Acidity	Total sugars	Starch	Sugar/ starch ratio	Sugar/ acid ratio	Ascorbic acid mg/100 g fresh
Stage 1.	135	70.5	0.0	0.10	0.0	14.6	0.00	0.00	1.2
Stage 2.	150	71.3	0.0	0.11	0.5	13.3	0.04	4.89	2.0
Stage 3.	165	72.2	2.2	0.13	1.1	11.3	0.10	9.09	4.5
Stage 4.	180	72.8	3.5	0.14	1.6	9.4	0.17	12.59	6.2
Stage 5.	195	73.3	5.8	0.14	3.0	5.8	0.52	21.01	8.0
Stage 6.	210	75.5	11.3	0.37	8.2	3.5	2.35	21.97	12.6

Source: Data adapted from Teaotia and Awasthi (1968). Values expressed as g per 100 g fresh material unless otherwise noted.

stage were slightly immature, and at the 195 day stage slightly over-mature. Fixation of an optimum maturity stage, which depends on variety and agro-climatic conditions is very important for the processing of bulbs.

COMPOSITION OF JACKFRUIT FROM DIFFERENT REGIONS

The chemical composition of fruits from different regions has been reported. Comparative tables giving the variation in composition are discussed in the following.

Composition of Tender Fruit

The moisture content of tender jackfruit varied from 80 to 84% and there were slight variations between region to region in the principal constituent. Tender jackfruit on average contain: moisture 80.7%; protein 2.3%; minerals 0.9%; fiber 3.6%; starch 8.8%; sugar 0.1% and give good quantities of calcium, phosphorus, iron, thiamine, riboflavin and vitamin C (Tables 11.21 and 11.22).

Composition of Ripe Jackfruit Bulbs

Ripe jackfruit bulbs are rich in sugars with a caloric value of about 90 calories per 100 g fresh weight. Bhatia *et al.* (1955A) reported that ripe bulbs contain 20.6% total sugars, with practically no starch. The bulbs are rich in calcium, potassium, iron, vitamins A and C, and some

TABLE 11.21. CHEMICAL COMPOSITION OF TENDER JACKFRUIT FROM DIFFERENT REGIONS OF SOUTH INDIA (G PER 100 G FRESH EDIBLE PORTION)

				Average	
Constituent	Mysore[1]	Bangalore[2]	Hyderabad[3]	Fresh	Dry
Moisture	84.2	84.0	84.0	80.7	—
Protein	1.7	2.6	2.6	2.3	11.9
Fat	—	0.3	0.3	0.3	1.6
Carbohydrates	7.5	9.4	12.2	11.2	58.0
Minerals (total)	0.9	0.9	0.9	0.9	4.7
Fiber	4.4	—	2.8	3.6	18.7
Ash	1.0	—	—	1.0	5.2
Acidity	0.2	—	—	0.2	1.0
Caloric value	32.0	51.0	61.0	50.0	250.0

[1]Source: Bhatia *et al.* (1955A).
[2]Source: Anon. (1979).
[3]Source: Aykroyd *et al.* (1966).

TABLE 11.22. MINERAL AND VITAMIN COMPOSITION OF TENDER JACKFRUIT FROM DIFFERENT REGIONS IN SOUTH INDIA (MG PER 100 G EDIBLE PORTION FRESH)

Constituent	Hyderabad[1]	Mysore[2]	Bangalore[3]	Average
Minerals				
Calcium	30.0	139.5	50.0	73.2
Phosphorus	40.0	34.7	97.0	57.2
Iron	1.7	2.5	1.5	1.9
Sodium	35.0	—[4]	—	35.0
Potassium	328.0	—	246.0	287.0
Vitamins				
Thiamine	0.05	—	0.25	0.15
Riboflavin	0.04	—	0.11	0.07
Vitamin C	14.0	—	11.0	12.5
Nicotinic acid	0.2	—	—	0.2

[1]Source: Aykroyd *et al.* (1966).
[2]Source: Bhatia *et al.* (1955A).
[3]Source: Anon. (1979).
[4](–) values not determined.

members of the B-group, viz., thiamine, riboflavin and nicotinic acid (Tables 11.23 and 11.24).

TABLE 11.23. CHEMICAL COMPOSITION OF RIPE JACKFRUITS FROM DIFFERENT REGIONS (G PER 100 G FRESH EDIBLE PORTION)

					Average	
Constituent	Hyderabad[1]	Mysore[2]	Unspecified[3]	Bangalore[4]	Fresh	Dry
Moisture	76.2	72.5	73.1	77.2	74.9	—
Protein	1.9	0.4	0.6	1.9	1.2	4.8
Fat	0.1	—	0.6	0.1	0.28	1.1
Minerals (total)	0.9	0.9	—	0.8	0.87	3.5
Fiber	1.1	0.9	1.8	1.1	1.23	4.9
Carbohydrates	19.8	24.8	23.4	18.9	20.7	82.5
Ash	—	1.0	0.5	—	0.75	2.0
Total sugars	—	20.6	—	—	20.6	82.4
Acidity	—	0.3	—	—	0.3	1.2
Starch	0	0	—	—	0	0
Caloric value	88	103	101	84	90	359

[1]Source: Aykroyd *et al.* (1966).
[2]Source: Bhatia *et al.* (1955A).
[3]Source: Purseglove (1968); region not specified.
[4]Source: Anon. (1979).

TABLE 11.24. MINERAL AND VITAMIN COMPOSITION OF RIPE JACKFRUIT (MG PER 100 G EDIBLE PORTION)

	India				
Constituent	Hyderabad[1]	Mysore[2]	Bangalore[3]	Unknown[4]	Australia[5]
Minerals					
Calcium	20.0	33.4	20.0	22.0	18.0
Phosphorus	41.0	18.9	30.0	38.0	—
Potassium	191.0	—[6]	—	407.0	350.0
Sodium	41.0	—	—	2.0	3.0
Iron	0.5	1.4	0.5	0.5	0.8
Copper	0.2	—	—	—	—
Sulfur	69.0	—	—	—	—
Magnesium	27.0	—	—	—	14.0
Zinc	—	—	—	—	0.8
Vitamins					
Vitamin A (I.U.)	292.0	700.0	540.0	540.0	—
Thiamine	0.03	—	—	0.03	0.4
Riboflavin	0.13	—	—	—	0.12
Nicotinic acid	0.40	—	—	4.0	0.6
Vitamin C	7.0	2.1	—	8-10	6.0

[1]Source: Aykroyd *et al.* (1966).
[2]Source: Bhatia (1953); vitamin A is expressed as mg β-carotene per 100 g fresh; converted by 1 I.U. = .6 μg β-carotene.
[3]Anon. (1979).
[4]Source: Morton (1987).
[5]Source: Wills *et al.* (1986).
[6](–) values not determined.

Jackfruit Seeds *Composition*

The approximate composition of jackfruit seeds is: water 64%; protein 7%; fat 0.4%; fiber 1.3%; carbohydrates 26.8% (Table 11.25). The principal minerals present in jackfruit seeds are calcium, potassium, magnesium, sulfur, phosphorus, and iron; also present are the B-complex vitamins (thiamine, riboflavin, nicotinic acid) and choline (Table 11.26).

The principal carbohydrate of seed is starch (25% of fresh weight, which has been isolated from the cotyledons and chemically analyzed by Bobbio *et al.* (1978). Their analyses showed an amylose content of 28.1% and hydrolysis of the starch yielded 99.1% d-glucose (Table 11.27). Jackfruit starch grains are round or bell-shaped and range in size from 7 to 11 μm.

The gel produced from jackfruit starch was odorless and was extremely firm without any sag value when compared to other common starches. The gels did not develop any syneresis during storage at 4° C

TABLE 11.25. CHEMICAL COMPOSITION OF JACKFRUIT SEEDS FROM DIFFERENT REGIONS (G PER 100 G FRESH MATERIAL)

					Average	
Constituent	S. Asia[1]	Mysore[2]	Hyderabad[3]	Brazil[4]	Fresh	Dry
Moisture	51.6	77.0	64.5	61.5	63.65	—
Protein	6.6	2.64	6.6	12.3	7.04	19.37
Fat	0.4	—	0.4	0.5	0.43	1.18
Fiber	1.5	1.0	1.5	—	1.33	3.66
Carbohydrates	38.4	17.9	25.8	25.1	26.80	73.73
Minerals	—	0.64	1.2	—	0.92	2.53
Calories	184.0	83.0	133.0	154.0	139.0	383.0

[1]Source: Purseglove (1968).
[2]Source: Bhatia *et al.* (1955A).
[3]Source: Aykroyd *et al.* (1966).
[4]Source: Bobbio *et al.* (1978).

for five days (Table 11.28). The paste characteristics of the starch, at 8 and 10% concentrations, indicated that the initial pasting temperature was 74.5° C without any change due to increase in concentration. At 8% concentration, the viscosity increased with an increase in temperature

TABLE 11.26. MINERAL AND VITAMIN COMPOSITION OF JACKFRUIT SEEDS FROM DIFFERENT REGIONS OF SOUTH INDIA (MG PER 100 G EDIBLE PORTION)

Constituents	Hyderabad[1]	Mysore[2]	Bangalore[3]
Minerals			
Calcium	50.0	87.2	21.0
Magnesium	54.0	—[4]	—
Phosphorus	97.0	40.9	28.0
Iron	1.5	1.5	—
Sodium	63.2	—	—
Potassium	246.0	—	—
Copper	0.19	—	—
Sulfur	356.0	—	—
Vitamin A (I.U.)	17.0	—	17.0
Thiamine	0.25	0.05	—
Riboflavin	0.11	—	—
Nicotinic acid	0.3	—	—
Vitamin C	11.0	—	—
Choline	52.0	—	—

[1]Source: Aykroyd *et al.* (1966).
[2]Source: Bhatia *et al.* (1955A).
[3]Source: Anon. (1979).
[4](–) values not determined.

TABLE 11.27. CHEMICAL ANALYSIS OF STARCH FROM JACKFRUIT COTYLEDONS (SEEDS)

Component	Percentage
Moisture	13.0
Fat	0.64
Protein	0.32
Ash	0.22
Amylose	28.1
Glucose (by hydrolysis)	99.1

Source: Bobbio *et al.* (1978).

from 74.5 to 95° C, and continued to increase even above 95° C, while at 10% concentration, the peak viscosity was reached at 94° C (Table 11.29). The starch granules are, therefore, considered to be resistant to breakdown by thermal or mechanical shear, and the bonding forces within the starch granules remain strong even after gelatinization and swelling to a high degree. In conclusion, jackfruit starch, because of its physicochemical characteristics, might be useful in the food industry where stable paste and rigid gel properties are desirable, as in pie fillings, canned soups, and baby foods (Bobbio *et al.* 1978).

FLAVOR OF JACKFRUIT

According to a report from the Philippines (Oliveros *et al.* 1971), the demand for ripe jackfruit has increased 100-fold, following the advance of food technology in the Philippines. The edible portion of jackfruit and also the other fleshy portions, e.g., the rind and unfertilized perianth, impart a strong, sweet aromatic odor; the bulbs possess a desirable tex-

TABLE 11.28. SAG AND APPARENT RIGIDITY VALUES OF JACKFRUIT STARCH (FROM COTYLEDONS) AS COMPARED TO OTHER STARCHES (AT 8% CONCENTRATION)

Starch[1]	Temperature of maximum viscosity (°C)	Sag value		Apparent rigidity
		t = 0 min	t = 10 min	
Jackfruit	95.0	0.0	0.0	*[2]
Rice	92.5	9.7	9.8	0.98
Potato	67.0	13.7	13.9	0.69
Arrowroot	79.5	15.3	15.7	0.39
Tapioca	70.0	25.5	26.2	0.37

Source: Bobbio *et al.* (1978).
[1]Determined by Brabender Visco/amylograph.
[2]Beyond maximum limits of apparatus.

TABLE 11.29. PASTE VISCOSITY CHARACTERISTICS OF JACKFRUIT STARCH (FROM COTYLEDONS) AT 8 AND 10% CONCENTRATIONS

Characteristic	Concentration (%)	
	8	10
Initial pasting temperature (°C)	74.5	74.5
Temperature of peak viscosity (°C)	95.0[1]	94.0
Peak viscosity (AU)[2]	420[3]	860
Viscosity (AU) at constant temperature (95°C)	450	840
Viscosity (AU) at 50°C (cooling cycle)	640	1000

Source: Bobbio *et al.* (1978).
[1]After 5 min of continuous heating.
[2]AU = Amylograph units.
[3]Viscosity at 94°C was 330 AU.

ture and a rich, appetizing taste. In shredded form, it may be eaten raw or it may be used as an ingredient of ice cream, candies and other forms of desserts. The local demand for the fruit is so great that it has become one of the most expensive fruits in the Philippines. Therefore, the short refrigerated storage life and the absence of any suitable flavor substitute for its use in the off-season led Oliveros *et al.* (1971) to explore the flavor concentrates of jackfruit as described below.

Using the left-over material after removal of the edible portion (bulbs and seeds), Oliveros *et al.* (1971) collected an essential oil from the rind by subjecting it to steam distillation, drying the essential oil over anhydrous sodium sulphate, and storing it in nitrogen-filled, amber-colored bottles. The yield of oil was low, about 0.03%. The essential oil was colorless, with a refractive index of $n_D^{25} = 1.4862$, and density of $d_{25}^{25} = 0.9124$. It possessed a strong odor of the fresh fruit. From this preliminary study, the authors concluded that with a more appropriate method of extraction under controlled conditions, commercial extraction of the essential oil would be profitable. They suggested its possible uses in food and perfumery industries.

Abraham (1971) also carried out some preliminary investigations on the flavor of jackfruit. He found that chloroform is a good solvent for the extraction of the flavor components of jackfruit; steam distillation destroys the aroma, and cannot be employed for the preparation of a flavor concentrate. Thin layer chromatography of the chloroform extract revealed the presence of four carbonyl compounds which were not identified.

Natarajan and Karunanithy (1974) have shown that the natural aroma could be reinforced (added back) in preserved products, viz., canned products and nectars, by the addition of a few milligrams per liter of two esters synthesized from gamma-butyrolactone, viz., the ethyl and butyl esters of 4-hydroxybutyric acid. Swords *et al.* (1978) reported the use of a combination of gas chromatography and mass spectrometry to identify from an extract of a distillate of jackfruit 20 major flavor components, including the aliphatic alcohols butanol, isoamyl alcohol, hexanol and octanol, of which the former two were in higher concentrations. The remaining identified components were esters. The predominant esters were: butyl isovalerate, butyl butyrate, isoamyl isovalerate, butyl acetate, isobutyl isovalerate and propyl isovalerate (Table 11.30). The au-

TABLE 11.30. VOLATILE CONSTITUENTS OF TWO VARIETIES (FROM BRAZIL AND MALAYSIA) OF JACKFRUIT (*ARTOCARPUS HETEROPHYLLUS*)

Compound	GC/MS identification[1]	% composition[2]
1-butanol	+	28.0
2-methyl-1-butanol (isoamyl alcohol)	+	26.6
Ethyl butyrate	+	0.5
Ethyl isovalerate	+	9.0
Propyl isovalerate	+	2.3
Butyl butyrate	+	0.4
Isobutyl isovalerate	+	0.6
Butyl isovalerate	+	5.5
Butyl hexanoate	+	0.2
Isoamyl hexanoate	+	0.2
Butyl acetate	+	0.7
1-propanol	—	1.6
2-methyl-1-propanol	—	3.0
2-methylbutyl butyrate (Isoamyl butyrate)	—	0.4
Methylbutyl isovalerate	—	3.8
2-2-methylbutyl acetate (Isoamyl acetate)	—	0.5
Octyl acetate	—	0.2
Methyl isovalerate	—	8.9
Methyl butyrate	+	—
Propyl butyrate	+	—
Butyl-2-methyl butyrate	+	—
Isobutyl butyrate	+	—
Isoamyl isobutyrate	+	—
Isoamyl isovalerate	+	—
Butyl valerate	+	—
Hexanol	+	—
Octanol	+	—

[1]Source: Swords *et al.* (1978); + detected, — not detected.

[2]Source: Rasmussen (1983); percent composition estimated from relative G.C. peak heights from two GC columns and totalling 100% for all identified peaks including 2 impurities from solvent not listed (6.5%).

thors concluded that the mild floral aroma of jackfruit was due to the predominance of esters over the four aliphatic alcohols. The total absence of terpenoids was considered a unique feature of jackfruit.

In a recent study, Rasmussen (1983) carried out identification of the volatile components of jackfruit by gas chromatography/mass spectrometry with two different columns, to overcome the difficulties in distinguishing the isomeric compounds, normally encountered when using a single column. Employing jackfruit obtained from the local markets of Kuala Lumpur, he separated the volatile components using vacuum-distillation with dichloromethane as solvent, and reduced the volume (100-200 μl) before analyzing by GC/MS.

By means of his advanced technique, he too was able to establish the presence of 20 flavor components, including eleven of the compounds identified earlier by Swords *et al.* (1978). Rasmussen (1983) was, however, able to isolate and identify eight additional compounds from Malaysian jackfruit (Table 11.30). He explained the absence of the nine previously established constituents by stating that they may either be present in small quantities, as small peaks for which mass spectra were not obtained, or as minor constituents of larger GC peaks.

Judging from the GC peak heights of Swords *et al.* (1978) and of Rasmussen (1983) (Table 11.30), the major constituents were generally the same, although there were some differences in the relative concentrations established by the two methods. Furthermore, the differences in the number of components isolated may be due to the differences in the jackfruit varieties used and, also, may be due to the variations occurring by seed propagation of the varieties.

PROCESSING

The seasonal nature of the fruit coupled with the short storage life even under low temperature conditions, as well as its exotic taste and flavor necessitate processing of the fruit in order to make it available in the off-season. A number of products have been developed from the raw, tender jackfruit, as well as from the ripe fruits and its seeds. There are at least five products prepared from tender jackfruit traditionally in India which have been standardized (Bhattacharjee 1981). Raw, tender jackfruit are (1) canned in brine, (2) canned in curried form as such (or in combination with cauliflower, beans, potatoes and tomatoes) (3) made into dehydrated products, (4) papadam or papad, (5) sweet or sour pickle in oil, and (6) sweet and sour pickle in vinegar.

The ripe jackfruit bulbs (excluding seed) and the rind of the ripe fruit (including the perianth of unfertilized flowers) have been used for processing into a number of products. Products from ripe jackfruit are (a) from bulbs and (b) from rind (Bhattacharjee 1981). Ripe jackfruit (bulbs) are (1) canned in syrup as such or with other fruits (mango, pineapple, banana and orange), (2) canned frozen fruit, (3) jams-pure or mixed (with pineapple, bananas and mangoes), (4) dehydrated bulbs, (5) chutney, (6) preserves, candy, and (7) concentrate and powder. Rinds of ripe jackfruit are made into (1) syrup, (2) pectin, (3) jelly, (4) pectin extract, and (5) papadam.

Like the raw, tender jackfruit, the seeds of ripe jackfruit are used as vegetables and are equally popular in the east, especially in India, Malaysia and Sri Lanka. Cooked with grated coconut, chilies, salt and spices, seeds form a very popular side dish with rice for the people of Kerala. The seeds, when roasted or fried like peanuts, are very tasty to eat. Stewed with meat, both the cubes of raw tender jackfruit, or the seeds, form a good side dish with rice (Thomas 1980). Therefore, besides the direct use in culinary preparations, there are a few products developed from seeds. Seeds are (1) canned in brine, (2) canned in curried style as such or in combination with other vegetables, (3) canned in tomato sauce, (4) dried for flour, and (5) roasted (Bhattacharjee 1981).

Canning Raw Jackfruit in Brine

Tender, raw jackfruit, after the outer rind and core were removed, were cut into 2 to 3 cm cubes. About 450 g of prepared cubes were filled into no. 2½ plain cans, covered with 340 g of boiling 2% brine, kept in boiling water until the center of the can reached 79.5° C (about 7 to 8 minutes), sealed and processed at 4.5 kg pressure for 75 minutes, and cooled in running water. The cans had a vacuum of 25-40 cm and head space of 8 mm. The drained weight was about 66%. The product had an attractive light buff color resembling roasted chicken, had good taste and flavor, and was quite soft in texture (Bhatia *et al.* 1956A).

Canning Raw Jackfruit Curry

Raw, tender jackfruit produced a good quality product when packed in curried style alone, or in combination with other common vegetables like potatoes, tomatoes, cauliflower, beans, etc. The vegetables were cut into pieces of desired size, packed in no. 2½ cans and covered with hot (80° C) spiced gravy. The cans were exhausted and processed for canning in brine. Each can contained 500 g of jackfruit pieces canned alone, or 350 g of jackfruit pieces and 115 g of other vegetables. When beans were used for mixed curry, they were blanched in 1% sodium bicarbonate for 5 minutes to remove bitterness and beany flavor. When onions were used, 30 g of onions and 425 g of jackfruit pieces were packed in sulfur-resistant lacquered cans. In each case 350 g of gravy was used as the covering liquid. The finished product in all these packs had adequate vacuum, head space and drained weight (Bhatia et al. 1955B, 1956A).

Canning Ripe Jackfruit Bulbs

Canning of jackfruit, if carried out at its original pH of 5.0 to 5.2 and processed at 0.70 kg/cm² pressure for 30 minutes, results in a very mashy product with a cooked taste and flavor (Bhatia *et al.* 1955A). Addition of extra acid in some form becomes necessary to bring the pH below 4.5 and allow safe processing in boiling water (Bhatia *et al.* 1956D). Jackfruit when canned alone has a strong flavor which is not liked by some people. Hence, canning jackfruit along with other fruits is desirable to make it mild and acceptable (Bhatia *et al.* 1956D). However, the aforementioned authors found that canning bulbs in syrup of 50° Brix containing 0.5% acid gave a good product with a storage life of 14 months at room temperature (21-35° C). The pH of the medium was 4.7 and vacuum was 23 cm. The effect of processing on the nutritive value has been studied in detail by others (Pruthi *et al.* 1955; Dhopeswarkar *et al.* 1956; Siddappa and Bhatia 1956; Siddappa 1957).

Canning Ripe Jackfruit Pulp

Jackfruit pulp is prepared by blending the bulbs with 25% water in a blender (on a large scale, the bulbs can be minced by passing through a mincer and later pulped by passing through a pulper with addition of 25% water to the minced material). The pulp with 1% citric acid added, when canned and processed in boiling water for 40 minutes, gave a product which looked attractive, and had good taste and flavor. When mixed with other fruit pulps like mango or pineapple, a 0.5% citric acid addition is adequate. The pulp can be used for preparation of ice cream or can be used for jams and squash making (Bhatia et al. 1955A, B, 1956C).

Canning Jackfruit Seeds

Jackfruit seeds canned in brine alone or in curried form, as in the case of raw, tender fruit, not only gave a good product but eliminated the trypsin inhibitor present in the seed that can be destroyed by heat (Siddappa 1957).

DEHYDRATION

Unripe as well as ripe jackfruit, and the seeds from mature ripe jackfruit, can be preserved for over one year by dehydration.

Dehydration of Jackfruit Bulbs

Jackfruit bulbs can be dried either in the sun or in a cabinet drier at 50-55° C after sulfuring them in a closed chamber. Sulfuring can be done by burning sulfur at 8 kg sulfur per ton of material in 1000 cubic feet of space. The treatment is necessary for the retention of bright yellow color and to avoid browning during storage. Most of the sulfur dioxide absorbed by the product is lost in the process of drying and during storage. A sample of product, after about 2 years of storage at room temperature (24-30° C) in an air-tight glass jar, contained only 280 ppm sulfur dioxide. It retained the yellow color for about one year, after which slight browning was noticeable. The product was quite satisfactory with regard to color and flavor but was slightly tough and hard to chew. Three kg of bulbs on drying would yield 1 kg of product (Bhatia *et al.* 1955C).

In a later study, Teaotia and Awasthi (1968) carried out detailed investigations on the dehydration of jackfruit bulbs. Their study covered the effect of fruit maturity, pretreatments with SO_2, dehydration and rehydration characteristics, storage and quality of the dried product, and cooking quality of the product. Fruits harvested between 165 to 195 days (from spike appearance) with 0.2% SO_2 treatment for 15 minutes, gave a good quality dehydrated product with a drying ratio around 11.5:1 (Table 11.31). At a drying temperature of 60-65° C, in a cross-flow drier with 12 aluminum trays, material spread in a single layer (with a loading of 900 g per square foot), the product took 20 hours to attain a moisture content of 5.3%. Dehydrated product preserved in glass jars with lids sealed with wax, maintained the flavor and original qualities for over 6 months at room temperature (24-30° C) (Table 11.32). Dried bulbs, when reconstituted in warm water (65° C) containing 1% sodium chloride for 30 minutes, gave a good quality product that could be used in culinary preparation (Table 11.33). The cooking quality of bulbs harvested between 165 and 195 days, with sulfite treatment during dehydration, and reconstituted as above gave a good quality cooked product, while control samples harvested at 180 days yielded a very good product (Table 11.34).

Dehydration of Jackfruit Pulp into Sheets

Jackfruit pulp, after careful experimentation and development of a patented method was successfully dried into sheets or slabs, as normal

TABLE 11.31. QUALITY ASSESSMENT OF JACKFRUIT BULBS DEHYDRATED AT SIX STAGES OF MATURITY FOR 20 HOURS AT 60-65°C

Stage of maturity (period in days from spike appearance)	Treatments	Color score[1]	Flavor score[1]	Texture score[1]	Total score[2]	Dehydration quality	Drying ratio—fresh fruit weight/dried product wt.
135	Plain water	0.0	0.0	0.0	0.0	Unacceptable	
	0.1% SO_2	0.5	0.4	0.3	1.2	Unacceptable	13.1:1
	0.2% SO_2	0.7	0.6	0.7	2.0	Unacceptable	
150	Plain water	3.2	3.3	3.1	9.6	Unacceptable	
	0.1% SO_2	3.5	3.4	3.4	10.3	Unacceptable	12.5:1
	0.2% SO_2	3.7	3.6	3.6	10.9	Unacceptable	
165	Plain water	4.0	4.2	4.1	12.3	Poor	
	0.1% SO_2	6.2	6.0	6.2	18.4	Fairly good	12.1:1
	0.2% SO_2	8.0	7.8	7.8	23.6	Good	
180	Plain water	5.6	5.8	5.5	16.9	Poor	
	0.1% SO_2	7.5	7.4	7.6	22.5	Fairly good	11.7:1
	0.2% SO_2	8.5	8.4	8.4	25.3	Fine	
195	Plain water	5.5	5.6	5.4	16.5	Poor	
	0.1% SO_2	7.3	7.0	7.2	21.5	Fairly good	11.5:1
	0.2% SO_2	8.2	8.1	7.1	23.0	Good	
210	Plain water	5.0	5.2	5.2	15.4	Poor	
	0.1% SO_2	6.0	5.8	5.8	17.6	Poor	11.1:1
	0.2% SO_2	6.2	6.2	6.2	18.6	Poor	

Source: Teaotia and Awasthi (1968).
[1]Maximum score = 10.
[2]Maximum total score = 20.

TABLE 11.32. QUALITY ASSESSMENT OF JACKFRUIT BULBS DEHYDRATED ON 165, 180 AND 195 DAYS (MATURITY) AND AFTER STORAGE FOR 6 MONTHS AT ROOM TEMPERATURE

Days	Treatments	Color score[1]	Flavor score[1]	Texture score[1]	Total score[2]	Quality
165	Control	3.0	3.2	3.0	9.2	Unacceptable
	0.2% SO_2	7.5	7.6	7.8	22.9	Good
180	Control	3.5	3.4	3.2	10.1	Unacceptable
	0.2% SO_2	8.4	8.0	8.2	24.6	Fine
195	Control	2.8	3.1	2.8	8.7	Unacceptable
	0.2% SO_2	7.8	7.5	7.6	22.9	Good

Source: Teaotia and Awasthi (1968).
[1]Maximum score = 10.
[2]Maximum total score = 30.

drying at 50° C gave a fibrous product. The method involves the addition of a sweetening agent, heating the pulp, sieving it, spreading on trays, drying in a cabinet drier, and then cutting the dried pulp into slabs which can be wrapped in cellophane paper and packed in cardboard boxes or glass jars. The product had a storage life of 15 months (Bhattacharjee 1981).

Dried Jackfruit Seeds, Roasted Nut, Seed Flour and Papads

The seeds of jackfruit are processed into different products, viz., roasted nuts, dried seeds, seed flour and papadam or papad.

TABLE 11.33. QUALITY ASSESSMENT OF REHYDRATED JACKFRUIT BULBS

Treatments	Texture[1]	Color[1]	Flavor[1]	Total[2]	Quality
Tap water (20°C) containing 1% NaCl	7.0	7.8	7.0	21.8	Fairly good
Tap water (20°C)	7.0	7.5	7.0	21.5	Fairly good
Warm water (65°C) containing 1% NaCl	8.2	8.5	8.3	25.0	Fine
Warm water (65°C)	7.8	8.2	8.0	24.0	Good
Boiling water (100°C) containing 1% NaCl	2.2	1.0	2.5	5.7	Unacceptable
Boiling water (20°C)	2.2	1.0	2.4	5.6	Unacceptable

Source: Teaotia and Awasthi (1968).
[1]Maximum score = 10.
[2]Maximum score = 30.

TABLE 11.34. COOKING QUALITIES OF DEHYDRATED JACKFRUIT BULBS

Treatments	Taste and flavor[1]	Cooking quality
165th day, with 0.2% SO_2	7.8	Good
180th day, with 0.2% SO_2	8.4	Very good
195th day, with 0.2% SO_2	7.5	Good
180th day, without SO_2	5.0	Poor

Source: Teaotia and Awasthi (1968).
[1]Maximum = 10.

Dehydration of Seeds

Jackfruit can be dried in the sun, or in a cabinet drier to a moisture content of 8% or less. These dried seeds can be preserved in airtight containers for over a year. Dried jackfruit seeds are white in color and can be reconstituted by soaking in 3 times their weight of water for 18 hours. They can be used like fresh seed for canning in different forms, or they can be cooked and consumed (Bhatia *et al.* 1956A).

Roasted Nuts

Good quality roasted nuts prepared from jackfruit seeds resemble chestnuts in nutritive value and flavor. The outer silvery white tough skin could be removed by roasting the fresh seeds at 200-300° C for 10 minutes in a coffee roasting machine. On further roasting for 30 minutes at 300° C, the seeds became soft enough to chew, and were very delicious (Bhatia et al. 1956B).

Jackfruit Seed Flour

The outer white coat can be easily removed by sun drying. The red cuticle of the seeds was removed mechanically using a potato peeling machine fitted with a revolving emery disc. The pieces with cuticle removed were then cut into 2 or 4 pieces and dried in the sun, or in a cabinet drier at 45° C, until they became hard and bony. They were ground in a flour mill to a flour that resembled wheat flour in appearance. The flour had a moisture content of 8.0% and was slightly bitter in taste. The jackfruit seed flour mix at a 25% level with wheat flour gave chapaties (local name for flat, unleavened bread, made from wheat flour) (Siddappa and Bhatia 1955).

Jackfruit Papad

Papad is a common household product made from jackfruit bulbs which are neither fully mature nor completely raw. Pale bulbs having a mild flavor, but little sweetness, are used for this preparation. They are ground into a pasty pulp in a stone grinder after steaming for 30 minutes at 4.5 kg pressure using about one-third water to facilitate grinding. The following recipe was found suitable: ground pulp, 1½ kg; common salt powder, 18 g; asafoetida, 1 g; chili powder, 2 g; caraway, 4 g; black pepper powder, 3 g; sodium bicarbonate, 1.8 g.

The pulp was ground further (after mixing in all the ingredients), made into round thin chapaties 7 to 10 cm in diam and dried in a home drier at 50-55° C for 16 hr. The drying ratio was about 3:1. The product is good in texture, has light brown color, is slightly sweetish in taste, and has a well balanced spicy flavor. These papads, when wrapped in paper, keep well at room temperature (24-30° C) for 4-6 months (Bhatia et al. 1956B). On frying in vegetable oil, papads swell slightly, have a crisp texture, light brown color, and a good spicy taste and flavor.

JACKFRUIT JAM AND JELLY

Jam

Good quality jackfruit jam can be prepared from jackfruit alone or in combination with pineapple, banana and mango. Bulbs of ripe fruit are cut into small bits, mixed with an equal weight of other fruits and boiled in a small quantity of water to soften the fruit pieces. Sugar and citric acid are then added at the rate of 75% and 0.3% respectively, of the weight of the fruit used. The contents are cooked with constant stirring until the solids (refractometer) are 68-70%. The jam is filled at a temperature of about 75° C in sterile glass jars, and cooled to room temperature; thereafter, the lids are closed tightly and waxed. The jam may also be packed in plain cans by filling hot, sealing and inverting the cans. Jams thus prepared have a predominant jackfruit flavor, and are of good texture. When stored at room temperature (24-30° C) these jams retain their texture and characteristic flavor for 1½ years (Bhatia *et al.* 1955B).

Jelly from Jackfruit Rind

Siddappa and Bhatia (1955) standardized a method for the preparation of jelly of good quality from the rind of jackfruit. The rind is rich in sugars and pectin, with the pleasant natural aroma of the bulbs. Since the rind is low in acidity, 1:1 ratio of extract to sugar with 0.6 to 0.8% added acidity was found suitable for preparation of a good jelly. Further details of preparation are also given in a technical leaflet (Anon 1953A).

JACKFRUIT SQUASH (NECTAR)

A suitable recipe, which has evolved after trying various blends, is: blanched pulp, 4.5 kg; sugar, 10 kg; water, 6 kg; citric acid, 200 g; potassium metabisulfite, 13 g.

A 60° Brix sugar syrup is prepared by dissolving sugar in hot water, to which citric acid is added, and then mixed with pulp after cooling. A squash of 52° Brix and 1% acidity is thus obtained (Anon 1953B). It is filtered and filled in bottles, leaving a headspace of 5 cm. The bottles are then closed air-tight with crown corks and stored in a cool dry place. The squash is diluted as desired (with 4-5 volumes of water) and may be taken with or without carbonation.

JACKFRUIT PRESERVE AND CANDY

Preserve

Bhatia *et al.* (1955C) reported the preparation of jackfruit preserve as follows: the bulbs together with 2½ times their weight of 40° Brix syrup containing 0.3% added citric acid are placed in glass flasks and are subjected to vacuum (66-69 cm) for about 2 hr (45° Brix stage) and then left over night. The next day the concentration is continued to 65% solids; the bulbs are then allowed to absorb sugar for about 2 hr until the finished product reaches 72% solids. The whole process is completed in 36 hr. The product obtained is firm in texture, has an attractive yellow color, and has full jackfruit flavor. The yield of the preserve is about 1½ times the processed fruit. The product can be packed in glass jars as a preserve, or the bulbs may be drained and dried to prepare good quality candy. The drained syrup can be marketed as a fruit syrup packed in glass bottles. It has the characteristic jackfruit flavor, and on diluting five times gives an excellent drink. These products (candy and preserve) have been found to retain their initial qualities for a period of 15 months when stored at room temperature (24-30° C).

JACKFRUIT CHUTNEY

For the preparation of sweet, spiced chutney from jackfruit bulbs, Bhatia *et al.* (1955C) evolved the following recipe: jackfruit pieces 750 g; sugar, 225 g; common salt, 20 g; spices (whole-mace, cinnamon, cardamom and black pepper) each 4.5 g, 18 g total; red chili powder, 5 g; onions (chopped), 30 g; garlic (ground), 7 g; acetic acid, 4 ml; citric acid (at 0.3% fruit wt), 2 g.

After adding citric acid, the fruit material is boiled with about a half liter of water for 15-20 minutes to soften it. Spices, onion, garlic and red chilies are loosely tied in a bag of mull cloth and cooked along with the fruit to a thick consistency. The sugar and salt are then added, and cooking continued until the material becomes thick again. Vinegar or acetic acid is added about five minutes before the termination of boiling, when solids (by refractometer) reach 68-70%. The bag of spices is removed, and the finished product is packed in freshly sterilized glass jars. After cooling, a metal lid is screwed on, placing an ordinary cork just below the metal cap to avoid blackening of the product during storage by the action of acetic acid on the metal lid. The product can be stored for over 15 months at room temperature (24-30° C).

JACKFRUIT TOFFEE

A recipe for the preparation of jackfruit toffee from jackfruit pulp has been standardized by Bhatia *et al.* (1956B). The pulp is prepared by blending the drained fruit from cans (the authors used in their study drained fruit from cut out cans) and passing through 16 mesh sieve. The recipe is: pulp, 175 g; drained syrup (32° Brix from the can), 85 g; skim milk powder, 50 g; sugar, 30 g; fat, 425 g; common salt, 1.5 g; vanillin, 5 drops. High melting point fat (m.p. 42° C) should be used.

The ingredients, except vanillin, are mixed and cooked in a shallow stainless steel pan with continuous stirring until the final weight is 200 g, when the product has attained a thick consistency. Vanillin is added and mixed uniformly and the thick mass is spread on a smooth stainless steel plant to which a little fat was previously smeared. When it sets (in 12 hr) it can be cut into toffees and packed in glass jars or cardboard boxes after wrapping each toffee in cellophane paper. The toffee thus prepared has a chocolate color, firm texture and mild jackfruit flavor. Even after a storage period of one year, the toffee retained its flavor and did not develop any rancid taste.

PICKLES

Curing of Jackfruit Slices

Slices of green, tender jackfruit were cured in an equal weight of salt solution by gradually raising the strength of the brine from 16° Salometer in the beginning to about 60° over a 2-week period. Production of lactic acid was, however, not satisfactory. The addition of 2% sugar to the brine helped in producing lactic acid to some extent. This could be maintained by further additions of sugar during curing. By this method the acidity could be raised to 0.7% (as lactic) (Bhatia *et al.* 1956C).

Sweet, Spiced, Vinegar-Jack Pickle

The cured slices were washed free from salt under running cold water, filled into glass pickle jars and kept covered with 8% acetic acid solution. After two days, the vinegar was drained off and sweet spiced vinegar of 50° Brix (Campbell 1950) having about 3% acetic acid was added as covering liquid. Any fall in Brix or acidity in the course of 3 or 4 days was adjusted by adding acetic acid or sugar. The final Brix of the liquid in the pickle was kept at 50° and the acidity at 3% (as acetic acid) (Bhatia *et al.* 1956C).

Sour Oil Pickle

The following recipe gave a good oil pickle: cured jackfruit slices, 1 kg; table salt, 55 g; chili powder, 17 g; caraway, 2.8 g; cloves (headless), 2.8 g; cardamon, 1.4 g; cinnamon (broken), 2.8 g; dried ginger (chopped), 14 g; onions (chopped), 30 g; gingelly oil, 750 g; garlic (ground), 2.8 g; vinegar (5-6% acetic acid), 12.5 g.

Chopped ginger, onions, garlic and mustard seeds were fried in a little oil until they turned brown in color. These, along with the remaining spices, were mixed thoroughly with the cured jackfruit slices and kept in the sun for 3-4 days in an earthenware jar. Vinegar was then added and the pickle was covered with gingelly oil and stored after covering the jar with a lid (Bhatia *et al.* 1956C).

Sweet Oil Pickle

The method of preparation of this pickle was similar to that of sour oil pickle except that 125 g of sugar was added along with the vinegar for the recipe given above. These different types of pickles kept well for more than two years at room temperature (24-30° C) (Bhatia *et al.* 1956C).

PECTIN FROM JACKFRUIT WASTE

Krishnamurthy and Giri (1949) studied the preparation and composition of pectin from the seeds, kernel and pericarp of jackfruit, and found the pericarp and kernel to be rich sources of pectin. Jain and Lal (1954, 1957) prepared high grade pectins from jackfruit waste. Bhatia *et al.* (1959) have prepared pectin and also pectin extract from jackfruit rind.

FUTURE CONSIDERATIONS

Commercial cultivation of jackfruit in tropical countries appears to be very profitable, not only for the yield—which is higher than for any other fruit crop—but also for the multiple uses of the tree and the fruit. The advances made, both in horticulture of jackfruit, and in technologies developed for the proper utilization of the fruit, combined with the growing demand for the fruit worldwide, make it necessary to step up the production of the fruit on a commerical scale. There is, however, a need for the selection and propagation of varieties suitable for processing, since the post-harvest life of the fruit even under refrigeration is limited to 4-6 weeks. There is also a need for updating and scaling up the technologies developed earlier for the utilization of the fruit to make its cultivation more profitable.

REFERENCES

ABRAHAM, K. O. 1971. Fruit flavors: Preliminary studies on jackfruit (*Artocarpus integrifolia*, L.) flavor. M.Sc. Dissertation, Mysore University, India.

ACHINEWHU, S. C. 1982. The nutritive qualities of plant foods: Chemical and nutritional composition of bread fruit (*Artocarpus utilis*) and climbing melon seed (*Colocyathis vulgaris*). Nutr. Rep. Int. 25, 643-647.

AMAR SINGH. 1980. Fruit Physiology and Production. Kalyani Publishers, Printed at R.K. Printers, 80-D, Kamla Nagar, Delhi-110 007.

ANON. 1951 (Mauritius). Annu. Rep. Dep. Agric. Mauritius (1949), 1951. Published by Govt. Printer, Port Louis, No. 25.

ANON. 1953A. Preparation of jackfruit jelly. Bull. CFTRI 3, 93.

ANON. 1953B. Preparation of jackfruit nectar. Bull. CFTRI 3, 70-73.

ANON. (Seychelles) 1964. Annu. Rep. Dep. Agric. for 1964.

ANON. 1979. Jackfruit. Indian Institute for Horticultural Research. Ext. Bull. 22, Bangalore.

ANON. 1986. Increase in average sale of jackfruits. A special article on jackfruits of Panruti. The Hindu, June 6, 20.

AYKROYD, W. R., GOPALAN, C. and BALASUBRAMANYAN, S. C. 1966. The nutritive value of Indian food and the planning of satisfaction diet. Spec. Rep. Ser. 42. Indian Council of Medical Research, New Delhi.

BARRAU, J. 1959. Investigation to extend season of breadfruit yield. South Pac. Comm. Quart. Bull. 4(4) 20-40.

BENNETT, F. D., and NOZZOLILLO, C. 1987. How many seeds in a seeded breadfruit, *Artocarpus altilis* (Moraceae). Econ. Bot. 41, 370-374.

BHATIA, B. S. 1953. Studies on the preservation of some jackfruit products. M.Sc. Thesis, University of Punjab, India.

BHATIA, B. S., SIDDAPPA, G. S. and LAL, G. 1955A. Composition and nutritive value of jackfruit. Indian J. Agric. Sci. 25, 303-306.

BHATIA, B. S., SIDDAPPA, G. S. and LAL, G. 1955B. Development of products from jackfruit - I. Canned jackfruit, frozen canned jackfruit and jackfruit jam. Indian Food Packer 9(11) 7.

BHATIA, B. S., SIDDAPPA, G. S. and LAL, G. 1955C. Development of products from jackfruit - II. Jackfruit preserve, candy, chutney and dried bulbs. Indian Food Packer 9(9) 7.

BHATIA, B. S., SIDDAPPA, G. S. and LAL, G. 1956A. Development of products from jackfruit. VI. Canned green jackfruit in brine and curry. Indian Food Packer 10, 7.

BHATIA, B. S., SIDDAPPA, G. S. and LAL, G. 1956B. Development of products from jackfruit. V. Dried jack seeds, flour, roasted nut and jack papad and toffee. Indian Food Packer 10, 9.

BHATIA, B. S., SIDDAPPA, G. S. and LAL, G. 1956C. Development of products from jackfruit. VII. Dried green jackfruit and jack pickle. Indian Food Packer 10(9) 13-14.

BHATIA, B. S., SIDDAPPA, G. S. and LAL, G. 1956D. Role of pH in the canning of jackfruit (*A. integrifolia*): Effect of adding acid or other fruits to the canned products. Indian J. Agric. Sci. 27, 133-135.

BHATIA, B. S., SIDDAPPA, G. S. and LAL, G. 1959. Preparation of pectin, pectin extract and syrup from jackfruit rind. Indian J. Agric. Sci. 29, 75.

BHATTACHARJEE, S. 1981. Chemistry and technology of some minor fruits. M.Sc. Dissertation, University of Mysore, India.

BIALE, J. B. and BARCUS, D. E. 1970. Respiratory patterns in tropical fruits of the Amazon Basin. Trop. Sci. 12, 93-104.

BOBBIO, F. O., EL-DASH, A. A., BOBBIO, P. A. and RODRIGUES, L. R. 1978. Isolation and characterization of the physicochemical properties of the starch of jackfruit seeds (*Artocarpus heterophyllus* Lam). Cereal Chem. 55, 505-511.

BOWERS, R. D. 1981. Breadfruits: A low energy requirement source of carbohydrate for wet tropics. Entwicklung Landlicher Raum. 15(2) 11-13.

BUTANI, D. K. 1978. Pests and diseases of jackfruit in India and their control (a review). Fruits 33(5) 351-357.

CAMPBELL, C. H. 1950. Canning, Preserving and Pickling. Vance Publishing Corp., 139 N. Clark St., Chicago.

CHATTERJEE, B. K. and MUKHERJEE, S. K. 1980A. Stooling potentiality of clones of jackfruit (*Artocarpus heterophyllus* Lam). Sci. Cult. 46(2) 55-56.

CHATTERJEE, B. K. and MUKHERJEE, S. K. 1980B. A note on the effect of leafy and non-leafy cuttings, rootings of jackfruit (*Artocarpus heterophyllus* Lam). Prog. Hortic. 11(4) 49-51.

CHATTERJEE, B. K. and MUKHERJEE, S. K. 1982. Studies on the efficacy of rates and methods of application of indole butyric acid (IBA) on rooting of cuttings of jackfruit (*A. heterophyllus* Lam). Sci. Cult. 48(2) 74-75.

COENEN, J. and BARRAU, J. 1961. The breadfruit tree in Micronesia. South Pac. Bull. 11, 37-67.

CORREA, A. M. N., NAKAMURA, T. and TOLMASQUIM, E. 1970. Study of the properties of breadfruit starch. Rev. Bras. Tecnol. 1, 27-31.

COURSEY, D. G., BURDEN, O. J. and RICHARD, J. E. 1976. Recent research on post-harvest handling of tropical and subtropical fruits. Acta Hortic. 57, 135-143.

COX, P. A. 1980. Two Samoan technologies for breadfruit and banana preservation. Econ. Bot. 34, 181-184.

CRUZ, S. R. 1949. Newspaper as protective wrapper for jackfruits (nangka) against fruitfly (*Bactrocera umbrosa* Fabr.). Philippine J. Agric. 14, 213-220.

DAS, G. C., RANA, S. and KONHAR, T. 1981. Effect of (2-chloro-ethyl) phosphonic acid (Ethephon) on induction of flowering and gynoecium in jackfruit (*Artocarpus heterophyllus* Lam). Orissa J. Hortic. 9(1) 49-52.

DE BRAVO, E. N., GRAHAM, H. D. and PADOVANI, M. 1983. Composition of the breadfruit (seeded breadfruit). Carib. J. Sci. 19, 27-32.

DHUA, R. S., SEN, S. K. and BOSE, T. K. 1983. Propagation of jackfruit (*Artocarpus heterophyllus* Lam.) by stem cuttings. Punjab Hortic. J. 23(1) 84-91.

DHUA, R. S. and SEN, S. K. 1984. Role of etiolation, auxinic and non-auxinic chemicals on root initiation of air-layers of jackfruit (*A. heterophyllus* Lam). Indian J. Hortic. 41(1) 116-119.

DHOPESHWARKAR, G. A., FAGAR, N. G., PRUTHI, J. S. and LAL, G. 1956. Effect of storage temperature on the nutritive value and acceptability of canned guava and jackfruit. Fruit and Vegetable Industry in India, 236-238, published by CFTRI, Mysore.

GOODMAN, R. A. 1972. Plants and man in Samoa. Pac. Discovery 25, 12-18.

GRAHAM, H. D. and DE BRAVO, E. N. 1981. Composition of the breadfruit. J. Food Sci. 46, 535-539.

GUNJATE, R. T., KOLEKAR, D. T. and SALVI, M. J. 1982. *Cited* by H. N. Samaddar, Proc. Natl. Sem. on Plant Propagation, Calcutta.

HABIB, A. T. 1973. Vivipary in *Artocarpus heterophyllus* Lam. Mysore J. Agric. Sci 7(1) 120-121.

HAMILTON, R. A., CRILEY, R. A. and CHIAC, L. 1982. Rooting of stem cuttings of breadfruit *(A. altilis* (Park) Fosb.) under intermittent mist. Proc. Int. Plant Propagators Soc. 32, 347-350.

HAYES, W. B. 1957. Fruit Growing in India. Ritabistan, Allahabad, India.

JAIN, N. L. and LAL, G. 1954. Preparation of high grade pectins from papaya, jackfruit and woodapple. Bull. Cent. Food Technol. Res. Inst. 4, 287.

JAIN, N. L. and LAL, G. 1957. Some studies on the utilization of jackfruit wastes as a source of pectin. Indian J. Hortic. 14, 1.

JARRET, F. M. 1959. Studies on *Artocarpus* and allied genera. I. General considerations. J. Arnold Arb. 40, 1-29.

JOHNSTON, A. 1964. Plant diseases in Samoa. FAO Plant Prot. Bull. 12, 79-81.

KANNAN, K. and NAIR, T. N. 1960. A note on the growth of jack on two root stocks. South Indian Hortic. 8(3) 26-27.

KAREEM, A. A., NACHIAPPAN, R. M., SUBRAMANIAM, T. R. and GOVINDARAJAN, V. 1971. Severe occurrence of spittle bug, *Cosmoscarta relata* Dist. (Cercopidae: Homoptera) on jack at Kolli hills in Tamilnadu and its control. South Indian Hortic. 19(1) 87-89.

KRIPAL SINGH, K. and MATHUR, P. B. 1954. A note on the investigations on the cold storage and freezing of jackfruit. Indian J. Hortic. 11, 149.

KRISHNAMURTHY, C. R. and GIRI, K. V. 1949. Preparation, purification and composition of pectins from Indian fruits and vegetables. Proc. Indian Acad. Sci. 29B, 155-167.

LANGE, N. A. 1944. Handbook of Chemistry, 14th Ed. Handbook Publishers Inc., Sandusky, Ohio.

LOOS, P. J., HOOD, L. F. and GRAHAM, H. D. 1981. Isolation and characterization of starch from breadfruit. Cereal Chem. 58, 282-286.

MACKENZIE, J. B., MCKNIGHT, R. K. and LAWRENCE, P. 1964. Breadfruit cultivation practices and benefits in the Trush Territory of the Pacific Islands. Office of AGANA, Guam. Anthropological working papers, No. 7 and 8.

MANJUNATH, B. L. 1948. Wealth of India, Vol. 1. CSIR Publications, New Delhi.

MARIOTT, J., PERKINS, C. and BEEN, B. O. 1979. Some factors affecting the storage of fresh breadfruit. Sci. Hortic. 10, 177-181.

MATHUR, P. B., SINGH, K. K. and KPUR, N. S. 1952. Cold storage of jackfruit. J. Indian Soc. Refrig. Engineers 3(1) 40-46.

McMILLAN, R. J. Jr. 1975. Rhizopus artocarpi rot of jackfruit (*Artocarpus heterophyllus*). Proc. Fla. State Hortic. Soc. 87, 392-393.

MORTON, J. F. 1987. Fruits of Warm Climates. J. F. Morton Publisher, Miami, FL.

MOTI, D. L. and CHATURVEDI, O. P. 1976. Propagating some tropical and subtropical fruits by budding. Punjab Hortic. J. 16(1) 33-38.

MUKHERJEE, S. K. and CHATTERJEE, B. K. 1978. Effect of invigoration, etiolation and IBA on the rootage of jackfruit (*Artocarpus heterophyllus* Lam) cuttings. Sci. Cult. 44(9) 414-415.

MUKHERJEE, S. K. and CHATTERJEE, B. K. 1979. Effects of forcing etiolation and indolebutyric acid on rooting of cuttings of *Artocarpus heterophyllus* Lam. Sci. Hortic. 10, 295-300.

MUKHERJEE, S. K., RAO, D. P. MUKHERJEE, S., RAY, M. and MOHANTY, B. B. 1983. Manurial treatment at planting on growth and development of jackfruit (*Artocarpus heterophyllus* Lam.) Indian J. Hortic. 40(3) 150-154.

MUZIK, T. J. 1948. Effect of hormone on root formation in *Artocarpus communis*. Science 107, 225.

NATARAJAN, P. N. and KARUNANITHY, R. 1974. Synthetic flavor enhancers for *Artocarpus integrifolia*. Flavour Ind. 5(11) 282-283.

NAIK, K. C. 1949. South Indian Fruits and Their Culture. P. Varadachary & Co., Madras.

NAIK, K. C. 1952. Jackfruit. ICAR Bull. 33, New Delhi.

OCHSE, J. J. 1931. Vegetables of Dutch East Indies. Published in collaboration with R. C. Bakhuizen van den Brink, by G. Holff and Company, Djakarta, Indonesia.

OLIVEROS, B. L., CARDENO, V. and PEREZ, P. 1971. Physical properties of some Philippine essential oils. Flavour Ind. 2(5) 305-309.

PANDEY, P. C., PURI, Y. N. and REHILL, P. S. 1981. Blossom and fruit blight of jackfruit and its control. Indian J. For. 4(1) 5-7.

PANGGABEAN, G. 1979. The effect of storage temperature on the germination of jackfruit seed (*Artocarpus heterophyllus*). Bull. Penelitian Hortikutura 7(4) 39-42.

PASSAM, H. C., MAHARAJ, D. S. and PASSAM, S. 1981. A note on freezing as a method of storage of breadfruit slices. Trop. Sci. 23, 67-74.

PRUTHI, J. S., LAL, G., DHOPESHWARKAR, G. A. and AGAR, N. G. 1955. Effect of processing operations on the nutritive value of canned guava and jackfruit. J. Indian Chem. Soc. 18, 236-238.

PURSEGLOVE, J. W. 1968. Tropical Crops. Longmans, London.

QUIJANO, J. and ARANGO, G. J. 1979. The breadfruit from Colombia—a detailed chemical analysis. Econ. Bot. 33, 199-202.

RAMAN, K. R. 1957. A note on the propagation of jack on individual jack tree selections as root stocks in grafting. South Indian Hortic. 5, 31-32.

RASMUSSEN, P. 1983. Identification of volatile components of jackfruit by gas chromatography/mass spectrometry with two different columns. Anal. Chem. 55, 1331-1339.

SAHADEVAN, P. C., ABDUL SAMAD, A. and MENON, K. C. P. 1985. Cited by H. N. Samaddar. Madras Agric. J. 37, 497-498.

SAMADDAR, H. N. 1980. Jackfruit. *In* Fruits of India—Tropical and Subtropical. T. K. Bose (Editor). Naya Prakash, 206 Bidhan Sarani, Calcutta.

SAMADDAR, H. N. and YADAV, P. S. 1970. A note on the vegetatie propagation of cashewnut, avocado, jackfruit and custard apples. South Indian Hortic. 18(1) 47-49.

SAMADDAR, H. N. and YADAV, P. S. 1982. Cited by H. N. Samaddar. The Hortic. J. 1, 54-60.

SAMBAMURTY, K. and RAMALINGAM, V. 1954. Preliminary studies in blossom biology of the jack (*Artocarpus heterophyllus* Lam.) and pollination effects. Indian J. Hortic. 11, 24-29.

SARAIVA, A. C. 1965. Fruit flies - Diptera, Tripetidae of Cape Verde Islands. Estud. Agron. Lisboa 6, 153-167.

SEN, P. K. and BOSE, T. K. 1959. Effect of growth substances on rooting of jackfruit (*Artocarpus integrifolia* Linn. F.) layerings. Indian Agric. 3, 43-47.

SHAMANTHAKA SASTRY, M. C. 1976. Studies on sunflower seed proteins. M.Sc. Thesis, University of Mysore, India.

SHANKAR, G. and SINGH, S. R. 1965. A note on the flowering and sex ratio in jackfruit (*Artocarpus heterophyllus* Lam.). Allahabad Farming 39, 230-231.

SHARMA, M. R. 1964. Morphological and anatomical investigations on *Artocarpus* Forst. IV. The fruit. Proc. Indian Acad. Sci., Sect. 8, 60, 380-393.

SIDDAPPA, G. S. 1957. Effect of processing on trypsin inhibitor in jackfruit seed. J. Sci. Ind. Res. 16c, 199-201.

SIDDAPPA, G. S. and BHATIA, B. S. 1955. Production and uses of jackfruit seed flour. Bull. Cent. Food Technol. Res. Inst. 5, 3.

SIDDAPA, G. S. and BHATIA, B. S. 1956. Effect of canning on the β-carotene content of mango, papaya and jackfruit. J. Sci. Ind. Res. 15c, 118-120.

SINGH, L. B. 1951. Propagating jackfruit (*Artocarpus integrifolia*) by air-layering with hormones. Curr. Sci. 20, 100-103.

SINGH, L. C. 1961. A new technique for inducing early bearing in jackfruit (*Artocarpus heterophyllus* Lam.). Annu. Rep. Hortic. Res. Inst. Saharanpur, 1966.

SINGH, S., KRISHNAMURTHY, S. and KATYAL, S. 1963. Fruit Culture in India. ICAR, New Delhi.

SINHA, M. M. 1969. Studies on blossom bud differentiation and ontogeny of flowers. Plant Sci. 1, 224-227.

SINHA, M. M. 1972. Studies on the sex-ratio in jackfruit (*Artocarpus heterophyllus* Lam.). Plant Sci. 4, 119-120.

SINHA, M. M. 1975. Studies on the floral biology of jackfruit (*Artocarpus heterophyllus* Lam.). Prog. Hortic. 7(3) 69-75.

SINHA, M. M. and SINGH, S. N. 1971. Growth studies (vegetative phase) in jackfruit (*Artocarpus heterophyllus* Lam.). Indian J. Hortic. 28(3) 203-205.

SRINIVASAN, K. 1970. "Mutton Varikka"—as promising jackfruit variety. Agric. Res. J., Kerala 8, 51-52.

SRIVASTAVA, R. P. 1961. Preliminary studies in sex distribution, growth behavior and the influence of α-naphthaleneacetic acid sprays on the size, seed number and weight of jackfruit (*Artocarpus heterophyllus* Lam.). J. Sci. Res. Banaras Hindu Res. 11, 213-221.

STONE, B. C. 1974. The correct botanical name for breadfruit. Polynesian Soc. 83, 92-93.

SWAMINATHAN, M. 1971. Nutritional aspects of oil seeds and their processed products. 79-89. *In* Proceedings of the workshop on the utilization of groundnut and other oil seed for edible purposes: Potential and prospects. Published by Association of Food Scientist and Technologists (India), CFTRI, Mysore.

SWORDS, G., BOBBIO, P. A. and HUNTER, G. L. K. 1978. Volatile constituents of jackfruit (*Artocarpus heterophyllus*). J. Food Sci. 43, 639-640.

TEAOTIA, S. S., and AWASTHI, R. K. 1968. Dehydration studies in jackfruit (*Artocarpus heterophyllus* Lam.). Indian Food Packer 22(5) 6-14.

TEAOTIA, S. S. and CHAUHAN, R. S. 1969. Studies on the bearing habits, floral biology and fruit development in jackfruit (*Artocarpus heterophyllus* Lam.). Hortic. Sci. 1(2) 29-37.

TEAOTIA, S. S., DAYAL, K. and ASTHANA, M. P. 1963. Propagation of jackfruit by budding. Sci. Cutl. 29, 46-47.

THOMAS, C. A. 1980. Jackfruit, *Artocarpus heterophyllus* (Moraceae), as a source of food and income. Econ. Bot. 34, 154-159.

THOMPSON, A. K., BEEN, B. O. and PERKINS, C. 1974. Storage of fresh breadfruit. Trop. Agric. 51, 407-415.

TRUJILLO, E. E. 1970. A Phytophthora fruit rot of breadfruit. Phytopathology 60, 1542.

TUMAALII, F. 1982. Influence of maturation on the composition of breadfruit (*Artocarpus altilis*). M.Sc. Thesis, The University of New South Wales, Kensington, Australia.

VAN WISSEN, H. L. M. 1978. The possibilities of breadfruit (*Artocarpus altilis* (Park) Forsberg) as a cash crop in western Samoa. Dept. of Agriculture and Forests, Apia, W. Samoa, Working Paper 8.

VENKATESWARA RAO, G. 1983. Studies on milling and utilization of high extraction flour and semolina from Indian wheats. Ph.D. Thesis, University of Mysore, India.

VENKATESWARA RAO, G., SHURPALEKAR, S. R. and NIERLE, W. 1986. Effect of wheat flour extraction rate on protein fraction. Lebensm. Wiss. Technol. 19, 184-186.

VENUGOPAL, S. 1961. Binomics and control of *Perina nuda* Fabr., a pest on Ficus sp. in South India. J. Madras Univ. Sect. B31(B) 1-5.

WILLS, R. B. H., LIN, J. S. K. and GREENFIELD, H. 1986. Composition of Australian foods. 31. Tropical and subtropical fruit. Food Technol. Aust. 38(3), 118-123.

WOOTTON, M. and TUMAALII, F. 1984A. Breadfruit production utilization and composition—a review. Food Technol. Aust. 36, 464-465.

WOOTTON, M. and TUMAALII, F. 1984B. Composition of flours from Samoan breadfruit. J. Food Sci. 49, 1396-1400.

ZAIGER, D. and ZENTMEYER, G. A. 1965. A new lethal disease of breadfruit in the Pacific Islands. Plant Dis. Rep. 50, 892-296.

ZAIGER, D. and ZENTMEYER, G. A. 1967. Epidemic decline of breadfruit in the Pacific Islands. FAO Plant Prot. Bull. 15(12) 25-28.

ACKNOWLEDGEMENTS

The author is grateful to Dr. B. L. Amla, Director, CFTRI for giving permission to take up this work, M/s. K. M. Dastur and K. A. Ranganath, Scientists, FOSTIS, CFTRI for going through manuscripts and to Mr. A. F. Fernandez for typing the manuscripts expeditiously and neatly.

The author acknowledges with thanks all the authors and their publishers whose references have been used in one form or another in the preparation of the manuscripts, especially Dr. N. E. De Bravo, Dr. H. D. Graham, Dr. M. Podvani, Dr. M. Wootton, Dr. F. Tumaalii, Dr. S. C. Achinewu, Dr. Teaotia, S.S.; Dr. J. Quijano, Dr. G. J. Arango, Dr. B. S. Bhatia, Dr. M. R. Sharma, Dr. G. S. Siddappa and Dr. Girdhari Lal.

12

Araza

Charles R. Clement

The araza (also called araca-boi in Brazil) is a semi-domesticated fruit species of western Amazonia that belongs to the myrtle family (Myrtaceae) (Clement 1983; Chavez and Clement 1984). The araza is classified as *Eugenia stipitata* McVaugh and is separated by McVaugh (1956, 1958) into two subspecies. The subspecies, *stipitata* (*E. stipitata* subsp. *stipitata* McVaugh) is a small tree of the forest understory that appears not to have been domesticated at all or which may have been ancestral to the domesticated subspecies.

The subspecies *sororia* (*E. stipitata* subsp. *sororia* McVaugh), is a shrub also of the forest understory that has larger fruit and contains a different number of stamens in the flower. *Sororia* is a medium to large shrub (McVaugh 1956) rarely a small tree (Cavalcante 1976), with dense dark green foilage (McVaugh 1956, 1958) (Fig. 12.1). The new leaves are red-tinted, changing to dark green with maturation (Chavez and Clement 1984). The 3 to 8 flowers are presented on small inflorescences (McVaugh 1956, 1958). Each has four white petals, 75 to 150 stamens and a single pistil with a multi-locular ovary (McVaugh 1956, 1958). The fruit is a small (30 g) to large (800 g) berry with a thin velvety canary yellow skin that exudes a delightful aroma (Chavez and Clement 1984) (Fig. 12.2). The very juicy pulp has the same canary yellow color, a delightful (but distinct) flavor and is very acidic (Chavez and Clement 1984). The fleshy seeds are small to medium in size and may number between 3 and 20 (Chavez and Clement 1984).

The first technical study of the araza was by Pinedo *et al.* (1981) who reported yields of 28 MT/ha/yr near Iquitos, Peru. Alfaia *et al.* (1988A) reported yields of 20 to 25 MT/ha/yr on a yellow Oxisol near Manaus,

Courtesy of W. B. Chavez F.

FIG. 12.1. FIVE YEAR OLD SEEDLING OF ARAZA (A) WITH FRUIT ATTACHED (B). RECOMMENDED MONOCULTURE SPACING IS 4 X 3 M

FIG. 12.1. (Continued)

Courtesy of W. B. Chavez F.

FIG. 12.2. MATURE FRUIT OF ARAZA. NOTE THE THIN SKIN, THICK MESOCARP AND FIBROUS PLACENTAL AREA

Brazil. The delightful flavor and easy acceptance of the araza, combined with precocity and high yields, make the araza a very promising crop for the lowland humid tropics (Chavez and Clement 1984; FA0 1986). The araza is currently receiving priority attention in Peru (INIPA—Iquitos) and Brazil (INPA—Manaus).

Many authors have mentioned the araza (Calzada 1980; Clement 1983) without discussing details. Several authors have presented results from agronomic studies; the most important being Pinedo *et al.* (1981) and Alfaia *et al.* (1988A,B). Pinedo *et al.* (1981) and Falcao *et al.* (1988) have presented phenological data. Swift and Prentice (1983) have reported on small-scale commercial operations in Ecuador and point out the difficulty of transporting araza as fresh fruit because it bruises easily. They also mention that fruit picked green do not ripen normally.

COMPOSITION

Three reports on araza mesocarp composition have been published to date and are presented in Table 12.1. Pinedo *et al.* (1981) present an average of three samples for most of the components analyzed, while other authors present data from only one sample. There appears to be considerable chemical variability in the species, which will surely be amenable to the selection process.

USES

The araza has traditionally been used as a juice, either in thick or dilute form. Pezo and Pezo (1984) studied the preparation of a nectar from the pulp and concluded that it has great potential, both locally and internationally. However, to date, no entrepreneur has taken avantage of this.

As with most fruits, the araza can be used to make jams and jellies. Pezo and Pezo (1984) reported on the preparation of a jam that appeared to be very similar to guava (*Psidium guajava* L.) jam, which is to say that it has a reddish color and an agreeable, though nonaraza flavor. Giselda Ranzani (personal communication) has also experimented with araza jam and found that the araza flavor disappeared within about 5 min during cooking. Ranzani was able to prepare a jam with the typical araza yellow color and the delightful flavor by reducing cooking time to about 2 min, just enough time to partially sterilize the mash. This kind of jam would be much more attractive on the market and should easily develop a demand, even in areas where the araza is currently unknown.

In Brazil, preparation of 'creams' is very popular and araza makes an excellent flavoring for these. The easiest recipes call for one can of

TABLE 12.1. CHEMICAL COMPOSITION OF ARAZA MESOCARP (PER 100 G MESOCARP)

Component[1]	Pinedo *et al.* (1981)	Aguiar (1983)	Pezo and Pezo (1984)
Energy (cal)	—	39.8	—
Water (% fresh wt)	—	90.0	94.3
Brix (% fresh wt)	—	—	4.0
pH	—	—	2.0
Protein (% dry wt)	9.54 ± 1.36	6.0	10.1
Carbohydrates (% dry wt)	70.21 ± 1.32	89.0	80.6
Fats (% dry wt)	3.15 ± 0.60	2.0	0.5
Ash (% dry wt)	—	3.0	2.5
Fiber (% dry wt)	6.07 ± 0.50	—	6.3
Pectin (x dry wt)	—	—	3.4
Nitrogen (% dry wt)	1.53 ± 0.22	—	—
Phosphorus (% dry wt)	0.09 ± 0.00	—	—
Potassium (% dry wt)	2.15 ± 0.32	—	—
Calcium (% dry wt)	0.19 ± 0.03	—	—
Magnesium (% dry wt)	0.10 ± 0.02	—	—
Manganese (ppm)	13	—	—
Copper (ppm)	5	—	—
Iron (ppm)	87		
Zinc (ppm)	11	—	—
Vitamin C (mg % fresh wt)	7.68	23.3	74.0
β-carotene (mg % fresh wt)	—	0.4	—
Vitamin A (μg % fresh wt)	7.75	—	—
Vitamin B_1 (μg % fresh wt)	9.84	—	—

[1]Units of measurement vary: % fresh wt = g/100 g fresh wt x 100; % dry wt = g/100 g dry wt x 100; mg % fresh wt = mg/100 g fresh wt x 100: μg % fresh wt = μg/100 g fresh wt x 100; ppm = mg/1000 g dry wt.

sweetened condensed milk, one can of cream and one can of araza pulp, all beaten in a blender (Ranzani, personal communication). This can be eaten as is or used as a pie filling, for example.

Other products remain to be developed. One alternative use that may be of interest is the extraction of the aromatic principles from the skin of the fruit. It is so delightful that it might find some use in the perfume industry, but it has not yet been studied. A single ripe fruit can aromatize a large closed room and delight all who arrive. The author believes this feature is well worth looking into.

FUTURE OF THE ARAZA

Falcao *et al.* (1988) reported on the first bearing at 18 months in the field, with year-round production divided into four major harvests. Given the high yields already observed and the precocity and frequency of bearing together with its delightful, distinctive flavor, the araza appears to have great potential. As with many new tropical fruits, it awaits an entrepreneur to take it from the forest to the world market. Hopefully this will not be delayed much longer.

REFERENCES

AGUIAR, J. P. L. 1983. Araca-boi (*Eugenia stipitata* McVaugh). Aspects and preliminary data concerning its chemical composition. Acta Amazonica 13, 953-954. (Portuguese)

ALFAIA, S. S., CHAVEZ F., W. B., FERREIRA, S. A. N. and CLEMENT, C. R. 1988A. Effects of spacing and mineral addition on araca-boi (*Eugenia stipitata* McVaugh). I. Production of fruit. An. Cong. Bras. Frutic. 9 (119-123). (Portuguese)

ALFAIA, S. S., CHAVEZ F., W. B., FERREIRA, S. A. N. and CLEMENT, C. R. 1988B. Effects of spacing and mineral addition on araca-boi (*Eugenia stipitata* McVaugh). II. Vegetative growth. An. Cong. Bras. Frutic. 9 (125-128). (Portuguese)

CALZADA B., J. 1980. 143 Native Fruits. Libreria "El Estudiante", Lima Peru. (Spanish)

CAVALCANTE, P. B. 1976. Eatable Fruits of the Amazonia. 3rd Edition. INPA, Belem, Brazil. (Portuguese)

CHAVEZ F., W. B. and CLEMENT, C. R. 1984. Considerations concerning the araca-boi (*Eugenia stipitata* McVaugh, Myrtaceae) in Brazilian Amazonia. An. Cong. Bras. Frutic. 7, 167-177. (Portuguese)

CLEMENT, C. R. 1983. Underexploited Amazonian fruits. Proc. Am. Soc. Hortic. Sci., Trop. Reg. 27A, 117-142.

FALCAO, M. A., CHAVEZ F., W. B., FERREIRA, S. A. N., CLEMENT, C. R., BARROS, M. J. B., BRITO, J. M. C. and SANTOS, T. C. T. 1988. Phenological and ecological aspects of the araca-boi (*Eugenia stipitata* McVaugh) in Central Amazonia. Acta Amazonica 18(3/4) 27-38. (Portuguese)

FAO. 1986. Food and Fruit-bearing Forest Species. 3. Examples from Latin America. FAO Forestry Paper 44/3, Rome, Italy.

McVAUGH, R. 1956. Tropical American Myrtaceae. Fieldiana Bot. 29(3) 143-228.

McVAUGH, R. 1958. Flora of Peru. IV. Field Mus. Nat. Hist.—Botany, Vol. XII, 736-737.

MORTON, J. L. 1987. Fruits of Warm Climates. J. L. Morton Publisher, Miami, FL.

PEZO A., A. and PEZO V., F. E. 1984. Assay and manufacture of the nectar and juice separated from the araza (*Eugenia stipitata* McVaugh). Chem. Eng. Thesis. National University, Pruvian Amazonia, Iquitos, Peru. (Spanish)

PINEDO P., M., RAMIREZ N., F. and BLASCO L., M. 1981. Preliminary notes concerning the araza (*Eugenia stipitata*), native fruit of the Peruvian Amazonia. M.A.A./INIA/IICA. Misc. Pub. 229, Lima, Peru. (Spanish)

SWIFT, J. F. and PRENTICE, W. E. 1983. Native fruit species of the Ecuadorian Amazon: Production techniques and processing requirements. Proc. Am. Soc. Hortic. Sci., Trop. Reg. 27A, 95-100.

13

Jaboticaba

M. Joseph Ahrens

The jaboticaba (*Myrciaria cauliflora* Berg.) is a tree fruit native to Brazil where it is widely planted and enjoys immense popularity. This highly nutritious grape-like fruit is borne in small aggregates or singularly along the trunk branches, and in leaf axils of an attractive diffuse-branched, evergreen tree. The berries are eaten fresh or processed into jellies, marmalades and wines. While considerable plantings are in dooryards, commercial orchards exist with a percentage of those being "U-Pick" operations. The jaboticaba occupies a prominent place in local Brazilian markets. While it prospers on a variety of soil types and under several cultivation regimens, it would appear ideally suited to many areas of the subtropics and tropics.

BOTANICAL DESCRIPTION AND NOMENCLATURE

The jaboticaba is a large (up to 12 m) lanceolate-leafed, hardwood tree. Its fruits (1 to 4 cm in diameter) are borne along the trunk, limbs, and in leaf axils, either sessile or with short pedicels. They tend to cluster but can be produced singularly (Fig. 13.1). Grape-like in appearance and flavor, the globose fruit has a tough, usually dark, skin surrounding a juicy, sweet, soft pulp, in which are imbedded 1-4 seeds.

The common name jaboticaba [English, French, Spanish: jaboticaba; Portuguese: boticaba, jaboticaba, jaboticaba sabara, jabuticaba, jabuti—from "jabitim" or "jabuti": the name given a small, indigenous Brazilian turtle (Soubihe Sobrinho and Gurgel 1952)] can refer to any of several closely related species found in the genus *Myrciaria* of the family Myrtaceae (Bailey 1920; Ochse *et al.* 1961; Wiltbank *et al.* 1983).

Courtesy of Claire Warshaw

FIG. 13.1. ARTISTRY OF JABOTICABA FRUIT, LEAVES AND BRANCHES

There currently exists differing opinions as to the proper classification of jaboticaba into genera and species. Previously, the plant was assigned to *Eugenia* (Hortus III 1976). Most botanists outside of Brazil (Chandler 1958; Hortus III 1976; Kennard and Winters 1960; Martin *et al.* 1987; Ochse *et al.* 1961; Popenoe 1939; Ruehle 1958) now classify all jaboticaba as *Myrciaria cauliflora* Berg. Wiltbank *et al.* (1983) suggest that this is probably due to the early importation of this species from Brazil. However Bailey (1920) listed 3 species in 1920, *M. cauliflora*, *M. jaboticaba*, and *M. trunciflora*, whereas in 1949, he recognized 2 species as jaboticaba, namely, *M. cauliflora* and *M. jaboticaba* (Bailey 1949). The former is described as having elliptical-lanceolate leaves with acute tip and base, sessile fruit or attached by a very short pedicel, and lanceolate calyx lobes. The latter is a smaller tree with smaller fruit, bearing acute-lanceolate leaves with acute tip and obtuse base, fruit attached via a longer pedicel, and acute calyx lobes.

Within Brazil, there is no concensus regarding classification or recognition of the species (Bailey 1920; Gomes 1983; Wiltbank *et al.* 1983). In addition to *M. cauliflora* and *M. jaboticaba*, there is some recognition of a third species, *M. trunciflora* (Correa 1969; Gomes 1983; Hoehne 1944A,B) which has smaller leaves and smaller fruit attached by a longer pedicel. Wiltbank *et al.* (1983) reviewed the conflicting classifications of major named cultivars. A list of major cultivars with descriptions and classifications appears in Table 13.1.

HORTICULTURE

Climate

The jaboticaba is encountered wild in southern Brazil from Rio Grande do Sul to Minas Geraes (Bailey 1920) There are reports of it succeeding at elevations up to 1400 m (Teixeira da Fonseca 1954). It is quite hardy in South Florida (Spielman 1948; Sturrock 1940) and has withstood freezes to –3° C (Bailey 1920; Ochse *et al.* 1961; Spielman 1948). It is a true subtropical tree, not thriving in areas of hot (Simao 1974) or freezing temperatures, but requiring a cool climate for optimum vigor (Bailey 1920). An even moist environment (Bailey 1920) with adequate water throughout the year is best (Chandler 1958), but an occasional drought is tolerated (Spielman 1948).

Nutrition and Soils

Very little information about nutrition or soil requirements of jaboticaba is available. A review of the literature would indicate that a

TABLE 13.1. DESCRIPTIONS AND CLASSIFICATIONS OF MAJOR NAMED CULTIVARS OF JABOTICABAS

Cultivar[1]	Description
Large trees with large leaves and large sessile fruit	
Acu, Gigante, Grande, Grauda Jaboticatuba, Olh de Boi	Largest fruited group, medium purple, thick skinned, less flavorful than other cultivar groups
Ponhema	Large fruited, coriaceous, astringent, for processing
do Mato, Mineira, Paulista	Medium large round fruit, dark purple, medium thick skin, juicy, sweet, highly flavored flesh
Rajada	Similar to Paulista group but striped bronze/green
Branca	Medium large, round, white fruit, less sweet than Paulista group
Smaller trees with smaller leaves, fruit are smaller with very short pedicel	
Miuda, Murta, Sabara	Small fruited, thin skinned early maturing, most highly flavored jaboticabas, group most cultivated
Small trees, fruit with longer pedicel	
de Cabinho	Similar to Paulista group

[1]Cultivars within groups are probably local names for the same cultivar. Major grouping suggested by Wiltbank *et al.* 1983.

wide variety of soil types are suitable to sustain healthy plants. It appears most adapted to deep, rich (Bailey 1920; Sturrock 1940), well-drained sandy loams (Ochse *et al.* 1961). Sturrock (1940) reported that application of organic manure and mulching were necessary for optimum growth. Seedlings do well in peat or acid-sand muck (Ackerman 1979, 1980). Reports of its ability to thrive in or tolerate sand, calcareous or rocky soils are conflicting (Bailey 1920; Ochse *et al.* 1961; Sturrock 1940).

Propagation

Most trees cultivated naturally from seed require at least 6 to 10 years to bear fruit (Menninger 1959; Spielman 1948), whereas grafted plants can bear in 4 to 5 years (Anderson and Gomes 1975; Ochse *et al.* 1961; Rizzini and Mors 1976; Soubihe Sobrinho and Gurgel 1952) or 3 years from date of grafting (Menninger 1959). Most jaboticaba are cultivated directly from seeds (Bailey 1920; Chandler 1958; Ochse *et al.* 1961; Wiltbank 1983) which do not store well (Aroeira 1962), but are available year-round. Reports on the success of vegetative propagation have been mixed. While jaboticaba has responded to propagation via cuttings

(Ochse and Reark 1950), cleft grafting (Wiltbank *et al.* 1983), grafting (Sampaio 1986) and inarching (Bailey 1920), as with other myrtaceae species success has been limited (Chandler 1958; McKenzie 1985). However, a report by Sampaio (1986) indicates that the season in which the graft is administered may contribute substantially to the take of the graft. Jaboticaba scions grafted on seedlings in the summer had a success rate of 30.5% as opposed to 80% for autumn-winter grafts. A promising development is the successful application of tissue culture to jaboticaba (Litz 1984).

Transplanting is difficult (Gomes 1975). Therefore, seeds or grafted stock should be sown or set in place. In Brazil, trees are planted on the recommended centers of 6 to 10 m (Chandler 1958; Gomes 1975). However, Bailey (1920) suggests that any distance less than 10 m is inadequate. Although slow growing, trees under optimum conditions can reach 13 m (Sturrock 1940) and fill this spacing. In less than ideal conditions, such as in south Florida, trees more often are closer to 5 m (Ochse *et al.* 1961). Chandler (1958) suggests that since the fruit is borne directly on the trunk and main branches, trees could successfully be maintained at a more manageable size and height of 5 to 7 m.

Pests and Control

Jaboticaba is mostly free of major pests. Wiltbank *et al.* (1983) reported most growers used no pest control program. However, there are reports of isolated damage by fruit-eating birds, insects and disease organisms. Rust occurs during warm, moist periods. Pruning for better light and air penetration, wide planting distances and fungicide treatments have been recommended for rust control (Galli *et al.* 1968; Simao 1974). Aphids, fruit flies, branch-girdling bees, scales and beetles have been known to attack jaboticaba (Abrahao and Orlando 1956; Junqueira 1962; Gallo *et al.* 1970).

HARVESTING AND HANDLING

The initiation, frequency, and density of flowering and fruit set vary by cultivar and are influenced by temperature and rainfall. Many cultivars flower and set fruit several times a year (e.g. Sabara), others only once (e.g. Paulista) (Fouque 1974; Popenoe 1939; Ruehle 1958; Wiltbank *et al.* 1983), resulting in flowers, immature fruit, and mature fruit being found on the same tree simultaneously (Fig. 13.2). Time of fruit development from fruit set to maturity averages 6 weeks on the tree. Because of the non-uniformity of crop ripening and a long development and ripening period, spot-harvesting by hand must be incorporated. For the

FIG. 13.2. YOUNG AND MATURING FRUIT WITH NEW BLOOM

dooryard or "U-Pick" operation, this presents no problem. However, for the mass market this increases harvesting costs, as pointed out by Chandler (1958). There are no reports of efforts to select and breed plants for uniformity in ripening.

Little work has been done on handling and storage of jaboticaba. The fruit are subject to moisture loss, chilling injury, and absorb some odors. Using the current recommendations (McGregor 1987) of room cooling and holding at 13° to 15° C at 90 to 95% relative humidity, storage life of ripe jaboticaba is 2 to 3 days. They are normally packed in fiberboard flats holding 4.5 kg. However, the fruits have a thick, tough, protective peel which should protect them during transit (Ochse *et al.* 1961) Response to modified or controlled atmospheres and ethylene emitters or absorbers is unknown.

FRUIT COMPOSITION

The jaboticaba is a nutritious fruit as evidenced by food composition data (Table 13.2). The major constituents are carbohydrates, and about

TABLE 13.2. COMPOSITION AND FOOD ENERGY VALUE OF JABOTICABA

	Edible portion (per 100 g)		
Constituent	INCAP (1961)	IBGE (1977)	Gutierrez *et al.* (1976)
Food Energy (cal)	46	43	—
Water (%)	87.1	87.7	—
Carbohydrates, total (g)	12.6	10.8	12.5
Sugars (g)	—	—	10.1
Sucrose (g)	—	—	5.4
Glucose (g)	—	—	1.6
Fructose (g)	—	—	3.1
Fiber (g)	0.1	0.3	—
Protein (g)	0.1	1.0	—
Lipids (g)	—	0.1	—
Ash (g)	0.2	0.4	—
Minerals			
Calcium (mg)	6	13	—
Phosphorus (mg)	9	14	—
Iron (mg)	0.5	19	—
Vitamins			
Thiamine (mg)	0.02	0.06	—
Riboflavin (mg)	0.02	0.16	—
Niacin (mg)	0.02	2.4	—
Ascorbic acid (mg)	23	—	—

80% of the total soluble carbohydrates are the sugars—glucose, fructose and sucrose (Trevisan *et al.* 1972; Gutierrez *et al.* 1976). As expected with most fruits, lipids and proteins comprise inconsequential levels in the jaboticaba. Amino acid composition data is sparse but IBGE (1977) reported values (mg per g of nitrogen) for lysine (316), methionine (178), threonine (219) and tryptophan (57).

The vitamins thiamine (B_1), riboflavin (B_2), niacin and ascorbic acid (vitamin C) have been determined. Most investigators reported ascorbic acid levels in the range of about 22 to 31 mg/100 g fruit (Soubihe Sobrinho *et al.* 1955; Malavolta *et al.* 1956; Brune *et al.* 1966A). Brune *et al.* (1966B) reported a drop in vitamin C contents of the flesh during ripening. Freezing whole fruits reduced vitamin C over a 2-week storage period (Brune *et al.* 1966A).

Acids of Sabara jaboticaba amount to about 1.2 g/100 ml of juice (Wiltbank *et al.* 1983) and are primarily citric and oxalic acids (Trevisan *et al.* 1972). Two pigments have been identified namely, peonidin and peonidin 3-glucoside (Trevisan *et al.* 1972).

FUTURE CONSIDERATIONS

Most workers reporting on jaboticaba's potential as a commercial crop have been enthusiastic, beginning with Liberty Hyde Bailey in 1920. However, it has received little attention outside its native area. With the introduction of successful propagation techniques to reduce juvenility, selection for uniform crops, and investigations into improved postharvest practices, the jaboticaba promises to be a fruit of true merit, with few pest problems, requiring little maintenance and adaptable to a wide variety of soil types and elevations within the tropics and subtropics.

REFERENCES

ABRAHAO, J. and ORLANDO, A. 1956. Notes on *Macrodactylus oumilio* Burm., about different cultural practices and its control with modern insecticides. O Biologico 22, 147-155. (Spanish)

ACKERMAN, E. D. 1979. Jaboticaba nutrition experiment. Proc. Fla. State Hortic. Soc. 91, 187-189.

ACKERMAN, E. D. 1980. Tropical fruit tree potting media experiment, Proc. Fla. State Hortic. Soc. 92, 273-275.

ANDERSON, O. and GOMES, F. R. 1975. Vegetative propagation of jaboticaba. Anais do III Congresso Brasileiro de Fruticultura II, 423-427. (Portuguese)

AROEIRA, J. S. 1962. Dormancy and storage of seeds of some fruit plants. Experientiae 2, 541-609. (Spanish)

BAILEY, L. H. 1920. Manual of Tropical and Subtropical Fruits. Macmillan and Co., New York.

BAILEY, L. H. 1949. Manual of Cultivated Plants. Macmillan and Co., New York.

BRUNE, W., BATISTA, C. M., SILVA, D. O., FORTES, J. M. and PINHEIRO, R. V. R. 1966A. Concerning vitamin C content in Myrtaceas. Rev. Ceres 13, 123-133. (Spanish)

BRUNE, W., SILVA, D. O. and MATTOS, J. R. 1966B. Concerning vitamin C content of Myrtacea. II. Rev. Ceres 13, 182-193. (Spanish)

CHANDLER, W. H. 1958. The jaboticaba. *In* Evergreen Orchards. Lea and Febinger, Philadelphia, PA.

CORREA, M. P. 1969. Plants utilized in Brazil. Vol. IV. Ministerio da Agricultura. Rio de Janeiro, Brazil. (Portuguese)

FOUQUE, A. 1974. Fruit species of the American tropics. Institut Francais de Recherches Fruitieres Outre-Mer, Paris, France. (French)

GALLI, F., TOKESHI, DE CARVALHO, P. C. T., BALMER, E., KIMATI, H., CARDOSO, C. O. N. and SALGADO, C. L. 1968. Diseases of goiabeira. *In* Manual de fitopatologia. Editora Agronomica Ceres. Sao Paulo, Brazil. (Portuguese)

GALLO, F., NAKANO, O., WIENDL, F. M., NETO, S. S. and CARVALHO, R. P. L. 1970. Jaboticaba. *In* Manual de entomologia. Editora Agronomica Ceres. Sao Paulo, Brazil. (Portuguese)

GOMES, R. P. 1975. The jaboticaba. *In* Fruticultura Brasileira. Livraria Nobel. Sao Paulo, Brazil. (Portuguese)

GOMES, R. P. 1983 Brazilian Fruit Culture. Livraria Nobel, Sao Paulo, Brazil. (Portuguese)

GUTIERREZ, L. E., CEZAR, W. P., FERRARI, S. E. and GUIMARAES, G. L. 1976. Soluble carbohydrates in fruits. I. Pomegranate, mango, banana, jaboticaba, lime, pineapple, orange and cabelluda. Anais da E.S.A. "Luiz de Queiroz" 33, 167-172. (Portuguese)

HOEHNE, F. C. 1944A. Myrtaceae. *In* Frutas Indigenas. Instituto de Botanca. Secretaria da Agric. Sao Paulo, Brazil. (Portuguese)

HOEHNE, F. C. 1944B. Indigenous fruits. Publicacao Serie D. Instituto de Botanica. Sao Paulo, Brazil. (Portuguese)

HORTUS III. 1976. The Liberty Hyde Bailey Herbareum of Cornell University. Macmillan and Co., New York.

IBGE. INSTITUTO BRAZILEIRO DE GEOGRAFIAE ESTATISTICA. 1977. Tables of Food Composition. Vol. 3. Special Publication. Rio de Janeiro, Brazil. (Portuguese)

INCAP. INSTITUTE OF NUTRITION OF CENTRAL AMERICA AND PANAMA. 1961. Composition of Foods Commonly Used in Latin American Countries. U.S. Govt. Printing Office, Washington, D.C.

JUNQUEIRA, G. M. 1962. A note on the yellow eucalyptus beetle. O Biologico 28, 326-328. (Spanish)

KENNARD, W. C. and WINTERS, H. F. 1960. Some fruit and nuts for the tropics. Agr. Res. Serv., U.S. Dept. Agric. Misc. Publ 801.

LITZ, R. E. 1984. Tissue culture for the improvement of tropical fruits. Fla. Agric. Res. 3(1) 26-28.

MALAVOLTA, E., LEME, J. J., GURGEL, J. T. A. and SOUBIHE SOBRINHO, J. 1956. Ascorbic acid content in fruits of *Myrciaria glomerata* Berg. Nature 178, 424.

MARTIN, F. W., CAMPBELL, C. W. and RUBERTE', R. M. 1987. Perennial edible fruits of the tropics: An inventory. U. S. Dept. Agric., Agric. Handbk. 642.

MCGREGOR, D. M. 1987. Tropical products transport handbook. U. S. Dept. Agric. Handbk. No. 668.

MCKENZIE, D. M. 1985. Grafting Australian native plants. Intl. Plant Propagators Soc. 34, 135-141.

MENNINGER, E. A. 1959. The cultivated eugenias in American gardens. Part 2. Nat. Hortic. Mag 38, 145-163.

OCHSE, J. J. and REARK, J. B. 1950. The propagation of subtropical fruit plants by cuttings, a progress report. Proc. Fla. State Hortic. Soc. 63, 248-251.

OCHSE, J. J., SOULE, JR., M. J., KIJKMAN, J. M. and WEHLBURG, C. 1961. Tropical and Subtropical Agriculture. Vol. I. Macmillan and Co., New York.

POPENOE, W. 1939. Manual of Tropical and Subtropical Fruits. Macmillan and Co., New York.

RIZZINI, C. T. and MORS, W. B. 1976. Jabuticaba. *In* Brazilian Economic Botany. University of Sao Paulo (Editor). Sao Paulo, Brazil. (Portuguese)

RUEHLE, G. D. 1958. Miscellaneous tropical and subtropical Florida fruits. Fla. Agric. Exp. Bul. 156A.

SAMPAIO, V. R. 1986. Propagation by cuttings of sabaro (*Myrciaria jaboticaba* Berg.). Anais da Escola Superior de Agricultura "Luiz de Queiroz" 41(1) 135-140. (Portuguese)

SIMAO, S. 1974. Cultivation of jaboticaba. Suplemento Agricola de Do Estado de Sao Paulo. 7 July 1974. (Portuguese)

SOUBIHE SOBRINHO, J. and GURGEL, J. T. A. 1952. Characteristics of the seeds of myrtacea fruit. Rev. Agric. (Piracicaba) 27, 83-90 (Portuguese)

SOUBIHE SOBRINHO, J. *et al.* 1955. Vitamin C in "cabeludinha" (*Myrciaria glomerata* Berg.). Bragantia 14, 193-201. (Spanish)

SPIELMAN, H. W. 1948. Jaboticaba—"grape of Brazil." Foreign Agric. 12, 18-19.

STURROCK, D. 1940. Tropical Fruits for Southern Florida and Cuba and Their Uses. The Arnold Arboretum of Harvard University. Jamaica Plains, MA.

TEIXEIRA DA FONSECA, E. 1954. Jabuticaba. *In* Frutas do Brasil. Minist. de Educ. e Cultura, Brazil. (Portuguese)

TREVISAN, L. M , BOBBIO, F. O. and BOBBIO, P. A. 1972. Carbohydrates, organic acids, and anthocyanins of *Myrciaria jaboticaba* Berg. J. Food Sci. 37, 818-819.

WILTBANK, W. J., CHALFUN, N. N. J., and ANDERSON, O. 1983 Proc. Amer. Soc. Hortic. Sci. Trop. Reg. 27A, 57-69.

14

Carambola and Bilimbi

Charles W. Wilson, III

The carambola (*Averrhoa carambola* L.), also known as star fruit and other names specific to geographic locations (Morton 1987), is thought to originate in Ceylon and the Moluccas and has been cultivated in Asia since ancient times (Campbell *et al.* 1985; Green 1987). It can be propagated in tropical and subtropical climates and is grown in Australia, the Philippines and other South Pacific islands, Central America, South America, the Caribbean islands, Africa, Israel, and subtropical areas of the United States (Banerjee *et al.* 1980; Morton 1987).

The carambola tree is relatively small and attains heights of 6-9 m with a spread of 6-10 m. Its dark green deciduous leaves are odd-pinnately compound, with 5-11 leaflets 3-9 cm long. The pink to purplish pink flowers, with a diameter of 6 mm, are borne in clusters laterally on old and young branches, and occasionally terminally on young twigs (Campbell *et al.* 1985; Morton 1987). The plants can be propagated from seed, grafting, air layering (marcottage), and tissue culture (Litz and Conover 1980). Seedling plants are variable in their characteristics and are used mainly for rootstock. Air root development is slow. Vegetative layering has been reported successful but propagation is necessary to produce cultivars true to type and veneer grafting and chip budding are the most popular propagation methods (Campbell *et al.* 1985; Morton 1987). Carambola fruit is oblong to ellipsoid, 6-15 cm long, with 4-6 longitudinal ribs. Cross-sections of the fruit are star shaped. The skin is translucent, smooth and waxy, and the color varies from whitish to a deep golden yellow. The flavor is variable and ranges from sour with little sugar to sweet with little acid (Wagner *et al.* 1975; Berry *et al.* 1977; Campbell *et al.* 1985; Morton 1987). The fruit contain up to 12 thin, flat seeds that range from 6-12.5 mm in length (Wagner *et al.* 1975; Siota 1984; Morton 1987) (Fig.14.1).

PRODUCTION

Recent estimates for carambola production are limited to those for Malaysia (peninsular Malaysia, Thailand, Singapore, and Borneo), Taiwan (The Republic of China), and Florida (USA) (Knight *et al.* 1984; Campbell *et al.* 1985; Campbell *et al.* 1987; Green 1987). In Malaysia, the state of Selanor produces the best quality carambola; some 200-300 acres of carambola have been planted there during the last 10 years. Trees are planted on hilly terrain on about 3.6 m centers, in what appears to be a haphazard fashion. After 4-5 years trees are growing together. The major cultivar, the B-10, is a large, high sugar, low acid, nonfibrous fruit with deep ribs. Thinning and bagging is commonly practiced to produce fruit weighing approximately 310 g each. Every fifth tree is grafted with the less sweet B-2 cultivar, which is similar to the Arkin, to serve as a pollinator. Flowering occurs during the drier months between June and January. Most of the carambola crop is harvested between April and December. Total production estimates are not available. However, production is considered to begin in the second year, and good trees can produce as many as 140 kg of fruit per year (Green 1987).

In Taiwan, carambola trees ranging from 1 to 10 years old are currently cultivated on 2792 hectares. Most of the fruit are grown on overhead trellises. The trees are planted on flat land, on 5.5-6 m centers, below the trellises. Tree canopies are pruned extensively to conform to the trellis. Fruit that hang down from the trellis are selectively culled and the better fruit are bagged with paper bags to prevent mechanical and insect damage. Only fruit that are straight and well formed with no black spots, and attached to a mature branch are bagged. The principle cultivars Meeshi, Er-lin, Soft Sih, and Cheng-Tsey begin fruiting in June and continue until March. One third of the Taiwan carambola crop is produced in Tainan County, where the Cheng-Tsey (Chun-Choi) is the principle cultivar. The trees produce a yellow-orange fruit with a deep, firm rib that is nonfibrous. The fruit has high sugar and low acid content, and weighs about 310 g after thinning. The Er-lin cultivar grafted on one limb of each tree serves as a pollinator. The Taiwan Agricultural Yearbook for 1986 cites yearly production as 34,717 MT, with yields of 13.9 MT per hectare (Green 1987).

The carambola, introduced into South Florida as early as 1935, currently enjoys an expanding market. In the late 1950's and early 1960's fruit from a few home gardens were sold in restaurants and stores, and

FIG. 14.1. CARAMBOLA FRUITS

marked the beginning of commercial interest in the carambola in the United States. By 1971, about 4 hectares of groves in Dade County supplied carambolas for commercial distribution. Improved commercial cultivars in addition to an increase in people from Asia resulted in increased commercial carambola plantings to about 24.3 hectares by 1985 and to over 121 hectares in 1987 (Campbell *et al.* 1985; Campbell *et al.* 1987). There are many cultivars in Florida but few have commercial importance. Cultivars planted commercially include Arkin, Fwang Tung, Golden Star, Maha, Newcomb, and Thayer. The Arkin is the most popular cultivar because of its excellent flavor and resistance to shipping damage. The Golden Star, Newcomb, and Thayer are also widely used because they are productive and readily retailed. The Maha and Fwang Tung are not popular because they are susceptible to shipping damage, cold damage in refrigerated storage, and have poor color. One cultivar not available in nurseries, B-10, has been rated as commercially desirable. In Florida it is best to plant young trees during the rainy season (June-July) so trees can be well established before they are subjected to the colder winter months. Trees should be planted in full sunlight for maximum yield. Tree spacing depends on whether the grower prefers hedgerows or trees separate from one another in a row. Most growers space trees about 7.5 m apart, but 6 m spacing is used with occasional hedgeing. For hedgerows, spacings of 3-3.5 m are used. Some producers plant at the 3-3.5 m spacing, then remove alternate trees to reduce crowding; others interplant with optional tree crops such as papaya (Campbell *et al.* 1985). Production estimates have not been reported for all Florida producers. However, one of Dade County's largest packing houses shipped 47.2 MT in 1983 and was expected to ship up to 68 MT for 1984. At that time carambola production was expected to exceed 450 MT within 5 years (Knight *et al.* 1984). In 1987 production estimates approached 500 MT (Campbell 1987).

HARVESTING AND STORAGE

In Malaysia approximatly 20% of the carambola crop is consumed as fresh fruit and the rest is exported to countries such as England, Austria Germany, Hong Kong, and Singapore. Fruit is hand harvested and packed 16-18 kg per cardboard carton. However, information on shipping and storage conditions is scarce (Green 1987).

Carambolas grown in Taiwan are generally consumed as fresh fruit within the country, and 10% or less is processed into juice. Some carambolas are exported to countries such as Singapore, Hong Kong, Japan, and Canada. Much of the fruit is hand harvested from family-operated farms by family members. Information on storage and shipping conditions of exported fruit is also limited (Green 1987).

In Florida, carambolas are hand harvested at color break in fiberboard boxes of about 125 fruit each. At the packinghouse they are sorted into three size categories and unmarketable fruit is discarded. No other postharvest treatment other than removal of surface debris is used. Fruit are wrapped individually in tissue and packed stem end down at a 45° angle with the flutes of one fruit fitting into the grooves of other fruit. Fruit are close packed with about 15-25 fruit per box (Campbell *et al.* 1987; Morton 1987). Initially carambolas were stored at 15°C until shipping. However, recent research indicates that carambolas retain moisture and nutrients better when stored at 5°C and 85-95% relative humidity (Campbell *et al.* 1987) rather than the higher storage temperatures reported earlier for these fruit (Grierson and Vines 1965; Vines and Grierson 1966; Kenny and Hull 1986).

Grierson and Vines (1965) reported carambolas remained palatable for 2 weeks at 21°C storage, 3 weeks at 10 and 15°C storage, and 4 weeks at 0 and 5°C storage and suggested that storage temperatures below 10°C were unnecessary for shipments within the continental U. S. and Canada. They reported changes in soluble solids, titratable acidity, and solids/acid ratio during storage and found an increase in ratio with storage time. The most dramatic increase occurred at 0°C where the ratio in unpalatable fruit increased from 8.4 to 14.1 during 5 weeks storage. They found that fruit with optimum eating quality had a light yellow color and a ratio of 12.6 (8.65% solids/0.69% acid). In a subsequent study with waxed and unwaxed carambolas stored for 4 weeks at 5 and 10°C, respectively, they reported the best storage conditions were with waxed fruit stored at 5°C. Waxing retarded color development but reduced moisture loss. Both waxed and unwaxed fruit stored at the higher temperature developed color more rapidly.

Kenny and Hull (1986) reported results of storage of carambolas (cvs. Fwang Tung and Arkin) harvested at color break and stored at 5, 8, 10 and 21°C for 8 weeks. Samples were stored in standard ventilated cardboard cartons and half of the samples of each cultivar were enclosed in sealed polyethylene bags. Fruit stored best at 8°C without enclosure in polyethylene bags. Storage at 5°C resulted in cold damage, and temperatures above 8°C resulted in quality loss. Rapid decay was noted at 21°C. The Fwang Tung cultivar was reported to be more susceptible to decay than the Arkin cultivar when stored in polyethylene bags.

Wan and Lam (1984) investigated storage of the B-10 cultivar with and without enclosure in polyethylene bags at 5, 10, 15, and 20°C. Fruit enclosed in perforated polyethylene bags did not keep as well as fruit stored unenclosed or enclosed in polyethylene bags at any storage temperature. Enclosure in polyethylene bags retarded spoilage of both green and ripe fruits and prevented coloration of green fruit. Green fruit colored normally after removal to the atmosphere. Ripe carambolas were not susceptible to chilling injury at 5°C, but green fruit could be stored only 5 weeks or less at 5°C without chilling injury. Other than changes in ethylene and carbon dioxide production, the effects of the higher storage temperatures were not discussed.

Campbell *et al.* (1987) studied the storage stability of Golden Star and Arkin cultivars that were obtained from a commercial packinghouse and Arkin harvested at color break from a commercial grove. Storage studies with Golden Star and Arkin cvs. were conducted with fruit packed in commercial cartons, 15-25 fruit per box, at temperatures of 10, 15, and 20°C and 85-95% relative humidity. Arkin fruit from a commercial grove were similarly stored at 5 and 10°. Both Golden Star and Arkin carambolas were seriously decayed when stored 30 days at 20°. Arkin exhibited acute fungal growth after 30 days at 15°C, but some Golden Star fruit were marginally acceptable. Desiccation was noted in all fruit stored at 15°C. Fruit stored at 10°C were considerably more turgid and less deteriorated than those stored at higher temperatures. Storage of tropical fruits at low, but nonfreezing temperatures, often results in chilling injury and is not commonly practiced. Carambolas did not exhibit chilling injury symptoms at 10°C storage; therefore, responses to lower storage temperature were investigated. Fruit stored for 44 days at 5°C showed considerably less necrosis and desiccation, and were more firm than those stored for the same time at 10°C. Approximately 90% of the carambolas stored at 5°C were commercially acceptable after 44 days storage, but only 50% of those stored at 10°C were commercially acceptable. Fresh weight loss was greater at 10°C (6.7%) than at 5°C (4.4% loss). Differences between the two treatments were more apparent after additional storage for six days at ambient (23°C) temperature. Losses in weight of 10 and 20% were obtained for fruit removed from 5 and 10°C storage, respectively, and kept at ambient (23°C) temperature for 6 days. Fruit stored for 44 days at both temperatures were less green than when harvested but less color development was apparent for fruit stored at 5°C. Chilling injury was not apparent in fruit kept at either temperature. Carambolas removed from 5°C storage and kept an additional 6 days continued to develop a normal yellow color and showed no evidence of chilling injury, although the fruit were not considered to be marketable.

Fruits are classified as climacteric or nonclimacteric according to their respiration pattern. Climacteric fruit show a sharp increase in respiration during ripening. After reaching a climacteric peak, respiration decreases and senescence and physiological breakdown occurs. Climacteric-type fruit also show autocatalytic production of ethylene. Carambolas were first reported to exhibit climacteric respiration (Vines and Grierson 1966) but later studies indicated carambolas undergo nonclimacteric respiration (Oslund and Davenport 1981, 1983). Lam and Wan (1983) studied the climacteric nature of carambolas and determined they were nonclimacteric. There was no characteristic increase in carbon dioxide that is exhibited by climacteric fruit. Ethrel-treated fruit showed no sharp increase in carbon dioxide and ethylene was absent.

PROCESSING

Carambolas are largely consumed as fresh fruit but they are also served in salads and used to garnish other foods. They are made into sherbets, beverages, stews, puddings, tarts, and jellies (Grierson and Vines 1965; Campbell and Malo 1967; Campbell *et al.* 1985; Morton 1987). They can be preserved by pickling or drying (Berry *et al.* 1977; Campbell and Campbell 1983).

CHEMICAL COMPOSITION

Organoleptic properties of the many carambola cultivars are variable. The fruit vary in texture from soft to firm with flavor attributes varying from very sour with little sugar to very sweet with little acidity (Campbell *et al.* 1985). Wagner *et al.* (1975) described the flavor of several cultivars as sweet, sour, and apple-like; sour, tart, and apple-like; and sweet, good, and mild. Another investigator has suggested carambolas have an apricot-like flavor (Harler 1983).

Although the entire fruit is edible, carambolas are reported to contain 72.5-80.7% edible portion and 17.5-27.5% skin (Stahl 1935; Wills *et al.* 1986; Abdulla and Ragab 1970). Wagner *et al.* (1975) investigated 18 different cultivars obtained from the Subtropical Horticulture Research Laboratory, USDA, Miami, Florida for potential commercial application. Among those 18 cultivars, length varied from 6.1-10.0 cm, and weight varied from 54-111 g. The average number of seeds per fruit varied from 2-9. The relative texture of carambola slices as determined by shear force ranged from 2.7 to greater than 45 kg with the majority of samples measuring between 11 and 18 kg. Organoleptic evaluation by a trained taste panel determined that selections numbered 17, 37, 42, and the Dah Pon and Tean Ma cvs. were preferred over the other selections and cultivars. Comparisons of juice characteristics from the 5 preferred carambolas did not indicate any juice characteristics that were responsible for their flavor preference.

Siota (1984) reported fruit characteristics and some compositional data for 13 cultivars and 2 numerical selections from the Housen Horticultural Research Station, Taiwan. Diameter varied from 4.5-5.6 cm, length varied from 7.6-9.2 cm., and fruit contained from 4-10 seeds each. Herrman (1981) reviewed the chemical composition and constituents of important exotic fruits and reported on carambolas.

The dietary fiber composition of carambolas (Tean Ma and selection number 44) was studied by Lund and Smoot (1982). Values for cellulose, hemicellulose, lignin, cutin, ash, neutral detergent residue (NDR), acid detergent residue (ADR), and enzymatic (soluble and insoluble) fractions were reported for results obtained by modified forms of the Van Soest detergent and Hellendoorn enzymatic methods. The NDR contained plant cell wall components without pectins (cellulose, hemicellulose, lignin, cutin, and insoluble minerals). The ADR contained cellulose, lignin, cutin, and insoluble minerals. As a check for anomalous ADR values, sequential and nonsequential extractions were performed in triplicate for each method. Use of the sequential method eliminated errors incurred from the presence of tannins and acid-insoluble pectins, which can appear in the ADR when samples are directly extracted with acid detergent solution. For the sequential method, the sample was extracted first with neutral detergent solution, and then the residue was extracted with acid detergent solution. In the nonsequential method, the neutral and acid detergent extractions were conducted separately on different portions of the sample. Sodium sulfite was added to the extracting solution as an antioxidant for improved protein extraction.

The fiber content (% fresh weight) determined by the NDR nonsequential and sequential methods was 0.89 and 1.10% for the Tean Ma cultivar. Slightly higher values, 1.04 and 1.21%, were obtained for selection 44. Values from the ADR nonsequential and sequential methods were 0.66 and 0.78% for Tean Ma, and 0.65 and 0.85% for selection 44. Sequential NDR and ADR values were greater than values obtained by the nonsequential method. Cellulose, hemicellulose, lignin, and ash values (% fresh weight) determined from sequential detergent analysis were 0.46, 0.32, 0.31, and 0.014% for the Tean Ma cultivar. For selection 44, cellulose, hemicellulose, lignin, and ash values were 0.44, 0.35, 0.33 and 0.012%, respectively. Values for insoluble and soluble fractions from enzymatic fiber determinations with Tean Ma were 1.30 and 0.44%, respectively. Cutin values of 0.011 and 0.085% were reported for Tean Ma and selection 44, respectively. Both lignin and cutin were concentrated in carambola. The high cutin value was probably derived from the peel since it was not found in the flesh. Selection 44 contained higher amounts of cutin than Tean Ma.

Nordby and Hall (1979) reported lipids in Tean Ma and selections 17, 37, 42, and 44. The lipids in the combined flesh and peel from the five cultivars, 0.12% as fresh weight, varied very little. Lipid fractions analyzed were fatty acids, sterols, and alkane and alkene hydrocarbons. Analysis revealed fatty acids from C-12 through C-26 were present but only seven C-14 through C-18 fatty acids were present in relative amounts greater than 1%. Tean Ma and selection 42 had more oleic acid than the other 3 selections and its amount was considerably higher than that reported for other fatty acids. Selection 17 and Tean Ma had the highest and lowests amounts of the C-14 acid among the carambolas studied. There were two linoleic acid isomers present in about a 3:2 ratio that are not present in the seeds. This acid may be a useful marker in future breeding and taxonomic studies. Sterols reported (in increasing amounts) were cholesterol, isofucosterol, campesterol, and sitosterol. Stigmasterol, a C-29 sterol, was not present in carambolas. However, isofucosterol, another C-29 sterol, was present in amounts comparable to the stigmasterol level reported in other plants. The predominant alkanes, C-21 to C-31, had carbon skeletons with odd numbers of carbon atoms and followed the same general pattern for higher plants. Branched chain alkanes were not observed. Of the even numbered alkanes commonly present, only the C-22 and C-24 analogs were reported in relative amounts greater than 1%. The predominant monounsaturated hydrocarbons were C-23 and C-25 alkenes. Differences in the alkene profiles for the five cultivars were small. However, differences occurred in the ratios of C-23:1 and C-25:1 alkenes to their respective C-23 and C-25 alkanes.

The proximate analysis of carambolas from several sources is shown in Tables 14.1A and 14.1B. Moisture content varied from 89-94%, crude protein ranged from 0.2-0.9 g/100 g, fat values varied from 0.1-0.9 g/100 g, dietary fiber values were from 0.6-1.5 g/100 g, and ash content ranged from 0.3-0.5 g/100 g. Acidity as reported in Table 14.1A ranged from 0.6-0.8%. However, other investigators (Wagner *et al.* 1975; Hou *et al.* 1978; Siota 1984; Benk 1985B) have reported a wider range of 0.2-1.4%. Carambola pH values have been reported within a range of 2.3-4.9 (Vines and Grierson 1966; Wagner *et al.* 1975). Energy values ranged from 36-57 cal/100 g.

The vitamin and mineral contents of carambolas are listed in Table 14.1B. Ascorbic acid (vitamin C) was the predominant vitamin and values reported were from 26-53 mg/100 g pulp. Vitamin C values (mg/100 g) reported by other investigators were: 25-37 (Fonesca *et al.* 1969), 14-50 (Wagner *et al.* 1975), and 22-27 (Baqar 1980). Schmandke and Guerra (1969) reported values of 12.5 and 27.5 mg/100 g pulp for ascorbic and dehydroascorbic acid, respectively. Wills *et al.* (1986) reported that the

TABLE 14.1A. PROXIMATE COMPOSITION AND FOOD ENERGY PER 100 G OF EDIBLE CARAMBOLA

	A								
		Hawaiian							
Edible portion	Philippine[1]	sweet	sour	B	C	D	E	F	G
Moisture (%)	91-94	92	91	92-94	91	90	90	91	89-91
Protein (g)	0.24-0.84	0.7	—	0.3-0.7	0.5	0.5	0.9	0.7	0.4
Fat (g)	0.7	0.8	0.3	0.7-0.8	0.1	0.3	0.9	0.2	0.1
Carbohydrate									
Total (g)	—	—	—	—	—	8.8	7.5	—	9.4
Fiber (g)	0.6	1.2	1.0	—	0.9	0.6	1.5	1.6	0.8-0.9
Ash (g)	0.4-0.5	0.4	0.4	0.3-0.4	0.4	0.4	0.5	0.4	0.3-0.4
Acidity (%)	—	—	—	0.6-0.8	—	—	—	—	—
Calories	—	—	—	—	—	36	37	57	36

Sources: A, Winton and Winton (1935); B, Stahl (1935); C, Munsell *et al.* (1950); D, Leung and Flores (1961); E, Wenkam and Miller (1965): F, Wills *et al.* (1986); G, Morton (1987).
[1]Combined results from two studies

TABLE 14.1B. VITAMIN AND MINERAL COMPOSITION OF CARAMBOLA (MG/100 G)

Constituent	A	B	C	D	E
Vitamins					
Vitamin A	0.36[1]	0.09	0.02	0.01	—
Thiamine	0.04	0.04	0.04	0.02	0.03-0.04
Riboflavin	0.02	0.02	0.04	0.04	0.02-0.03
Niacin	0.3	0.3	0.7	0.5	0.3-0.4
Ascorbic acid (vit. C)	35	35	35	35	26-53
Minerals					
Calcium	4	5	0.9	3	4-6
Phosphorus	17	18	11	—	15-21
Iron	1.5	0.4	0.1	0.3	0.3-1.7
Sodium	—	—	—	3	—
Magnesium	—	—	—	3	—
Zinc	—	—	—	0.2	—
Potassium	—	—	—	110	—

Sources: A, Leung *et al.* (1952); B, Leung and Flores (1961); C, Wenkam and Miller (1965); D, Wills *et al.* (1986); E, Morton (1987).
[1]Vitamin A conversion: 0.3 μg vitamin A = 1 IU vitamin A.

35 mg of ascorbic acid per 100 g pulp provided the 30 mg minimum daily requirement (MDR) for the Australian dietary allowance. The highest reported value for vitamin C (53 mg/100 g pulp) provides about 90% of the U. S. Recommended Daily Allowance (U. S. RDA). Gross (1987) reported the vitamin A value to be relatively high, namely 680 μg retinol

equivalents per kg; which was derived from a rather uncommon provitamin A (β-cryptoflavin). Other vitamins listed were in the range reported for several other tropical fruits (Nagy and Shaw 1980).

The mineral contents of carambolas are also listed in Table 14.1B. Wills *et al.* (1986) carried out the most extensive mineral analyses reported to date. Their values for calcium and iron, 3 and 0.3 mg/100 gm pulp, respectively, were within the ranges reported previously. They reported values of 3, 3, and 0.2 mg/100 gm pulp for sodium, magnesium and zinc, respectively. Phosphorus was the second most abundant mineral in carambolas, ranging from 15.0 to 21.0 mg per 100 g of pulp.

Sugars and Carbohydrates

Values from several sources for sugars, total soluble solids, total carbohydrates and starch are shown in Table 14.2. Fructose values varied from 1.0-3.7% and glucose varied from 0.6-3.3%. Sucrose values were the most variable and ranged from a trace to 1.0%. With one exception, values for fructose and glucose were generally similar. Wills *et al.* (1986) reported more than 3 times as much fructose as glucose for carambolas obtained from a Sydney, Australia retail market. They also reported about 8 times more sucrose than for other reported values. Shaw and Wilson (1983) reported results for HPLC analysis of sugars from three carambola cultivars, Robert Newcomb, Arkin, and selection number M18960 grown in Florida. Fructose and glucose values were about 3% for both sugars, or about 2-2.5 times greater than values reported for other cultivars. Sucrose values (trace to 1.0%) were similar to values (0.4%) in three other studies (Chan and Heu 1975; Benk 1985A,B). Campbell *et al.* (1987) reported sugar levels for the Arkin cultivar after 44 days storage at 5 and 10°C followed by 6 days storage at ambient temperature (23°C). Their results obtained by a similar HPLC method were close to those reported for the same cultivar by Shaw and Wilson (1983). Sugar levels showed relatively little change in fruit stored at 5 and 10°C for 44 days. Sugar levels in fruit stored at 5°C and held for an additional 6 days at ambient temperature did not change significantly. However, for fruit from 10°C storage, there was a 5-fold decrease in both fructose and glucose during storage. Campbell and Koch (1989) reported higher levels of glucose and fructose for Arkin than for the Golden Star cultivar. They reported steady increases in glucose, fructose and sucrose from fruit set to commercial maturity.

Reducing sugars (Table 14.2) ranged from 2.6-5.3%. Winton and Winton (1935) reported values of 3.4 and 5.3% for sweet and sour Hawaiian cultivars, respectively. Shaw and Wilson (1983) reported 5.1% reducing

TABLE 14.2 QUANTITATIVE VALUES FOR SUGARS AND TOTAL SOLUBLE SOLIDS IN CARAMBOLA FRUIT

	Composition %					
Source	Fructose	Glucose	Sucrose	Reducing Sugars	Total Sugars	TSS
Winton and Winton (1935)						
Philippine	—	—	0	2.6	6.1-9.4[1]	—
Hawaiian						
Sweet	—	—	—	3.4	—	8.2
Sour	—	—	0.2	5.3	—	8.7
Stahl (1935)	—	—	—	4.0	4.0	—
Chan and Heu (1975)	1.2	1.6	0.4	—	3.2	—
Wagner *et al.* 1975)	—	—	—	—	—	5.0-9.9[2]
Hou *et al.* (1978)	—	—	—	—	—	8.3
Siota (1984)	—	—	—	—	—	3.9.6.8[3]
Wan and Lam (1984)	—	—	—	—	7.1	8.6
Benk (1985B)	1.2	1.6	0.4	2.8	3.2	—
Shaw and Wilson 1983)[4]						
M-18960	3.7	3.1	trace	—	6.8	8.4
Robert Newcomb	2.7	2.9	1.0	5.1[5]	6.6[5]	7.0
Arkin	3.1	3.3	0.6	—	6.9	7.9
Campbell *et al.* (1987)						
Arkin[6]	1.5	1.3	0.4	—	—	—
Arkin[7]						
5°C	1.3	1.2	0.3	—	—	—
10°C	1.3	1.2	0.3	—	—	—
Arkin[8]						
5°C	1.3	1.2	0.2	—	—	—
10°C	1.0	0.6	0.1	—	—	—
Campbell and Koch (1989)						
Arkin[9]	1.3	1.2	0.4	—	—	—
Golden Star[10]	1.0	0.9	0.4	—	—	—

[1]Range of values from two studies.
[2]Range of values for 18 cultivars from USDA Subtropical Horticulture Research Lab, Miami, FL.
[3]Range of values for 15 cultivars.
[4]Three of the same cultivars used by Wagner *et al.* 1975.
[5]Total sugars and reducing sugars determined by a method independent from using HPLC. Total sugars = 6.1, with reducing sugars = 5.1 and sucrose = 0.97 by using the Lane-Eynon method.
[6]Arkin cv. harvested at color break.
[7]Arkin cv. after 44 days storage at 5 and 10°C.
[8]Arkin cv. held 6 days at ambient after 44 days storage at 5 and 10°C.
[9]Determined at commercial maturity stage (65 days from fruit set).
[10]Determined at commercial maturity stage (60 days from fruit set).

sugars for the Robert Newcomb cultivar that was determined independently from HPLC. Total fructose and glucose values by HPLC (5.6%) were close to the values obtained by a different method.

Few values are available for total sugars (Table 14.2). Shaw and Wilson (1983) reported an average value of 6.7% from HPLC analysis (6.1% by an independent method) for the cultivars grown in Florida, and Chan and Heu (1975) reported 3.2% total sugars for a Hawaiian cultivar. The Florida cultivars contained more than twice as much total sugars as the Hawaiian cultivar. Winton and Winton (1935) reported a range of total soluble solids of 6.1-9.4% from two studies with Philippine cultivars (Table 14.2). The average value for the majority of samples reported was about 8.3%. Total soluble solids content for two cultivars from Florida and the sweet and sour Hawaiian varieties were similar, but values for Robert Newcomb were slightly lower. Wagner *et al.* (1975) and Siota (1984) (Table 14.2) reported the degrees brix in the cultivars they investigated and ranges were 5.0-9.9 and 3.9-6.8%, respectively. Few values for total carbohydrates and starch are available. Reported total carbohydrate values of 8.8 and 9.4% were close to average total soluble solids values. Carambolas apparently have little starch; the only value reported was 0.5%.

Organic Acids

Quantitative values for nonvolatile organic acids are shown in Table 14.3. The most detailed report was by Vines and Grierson (1966) for the Golden Star carambola, which lists the identity and quantitative amounts for seven acids present in green-mature and yellow-mature fruit. Oxalic and tartaric acids were the major acids in green mature fruit, but oxalic and α-ketoglutaric acids were major acids in the yellow-mature fruit. Oxalic acid increased and tartaric acid decreased as the fruit matured. Wills *et al.* (1986) reported higher values for malic and α-ketoglutaric acids (0.22 and 0.77 g/100 g pulp, respectively) than either Vines and Grierson (1966) or Campbell *et al.* (1987). In addition to quantifying sugars, Campbell *et al.* (1987) reported oxalic and malic acid values obtained during a storage study with Arkin carambolas. Neither oxalic nor malic acid decreased in fruit stored at 5°C for 44 days or during an extended 6-day storage period, but there was a 5-fold decrease in both acids for fruit held at 10°C for both storage periods.

Wilson *et al.* (1982) reported oxalic acid values obtained by an HPLC method for 7 named cultivars and 7 numerical selections that were obtained from the USDA Subtropical Horticultural Research Laboratory, Miami, FL (Table 14.3). Five cultivars used in this study were also used

TABLE 14.3. QUANTITATIVE VALUES FOR ORGANIC ACIDS IN CARAMBOLA FRUIT

	Acid (g/100 g)							
Source	Oxalic	Tartaric	Citric	Malic	α-Ketoglutaric	Succinic	Fumaric	% Total
Winton and Winton (1935)	0.31[1]	—	—	—	—	—	—	—
Vines and Grierson (1966)								
Green mature	0.50	0.44	0.13	0.12	0.4	0.02	tr	1.2
Yellow mature	0.96	0.09	—	—	0.22	—	0.03	1.3
Singh (1973)	0.40	—	—	—	—	—	—	—
Wagner *et al.* (1975)[2]	0.04-0.68	—	—	—	—	—	—	—
Wilson *et al.* (1982)[2]	0.08-0.73	—	—	—	—	—	—	—
Wan and Lam (1984)	0.29	—	—	—	—	—	—	—
Benk (1985B)	0.04-0.68	—	—	—	—	—	—	1.0
Wills *et al.* (1986)	—	—	—	0.21	0.77	—	—	—
Campbell *et al.* (1987)								
Arkin[3]	0.15	—	—	0.10	—	—	—	—
Arkin[4]								
5°C	0.15	—	—	0.1	—	—	—	—
10°C	0.08	—	—	0.05	—	—	—	—
Arkin[5]								
5°C	0.15	—	—	0.1	—	—	—	—
10°C	0.08	—	—	0.05	—	—	—	—
Campbell and Koch (1989)								
Arkin[6]	0.08	—	—	0.1	—	—	—	—
Golden Star[7]	0.65	—	—	0.06	—	—	—	—

[1]Acids reported as oxalic.
[2]Carambolas obtained from USDA Subtropical Horticulture Research Lab, Miami, FL. Some of same cultivars used in both studies.
[3]Arkin cv. harvested at color break.
[4]Arkin cv. after 44 days at 5 and 10°C.
[5]Arkin cv. held 6 days at ambient temperature (23°C) after 44 days storage at 5 and 10°C.
[6]Arkin cv. 65 days after fruit set (commercial maturity stage).
[7]Golden Star cv. 60 days after fruit set (commercial maturity stage).

by Wagner *et al.* (1975). Values obtained by HPLC were in agreement to those obtained by an independent titrimetric method. Oxalic acid content varied from 0.08-0.73 g/100 g pulp and with two exceptions results were close to previously reported values (Wagner *et al.* 1975; Campbell *et al.* 1987; Vines and Grierson 1966). Shaw and Wilson (1983) reported oxalic acid values for numerical selections M18960 and M8921 were 0.26 and 0.28 g/100 g pulp, respectively, whereas Wagner *et al.* (1975) reported 0.04 and 0.68 g/100 g pulp for these same selections. Campbell and Koch (1989) measured the oxalic and malic acid levels in the sweet Arkin and the tart Golden star cultivars, and found the levels of both acids decreased steadily with maturity in the sweet cultivar, but remained relatively constant with maturity in the tart cultivar.

Carotenoids

The identification and quantitative estimation for 13 carotenoids from unripe and ripe Golden Star carambolas is shown Table 14.4 (Gross *et al.* 1983). Total carotenoids for the unripe and ripe fruit amounted to 1.5 and 2.2 mg/100 g pulp, respectively. β-Cryptoflavin was the most abundant carotenoid for both ripe and unripe fruit, but it was higher in ripe fruit. The structure of cryptoflavin as determined by mass spectrometry was 5,8-epoxy-5,8-dihydro-β,β-caroten-3-ol with the epoxy group at the hy-

TABLE 14.4. CAROTENOIDS IN CARAMBOLA FRUIT

	Concentration (mg/100 g)		
	Gross *et al.* (1983)[1]		
Carotenoid	unripe	ripe	Wills *et al.* (1986)
phytofluene	0.12	0.37	
β-carotene	0.01	0.01	0.02
ζ-carotene	0.22	0.56	
neurosporene	—	0.004	
β-apo-8′-carotenal	trace	0.02	
β-apo-8′-carotenol	—	—	
β-cryptoxanthin	0.02	0.03	0.01
β-cryptoflavin	0.56	0.75	
β-cryptochrome	0.04	0.06	
unidentified mixture	0.08	—	
leutin	0.04	0.03	
mutatoxanthin	0.32	0.31	
unidentified mixture	0.09	0.06	
total carotenoids	1.5	2.2	

[1]cv. Golden star

droxylated ring. Cryptoflavin is a provitamin A (Gross 1987). Substantial amounts of phytofluene, ζ-carotene, and mutatoxanthin were also present. ζ-Carotene was higher in ripe fruit.

Amino Acids

Vines and Grierson (1966) quantified 17 amino acids in green mature and yellow mature Golden Star carambolas. Seven essential amino acids were quantified with threonine the most abundant. Methionine, leucine phenylalanine, lysine and histidine were found in trace amounts. The major amino acids were alanine, serine, and glutamic acid. All amino acids were lower in the yellow mature fruit.

Hall *et al.* (1980) reported protein and amino acid composition for Tean Ma carambolas. They found (per 100 g pulp) 0.38 g crude protein, 362 mg total amino acids, and 149 mg required amino acids. Leucine was the most abundant and cysteine-cystine and histidine were the least abundant essential amino acids quantified. Glutamic and aspartic acids were the major nonessential amino acids and serine and alanine were present in substantial amounts. Arginine was the least abundant nonessential amino acid. Values for amino nitrogen, ammonia nitrogen, and total nitrogen reported for nine Taiwanese carambola cultivars were 10.0, 4.5 and 69.6 mg/100 ml, respectively (Hou *et al.* 1978).

Volatile Flavor Constituents

Very few studies on the indentity and quantitative estimation of volatile flavor components of carambolas (Table 14.5) have been reported. Stahl (1935) reported 0.78% oil in carambolas. Belardo *et al.* (1971) described physical properties of some Philippine essential oils and reported carambolas contained a trace amount of yellow oil with no semblance of fruit odor. Toulemonde and Beauverd (1984) reported the most comprehensive list of volatile flavor components found in the headspace above carambola puree but they did not report any quantitative data. They reported the identity of 104 compounds (5 alcohols, 6 aldehydes, 7 ketones, 52 esters, 4 acids, 21 hydrocarbons, and 9 miscellaneous compounds). Siota (1984) and Wilson *et al.* (1985) reported the GLC area percent for volatile components in a solvent extract of carambolas. Siota (1984) reported estimations for 41 compounds (5 alcohols, 3 aldehydes, 4 ketones, 19 esters, 9 hydrocarbons, and 1 miscellaneous compound). Wilson *et al.* (1985) reported amounts for 13 alcohols, 4 aldehydes, 6 ketones, 13 esters, 3 hydrocarbons, and 4 compounds classified as miscellaneous. They reported methyl anthranilate (27.2%) as the major component, whereas Siota (1984) reported ethyl anthranilate (11.07%) as the major

TABLE 14.5. CARAMBOLA VOLATILES

Compound	GLC Area % Solvent Extract		GLC Headspace Volatiles
	Wilson *et al.* (1985)[1]	Siota (1984)	Toulemonde and Beauverd (1984)
Alcohols			
Aliphatic			
2-methyl-1-propanol	0.20	—	—
1-butanol	—	—	x[7]
2-butanol	—	—	x
2-methyl-1-butanol	—	—	x
3-methyl-1-butanol	0.20	—	x
1-pentanol	0.05	—	—
1-penten-3-ol	0.04	—	—
1-hexanol	0.60	—	—
cis-3-hexen-1-ol	0.80	tr[2]	—
trans-2-hexen-1-ol	—	tr	—
trans-3-hexen-1-ol	1.20	—	—
6-methyl-hepten-2-ol	5.00	—	—
1-octanol	x[3]	0.98	—
1-nonanol	—	0.35	—
1-decanol	—	0.64	—
Aromatic			
benzyl alcohol	0.60	—	—
phenylethyl alcohol	3.41	—	—
Terpene			
borneol	3.20	—	—
terpinen-4-ol	0.07	—	—
Aldehydes			
Aliphatic			
acetaldehyde	1.60	—	—
hexanal	x[4]	—	x
cis-3-hexenal	—	—	x
trans-2-hexenal	—	—	x
trans-2,4-hexadienal	—	—	x
heptanal	—	tr	—
octanal	—	tr	—
nonanal	—	3.52	—
Aromatic			
benzaldehyde	0.41	—	—
phenyl acetaldehyde	—	—	x
cinnamyl aldehyde	0.80	—	x
Ketones			
Aliphatic			
2-butanone	—	—	x

TABLE 14.5. CARAMBOLA VOLATILES (Continued)

	GLC Area % Solvent Extract		GLC Headspace Volatiles
Compound	Wilson *et al.* (1985)[1]	Siota (1984)	Toulemonde and Beauverd (1984)
2,3-butanedione	—	—	x
3-methyl-2-butanone	0.20	—	—
2-pentanone	—	—	x
6-methyl-5-hepten-2-one	5.28	—	—
Alicyclic			
3,4-dihydro-β-ionone	—	2.25	x
β-ionone	1.01	0.40	—
isophorone	—	tr	—
Aromatic			
Acetophenone	x[3]	tr	—
O-methyl acetophenone	2.55	—	—
Terpene			
carvone	0.80	—	—
menthone	—	—	x
isomenthone	—	—	x
piperitone	—	—	x
Esters			
Aliphatic			
ethyl acetate	1.20	—	x
propyl acetate	—	—	x
2-methyl propyl acetate	—	—	x
butyl acetate	—	—	x
3-methyl butyl acetate	—	tr	x
sec-3-methyl butyl acetate	—	—	x
hexyl acetate	—	—	x
cis-3-hexenyl acetate	—	—	x
trans-2-hexenyl acetate	—	0.35	—
octyl acetate	—	—	x
methyl propanoate	—	—	x
methyl-2-methyl propanoate	—	—	x
ethyl-2-methyl propanoate	—	—	x
ethyl-2-methyl prop-2-enoate	—	—	x
butyl propanoate	—	—	x
butyl-2-methyl propanoate	—	—	x
methyl butanoate	—	—	x
methyl-2-butenoate	—	—	x
methyl-2-methyl butanoate	—	—	x
ethyl butanoate	x[4]	1.67	x
ethyl-2-butenoate	—	tr	x
ethyl-2-methyl butenoate	—	—	x

TABLE 14.5. CARAMBOLA VOLATILES (Continued)

Compound	GLC Area % Solvent Extract		GLC Headspace Volatiles
	Wilson *et al.* (1985)[1]	Siota (1984)	Toulemonde and Beauverd (1984)
propyl butanoate	—	—	x
butyl-2-methyl butanoate	—	—	x
hexyl butanoate	—	—	x
cis-3-hexenyl butanoate	—	—	x
methyl pentanoate	—	—	x
ethyl pentanoate	—	—	x
methyl hexanoate	—	tr	x
methyl-*cis*-3-hexenoate	—	—	x
methyl-*trans*-3-hexenoate	—	—	x
ethyl hexanoate	1.01	3.67	x
ethyl-*trans*-2-hexenoate	—	—	x
ethyl-2,4-hexadienoate	14.51	1.28	x
propyl hexanoate	—	—	x
isopropyl hexanoate	—	—	x
3-methyl butyl hexanoate	—	—	x
ethyl heptanoate	—	—	x
methyl octanoate	—	—	x
ethyl octanoate	—	x[5]	x
ethyl-*trans*-2-octenoate	—	—	x
methyl nonanoate	—	—	x
ethyl nonanoate	—	tr	x
methyl decanoate	—	—	x
ethyl decanoate	—	0.58	x
methyl tiglate	—	—	x
ethyl tiglate	—	—	x
Dicarboxylic acid esters			
diethyl succinate	1.20	x[6]	x
diethyl glutarate	0.05	—	—
Aromatic			
methyl benzoate	0.60	2.65	x
ethyl benzoate	0.60	x[6]	x
methyl salicylate	0.80	x[5]	—
phenyl ethyl acetate	4.61	—	—
cinnamyl acetate	0.41	—	—
methyl cinnamate	—	—	x
ethyl cinnamate	—	6.73	x
methyl anthranilate	27.20	1.22	—
ethyl anthranilate	—	11.07	—
methyl-N-methyl anthranilate	—	1.45	x

TABLE 14.5. CARAMBOLA VOLATILES (Continued)

Compound	GLC Area % Solvent Extract		GLC Headspace Volatiles
	Wilson *et al.* (1985)[1]	Siota (1984)	Toulemonde and Beauverd (1984)
ethyl-N-methyl anthranilate	—	4.57	—
ethyl nicotinate	0.80	—	—
Acids			
butanoic	—	—	x
2-methyl butanoic	—	—	x
hexanoic	—	—	x
heptanoic	—	—	x
Hydrocarbons			
Aliphatic			
undecane	—	—	x
dodecane	—	—	x
tridecane	—	tr	x
tetradecane	—	—	x
pentadecane	—	0.49	x
hexadecane	—	—	x
heptadecane	—	0.32	—
nonadecane	—	0.51	—
eicosane	—	0.23	—
tricosane	—	4.03	—
pentacosane	—	2.15	—
Aromatic			
benzene	—	—	x
o-xylene	—	—	x
m-xylene	—	—	x
p-xylene	—	—	x
trimethyl benzene	—	—	x
styrene	—	—	x
naphthalene	—	—	x
Terpene			
p-cymene	—	—	x
limonene	5.40	—	x
trans-ocimene	—	tr	—
allo-ocimene	—	—	x
α-pinene	0.20	—	x
β-pinene	1.40	—	x
α-terpinene	—	—	x
γ-terpinene	—	—	x
terpinolene	—	—	x
thespirine	—	tr	—
Sesquiterpene			
β-caryophyllene	—	—	x

TABLE 14.5. CARAMBOLA VOLATILES (Continued)

Compound	GLC Area % Solvent Extract		GLC Headspace Volatiles
	Wilson *et al.* (1985)[1]	Siota (1984)	Toulemonde and Beauverd (1984)
ar-curcumene	—	—	x
humulene	—	—	x
Miscellaneous			
diethyl carbonate	—	—	x
1,8-cineole	1.40	—	—
veratrole	0.80	x	—
benzothiazole	0.80	—	—
quinoline	0.50	—	—

[1]Solvent extract of macerated fruit.
[2]tr = trace.
[3]Octanol and acetophenone coelute from column, GLC area % = 4.10.
[4]Ethyl butanoate and hexanal coelute from column, GLC area % = 6.41.
[5]Ethyl octanoate and methyl salicylate coelute from column, GLC area % = 7.54.
[6]Diethyl succinate and ethyl benozate coelute from column, GLC area % = 1.74.
[7]Nonquantitative data.

component. Siota (1984) also found substantial amounts of methyl and ethyl-N-methyl anthranilate. Wilson *et al.* (1985) described the most prominent aroma as grape-like and the presence of anthranilate esters helps substantiate this observation. They also attributed the strong fruity aroma of the extract to the presence of esters and ketones; the majority of components reported by Toulemonde and Beauverd (1984) were esters.

Enzymes

The enzymes studied in carambola are those affecting color and texture of cut or processed fruit during storage. Adnan and Ghazali (1986) studied polyphenol oxidase (PPO) in an unidentified carambola cultivar from Malaysia. The enzyme was isolated by acetone precipitation and purified by dialysis. Specific activities for the crude extract, acetone precipitate, and dialyzed enzyme were 2.6, 24.7, and 56.7 U/mg, respectively The optimum pH for the acetone powder PPO catalyzed oxidation of catechol was 7.2. The Michaelis constant (Km) and Vmax for catechol with PPO were 0.061M and 20 U/ml, respectively. The enzyme was heat labile and could be inactivated at temperatures of 70 to 90°C for short durations. Ghazali and Leong (1987) reported polygalacturonase (PG) activity in an unidentified Malaysian carambola cultivar. The enzyme was isolated by ammonium sulfate precipitation and purified by ultracentrifugation and gel filtration. There was no detectable specific activity in the crude extract. However, specific activities for ammonium sulfate precipitate, ultracentrafugate, and gel filtrate were 1.61, 4.53, and 4.62 U/mg, respectively. The optimum pH of PG for the purified enzymes was 5.2. Viscometric and reductometric analysis on PG in the gel filtrate and paper chromatographic analysis of the reaction mixture was conducted to determine if the mode of action was from indo- or exo-PG. Results indicated that only exo-PG was present in carambolas. Baldini *et al.* (1982) reported the biochemical evaluation of an unidentified cultivar from Brazil. Qualitative tests indicated the presence of pectin esterase (PE), polygalacturonase (PG), amylase, invertase, catalase, and peroxidase. Specific activities (U/mg) for PE, PG, and peroxidase were 0.08, 22.2, and 34.8,respectively. They reported carambolas contained 0.63% pectin that was 81.5% esterified.

Seeds

There is very little information available on the analysis of carambola seeds. Berry (1978) reported the analysis of seeds in a unidentified carambola cultivar from Malaysia. The seeds contain 4.0% moisture, 73.87% oil, and 14.0% protein and had the following characteristics: refractive index (25°C) 1.4658, iodine number (Wijs) 81.64, saponification number 197.68, unsaponifiable matter 2.96%, and acid value 0.6. The trace amount of oil isolated was clear yellow and its composition was similar to that for groundnut oil. The unsaturated fatty acids, oleic (45.8%) and linoleic (22.3%), were the major fatty acids. Palmitic acid (21.3%) and stearic acid (8.15%) were the major saturated acids. Although the seeds are high in oil content they could hardly be considered a commercial product because of the limited supply.

BILIMBI

The bilimbi (*Averrhoa bilimbi* L.) is closely related to the carambola, and is probably a native of the Malayan region. It is widely distributed throughout the tropics and is common to Indonesia, the Philippines, Ceylon, Burma, Thailand, Malaysia, Singapore, India, Australia, the Caribbean Islands, and Central and South America. It is also found in gardens in warmer sections of southern Florida. The bilimbi has numerous different names. Cucumber tree and tree sorrel are common English names, and variations of bilimbi are used according to specific geographic areas (Mowry *et al.* 1967; Morton 1987).

The tree is long lived, has a short multibranched trunk, and reaches heights of 5 to 10 m. It is differentiated from the carambola by its larger leaves, which are densely clustered near the fragile branch tips. The leaves have 11 to 37 oblong leaflets, and the crimson flowers borne in small panicles arising from the trunk and oldest branches (Sturrock 1959; Mowry *et al.* 1967; Morton 1987). The tree bears its fruits in clusters on the trunk and at the base of the branches. Bilimbi is cultivated mainly by seed but shield-budding and air-layering has been successfully used (Mowry *et al.* 1967; Morton 1987).

The translucent greenish yellow fruit, 4-10 cm long, is cylindrical or obscurely five-angled with 6 or 7 flat disc-like seeds (Sturrock 1959; Mowry *et al.* 1967; Morton 1987). The fruit is very acid with firm juicy flesh that is generally not eaten raw but is made into tart jellies, pickled, stewed, or made into relishes for additon to vegetables and meats (Sturrock 1959; Mowry *et al.* 1967; Morton 1987). The juice (about 44.2% yield, pH 4.47) is made into cooling beverages and ades (Morton 1987). Its main use is in place of mango for making chutney (Morton 1987).

The chemical analysis of bilimbi is shown in Table 14.6, and amounts listed are a range of values from several sources (Winton and Winton 1935; Sturrock 1959; Morton 1987). The range of values for moisture and protein was 94.2-95.0 and 0.61-0.68 g/100 g pulp, respectively. Fiber content was 0.61 g/100 g pulp and ash varied from 0.31-0.41 g/100 g pulp. Calcium (5.4), phosphorus (11.1) and iron (1.01 mg/100 g pulp) levels were similar to the amounts of these three elements reported in carambola. Carotene, thiamine, riboflavin and niacin were present in amounts

TABLE 14.6. COMPOSITION OF BILIMBI

Component	Amount per 100 g edible portion
Water[1] (g)	94.2-95.0
Protein (g)	0.61-0.68
Fiber (g)	0.61
Ash (g)	0.31-0.41
Calcium[2] (mg)	3.4
Phosphorus (mg)	11.1
Iron (mg)	1.01
Carotene (mg)	0.035
Thiamine (mg)	0.010
Riboflavin (mg)	0.026
Niacin (mg)	0.302
Ascorbic acid (mg)	15.5

[1]Sources: Winton and Winton (1935); Sturrock (1959); Morton (1987).
[2]Sources: Sturrock (1959); Morton (1987).

of 0.0335, 0.010, 0.026, and 0.30 mg/100 g pulp, respectively. Bilimbi is not an exceptional source of vitamin C (15.5 mg/100 g pulp) but it would supply half of the Australian minimum daily requirement (MDR) and about one fourth of the U. S. RDA. Bilimbi is reported to contain 1.5 g/100 g acid (as oxalic), 2.2 g/100 g reducing sugars, and 1.39 g/100 g fat (Winton and Winton 1935).

REFERENCES

ABDULLAH, F. and RAGAB, M. H. H. 1970. Vitamin C content, pH value and titratable acidity of 26 types of Malayan fresh fruits grown in Selangor. Malaysian Agric. J. 47, 416-428.

ADNAN, T. A. B. T. and GHAZALI, H. M. 1986. Polyphenoloxidase from starfruit (*Averrhoa carambola*, L.). Pertanika 9, 219-224.

BALDINI, V. L., DRAETTA, I. D. and NORUMA, E. H. 1982. Biochemical evaluation of carambola (*Averrhoa carambola*). Colet. Inst. Technol. Aliment. Campina, Brazil, CITAC 7,12, 283-291.

BANERJEE, D. P., CHATTERJEE, B. K. and RAO, D. P. 1980. Propagating carambola (*Averrhoa carambola* L.) by air-layering with different concentrations of growth regulators. Indian Agric. 24, 165-168.

BAQAR, M. R. 1980. Vitamin C content of some Papua New Guinean fruits. J. Food Technol. 15, 459-461.

BELARDO, O. L., CARDENO, V. and PEREZ, P. 1971. Physical properties of some Philippine essential oils. Flavour Ind. 2, 305-309.

BENK, E. 1985A. Tropical and subtropical fruits for production of jams and jellies. Gordian 85, 122-125.

BENK, E. 1985B. Tropical and subtropical fruit and fruit juices used for production of fruit nectars and fruit-juice beverages. Fluess. Obst. 52, 358-364.

BERRY, R. E., WAGNER, C. J., Jr. and SHAW, P. E. 1977. Promising products from tropical fruits. Food Prod. Dev. 7, 109-112.

BERRY, S. K. 1978. The composition of the oil of starfruit (*Averrhoa carambola* L.) seeds. J Amer. Oil Chem. Soc. 55, 340-341.

CAMPBELL, B. A. and CAMPBELL, C. W. 1983. Preservation of tropical fruits by drying. Proc. Fla. State Hortic. Soc. 96, 229-231.

CAMPBELL, C. A. 1987. Personal communication.

CAMPBELL, C. A., HUBER, D. J. and KOCH, K. E. 1987. Postharvest response of carambolas to storage at low temperatures. Proc. Fla State Hortic. Soc. 100, 272-275.

CAMPBELL, C. A. and KOCH, K. E. 1989. Sugar/acid composition and development of sweet and tart carambola fruit. J. Amer. Soc. Hortic. Sci. 114:445-457.

CAMPBELL, C. W., KNIGHT, R. J., Jr. and OLSZACK, R. 1985. Carambola production in Florida. Proc. Fla. State Hortic. Soc. 98, 145-149.

CAMPBELL, C. W. and MALO, S. E. 1967. Utilization of Florida grown fruits in ice cream and sherbets. Fla. Agric. Exp. Stn. Ann. Report.

CHAN, H. T., Jr. and HEU, R. A. 1975. Identification and determination of sugars in starfruit, sweetsop, green sapote, jack fruit and wi apple. J. Food Sci. 40, 1329-1330.

FONSECA, H., NOGUEIRA, J. N. and MARCONDES, A. M. S. 1969. Ascorbic acid and beta carotene contents of Brazilian fruits and vegetables Arch. Latinoam. Nutr. 19, 9-16.

GHAZALI, H. M. N. and LEONG, C. K. 1987. Polygalacturonase activity in starfruit. Food Chem. 24, 147-157.

GREEN, J. G. 1987. Carambola production in Malaysia and Taiwan. Proc. Fla. State Hortic. Soc. 100, 275-278.

GRIERSON, W. and VINES, H. M. 1965. Carambolas for potential use in gift fruit shipments. Proc. Fla. State Hortic. Soc. 78, 349-353.

GROSS, J. 1987. Pigments in Fruits. Academic Press, New York.

GROSS, J., IKAN, R. and ECKHARDT, G. 1983. Carotenoids of the fruit of *Averrhoa carambola*. Photychemistry 22, 1479-1481.

HALL, N. T., SMOOT, J. M., KNIGHT, R. J., Jr. and NAGY, S. 1980. Protein and amino acid composition of ten tropical fruits by gas-liquid chromatography. J. Agric. Food Chem. 28, 1217-1221.

HARLER, C. 1983. What insiders say about carambola. Fla. Grower and Rancher 76, 9.

HERRMAN, K. 1981. Review on chemical composition and constituents of some important exotic fruits. Z. Lebensm-Unters. Forsch 173, 47-60.

HOU, T. Y., TSIANG, J. M., CHEN, H. C., CHEN, K. Y., CHEN, C. K. and HUANG, T. M. 1978. Study of the effect of different varieties and maturities of the raw material upon the chemical composition of the fruit juices J. Chinese Agric. Chem. Soc. 16, 14-23.

KENNEY, P. and HULL, L. 1986. Effects of storage condition on carambola quality. Proc. Fla. State Hortic. Soc. 99, 222-224.

KNIGHT, R. J., Jr., LAMBERTS, M. and BUNCH, J. S. 1984. World and local importance of some tropical fruit crops grown in Florida. Proc. Fla. State Hortic. Soc. 87, 351-354.

LAM, P. F. and WAN, C. K. 1983. Climacteric nature of carambola (*Averrhoa carambola* L.) fruit. Pertanika 6, 44-47.

LEUNG, W. T. W. and FLORES, M. 1961. Food composition table for use in Latin America. The Institute of Nutrition of Central America and Panama, Guatemala City, C.A. and the Interdepartmental Committee on Nutrition for National Defense, NIH., Bethesda, Md.

LEUNG, W. T. W., PECOT, R. K. and WATT, B. K. 1952. Composition of foods used in far eastern countries. USDA Handbook 34.

LITZ, R. E. and CONOVER, R. A. 1980. Partial organogensis in tissue cultures of *Averrhoa carambola*. Hortic. Sci. 15, 735.

LOPEZ, H., CIMADEVILLA, M., FERNANDEZ, E., DUBRUTHY, C., NAVIA, J. M., VALIENTE, A., CLEMENT, I. D. and HARRIS, R. S. 1956. Provisional Table of Nutrient Composition of Cuban Foods. Lab. FIM de Nutricion. Pub. #3. Comite Cubano Amer. pro Fundacion Invest. Med., Havana, Cuba.

LUND, E. D. and SMOOT, J. M. 1982. Dietary fiber content of some tropical fruits and vegetables. J. Agric. Food Chem. 30, 1123-1127.

MORTON, J. F. 1987. Carambola. *In* Fruits of Warm Climates. Published by J. F. Morton, Miami, FL.

MOWRY, H., TOY, L. R. and WOLFE, H. S. 1967. Miscellaneous tropical and subtropical Florida fruits. Rev. by Ruehle. G. D., Agric. Ext. Serv., Inst. Food and Agric. Sci., Univ. Fl., Gainesville, Bull. 156A.

MUNSELL, H. E., WILLIAMS, L. O., GUILD, L. P., KELLEY, L. T. AND HARRIS, R. S. 1950. Composition of food plants of Central America. VII. Food Res. 15, 421-438.

NAGY, S. and SHAW, P. E. 1980. Tropical and Subtropical Fruits. Avi Publishing Co., Westport, CT.

NORDBY, H. E. and HALL, N. T. 1979. Lipid markers in chemotaxonomy of tropical fruits: preliminary studies with carambola and loquat. Proc. Fla. State Hortic. Soc. 92, 298-300.

OSLUND, C. R. and DAVENPORT, T. L. 1981. No climacteric in the starfruit (*Averrhoa carambola*). Hortic. Sci. 16, 424.

OSLUND, C. R. and DAVENPORT, T. L. 1983. Ethylene and carbon dioxide in ripening fruit of *Averrhoa carambola*. Hortic. Sci., 18, 229-230.

SCHMANDKE, H. and GUERRA, O. 1969. The carotene, L-ascorbic acid, dehydroascorbic acid and tocopherol content of cuban vegetable products. Nahrung 13, 523-530.

SHAW, P. E. and WILSON, C. W., III. 1983. Separation of fructose, glucose, and sucrose in fruit by high performance liquid chromatography using UV detection at 190 nm. J. Sci. Food Agric. 34, 109-112.

SINGH, P. P. 1973. Oxalic acid content of Indian foods. Qual. Plant Matter 22, 335-347.

SIOTA, H. 1984. The botanical features of starfruit. Koryo 143, 37-43. (Japanese).

STAHL, A. L. 1935. Composition of miscellaneous tropical and subtropical Florida fruits. Univ. of Fla. Agric. Exp. Stn. Bull. 283.

STURROCK, D. 1959. The carambola-*Averroha carambola* L. *In* Fruits for Southern Florida. Southeastern Printing Co., Stuart, Fl.

TOULEMONDE, D. and BEAUVERD, D. 1984. Headspace analysis: Trap desorption by microwave energy application to the volatile components of some tropical fruit. Dev. Food Sci. 10, 533-548.

VINES, H. M. and GRIERSON, W. 1966. Handling and physiological studies with the carambola. Proc. Fla. State Hortic. Soc. 79, 350-355.

WAGNER, C. J., Jr., BRYAN, W. L., BERRY, R. E. and KNIGHT, R. J., Jr. 1975. Carambola selection for commercial production. Proc. Fla. State Hortic. Soc. 88, 466-469.

WAN, C. K. and LAM, P. F. 1984. Biochemical changes, use of polyethylene bags and chilling injury of carambola (*Averrhoa carambola* L.) stored at various temperatures. Pertanika 7, 39-46.

WENKAM, N. S. and MILLER, C. D. 1965. Composition of Hawaiian fruits. Hawaii Agric. Exp. Stn. Bull. 135.

WILLS, R. B. H., LIM, J. S. K. and GREENFIELD, H. 1986. Composition of Australian foods. 31. Tropical and subtropical fruit. Food Technol. Australia 38, 118-123.

WILSON, C.W., III, SHAW, P. E. and KNIGHT, R. J., Jr. 1982. Analysis of oxalic acid in carambolas (*Averrhoa carambola* L.) and spinach by high-performance liquid chromatography. J. Agric. Food Chem. 30, 1106-1108.

WILSON, C. W., III, SHAW, P. E., KNIGHT, R. J., Jr., NAGY, S. and KLIM, M. 1985. Volatile constituents of carambola (*Averrhoa carambola* L.). J. Agric. Food Chem. 33, 199-201.

WINTON, A. L. and WINTON, K. B. 1935. Carambola. *In* The Structure and Composition of Foods. John Wiley and Sons, Inc., New York.

15

Pejibaye

Charles R. Clement

The pejibaye (*Bactris gasipaes* H.B.K.) is a caespitose palm that may attain 20+ m in height. Stem diameter varies from 15 to 30 cm and internode length from 2 to 30 cm, but becomes reduced with age after 5 years. The internodes are armed with numerous black, brittle spines, although spineless mutants occur and have been selected for planting in several areas. The stem is topped by a crown of 15 to 25 pinnate fronds (Fig. 15.1), with the leaflets inserted at different angles. Among the axils of the senescent fronds the inflorescences may develop. After pollinization the bunch (Fig. 15.2) may contain between 50 and 1000 fruit and weigh between one and 25 kg. Numerous factors may cause fruit drop: poor pollinization, poor nutrition, drought, crowding, insects and diseases. The fruit (Fig. 15.3) that ripen have a starch/oily moist mesocarp, a fibrous red, orange or yellow epicarp, and a single endocarp with a fibrous/oily white endosperm. Individual fruit may weigh between 10 and 250 g.

The pejibaye was domesticated in tropical America, probably in western Amazonia (Clement 1988). The domestication process enormously increased the genetic variability of the species and will be described below with respect to fruit quality and uses. Most Amerindian groups that used it, selected the pejibaye as a starchy edible fruit, while others appear to have selected for starch to make meal, and fermented to make alcoholic beverages (Clement 1988). In some areas the pejibaye may have been as important a staple as cassava (*Manihot esculenta* Krantz). Pejibaye is much more nutritious and can produce considerable useable product year after year, whereas cassava is generally only harvested once or twice from each area.

FIG. 15.1. A SEVEN-YEAR-OLD PEJIBAYE PALM. NOTE THE MULTIPLE OFF-SHOOTS ARISING FROM THE BASE OF THE PRINCIPAL STEM.

Courtesy of S.A.N. Ferreira

FIG. 15.2 A MESOCARPA LANDRACE FRUIT BUNCH WEIGHING ABOUT 20 KG FROM YURIMAGUAS, PERU.

From an initial domestication in western Amazonia the Amerindians distributed the pejibaye throughout the tropical lowlands of South America and as far north as Honduras (Mora Urpi 1984). Figure 15.4 represents its current distribution.

Mora Urpi and Clement (1985) described eight landraces of pejibaye found in Amazonia by germplasm collection expeditions sponsored by the U.S. Agency for International Development. Their classification is based principally on fruit size, because this appears to have been most modified by selection during domestication. Fruit composition also varies considerably but has not yet been clearly defined. According to Mora Urpi and Clement (1985) there are three groups of landraces, based exclusively on fruit size: 'microcarpa', with small, generally more oily, fibrous fruit, averaging 20-25 g (2 landraces described); 'mesocarpa', with medium, rather starchier, less oily and fibrous fruit, averaging 25-70 g (4 landraces described); and 'macrocarpa', with large, very starchy, low oil and fiberous fruit, averaging more than 70 g (2 landraces described). Clement (1986A) defined a ninth landrace in the mesocarpa group and suggested that Mora Urpi's (1984) Occidental group of races was, in fact, one landrace within the mesocarpa group. As will be made clear in the following, different landraces appear ideally suited for different uses.

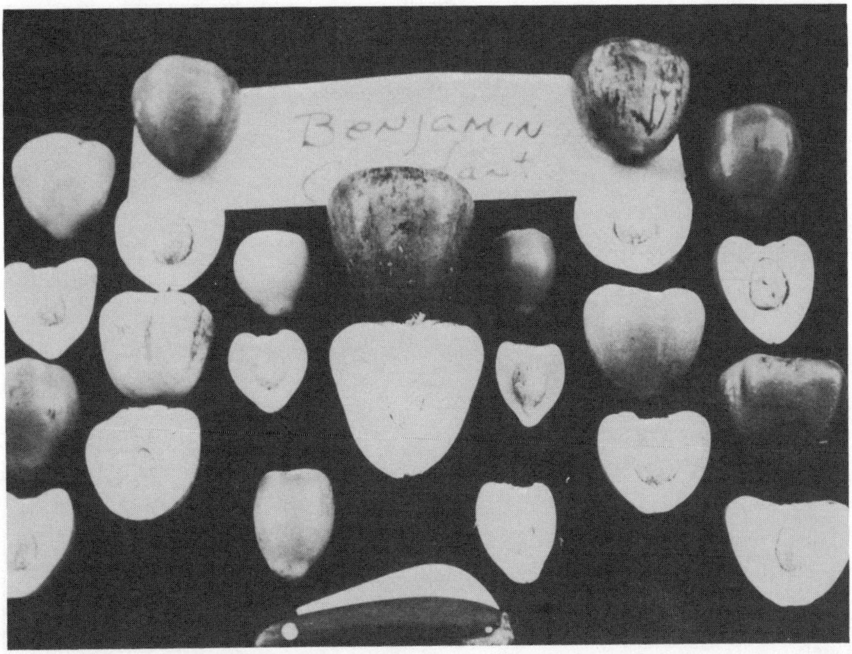

FIG. 15.3 FRUIT SIZE AND SHAPE DIVERSITY FROM THE PUTUMAYO MACROCARPA LANDRACE, COLLECTED FROM THE BENJAMIN CONSTANT, AMAZONAS, BRAZIL POPULATION MAINTAINED IN THE UNIVERSITY OF COSTA RICA GERMPLASM COLLECTION.

Clement and Mora Urpi (1987) outlined five possible uses for pejibaye that appear to have considerable economic potential: fruit for (1) direct human consumption; (2) animal ration; (3) flour and meal; (4) oil; and (5) heart of palm (palmito). Those authors, and Clement (1988) discussed these products from a breeder's point of view and presented initial ideotypes for each product. This chapter will outline the currently available results of interest to food technology and agrobusiness.

COMPOSITION

Bunch Components

Recent work has shown that bunch components, especially bunch weight and fruit number (Table 15.1), vary considerably from population to population and landrace to landrace. Piedrahita and Velez (1982), working on the Colombian Pacific (mesocarpa group), found surprisingly low bunch component values and a very low fruit to bunch ratio. They considered these to be due to bunch phytosanitary conditions. Arkcoll

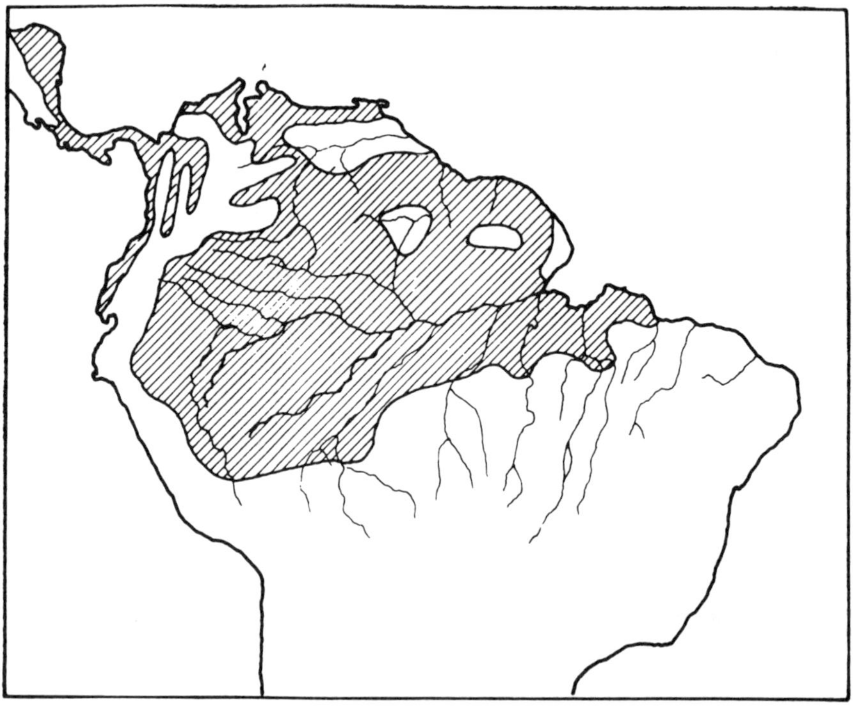

FIG. 15.4. APPROXIMATE MODERN DISTRIBUTION OF THE PEJIBAYE IN LATIN AMERICA. RECENT INTRODUCTIONS TO NEW AREAS HAVE NOT BEEN INCLUDED.

and Aguiar (1984) studied more than 100 bunches near Manaus, AM, Brazil with samples from three different landraces and many populations. As their work was prior to Mora Urpi and Clement's (1985) definition of Amazonian landraces, they were not able to distinguish them. Their data show low bunch weight, but high fruit number and fruit to bunch ratio, suggesting that most of their sample was from the Para microcarpa landrace.

TABLE 15.1. AVERAGE BUNCH COMPOSITION DATA FROM SEVERAL SOURCES

Reference		Bunch weight (kg)	Fruit number	Fruit/bunch ratio (%)
Piedrahita and Velez	(1982)	3.3	61	87
Arkcoll and Aguiar	(1984)	3.6	96	93
Clement and Mora Urpi	(1985)	7.9	149	96
Clement	(1986A)	4.8	106	91

Clement and Mora Urpi (1985) found significant differencs in bunch component values among Amazonian landraces, with the microcarpa landraces presenting low weights and high fruit numbers, whereas the macrocarpa landraces presented moderate to high bunch weights, lower fruit numbers and about equivalent fruit to bunch ratios. Mesocarpa landraces were generally intermediate in bunch weights and fruit numbers, although the largest bunches found were from mesocarpa populations. Clement (1986A) found significant differences among three mesocarpa populations in Costa Rica, both *in situ* and *ex situ*, for fruit number and bunch weight.

Piedrahita and Velez's (1982) observations in Colombia and Arkcoll and Aguiar's (1984) comments about fruit drop in Manaus show that there are already serious phytosanitary problems that must be resolved before monoculture plantations can be recommended. Pejibaye was domesticated and grown in a controlled environment (Clement 1988), which kept phytosanitary problems at acceptable levels. A change to monoculture, at least in its natural range, will always be threatened by these problems.

Fruit Composition

Piedrahita and Velez (1982) report typical mesocarpa fruit data, with an especially high pulp to fruit ratio. Arkcoll and Aguiar (1984) report intermediate data, due to the mixture of landraces in their sample. Their slightly lower mesocarp and pulp percent data indicate the lower values typical of microcarpa landraces. Clement and Mora Urpi (1985) found significant differences among Amazonian landraces for the three components studied (Table 15.2). Clement (1986A) found similar differences among three Costa Rican mesocarpa populations.

Mesocarp Composition

The literature on pejibaye mesocarp composition is relatively extensive, but most early reports are based on only one or two samples, with-

TABLE 15.2. AVERAGE FRUIT COMPOSITION DATA FROM SEVERAL SOURCES

Referance		Fruit weight (g)	Seed weight (g)	Mesocarp (%)	Pulp (%)
Piedrahita and Velez	(1982)	50	3.6	80.8	92.8
Arkcoll and Aguiar	(1984)	35	2.9	76.1	90.3
Clement and Mora Urpi	(1985)	58	4.1	—	91.7
Clement	(1986A)	42	4.6	—	88.5

out information on collection locality or population characteristics. [See Clement and Arkcoll (1985) for a review].

The three studies reported in Table 15.3 did not take into account Mora Urpi and Clement's (1985) landrace classification, but did find enormous variability in their different samples. While Piedrahita and Velez (1982) and Arkcoll and Aguiar (1984) worked with *in situ* samples, CIPRONA (1986) worked with *ex situ* samples from the Universidad de Costa Rica (UCR) and the Centro Agronomico Tropical para Investigacion y Ensenanza (CATIE) germplasm banks. Arkcoll and Aguiar (1984) were looking for plants with high oil levels; they found up to 61.7% in the dry mesocarp of an apparent microcarpa plant. CIPRONA (1986) found up to 17.5% protein. Both of these maxima show the available genetic potential for exploitation in a breeding program.

Arkcoll and Aguiar (1984) and CIPRONA (1986) present data on kernel composition. As with other palms this is a potential source of lauric oils, but it is also extremely rich in fiber. It is, however, a potentially important byproduct.

Proteolytic Enzyme Inhibitor

The fruit of the pejibaye is always eaten cooked. Fonseca (1927) reported the presence of an alkaloid, pupunhadine, as a probable reason for cooking. Murillo *et al.* (1983) reported a trypsin inhibitor. Arkcoll and Aguiar (1984) reported the presence of oxalate-like crystals in and under the exocarp. Most research has concentrated on the trypsin inhibitor, since this can have a serious negative effect on growth, especially with young birds and pigs.

CIPRONA (1986) reported a method to quantify the presence of this inhibitor and found considerable variation among samples, suggesting that there is an exploitable genetic variability. CIPRONA is continuing their research and hope to describe the inhibitor by the end of 1990. Early indications suggest that it is very similar to the inhibitor found in

TABLE 15.3. AVERAGE MESOCARP COMPOSITION DATA FROM SEVERAL SOURCES

Reference		% wet wt Water	% dry wt Protein	Oil	N-free extract[1]	Fiber	Ash
Piedrahita and Velez	(1982)	49.9	9.8	11.5	73.7	2.8	2.4
Arkcoll and Aguiar	(1984)	55.7	6.9	23.0	59.5	9.3	1.3
CIPRONA	(1986)	56.7	6.1	8.3	79.9	3.6	2.1

[1]Fraction is largely starch and sugars

the common bean (*Phaseolus vulgaris* L.) (Calzada, personal communication). Heat, as in boiling water, eliminates this inhibitor. The economic application of this is the subject of considerable research in using pejibaye as an animal ration, as will be reviewed below.

Protein Quality

Very little work has been done on the amino acid composition of pejibaye mesocarp protein, although some comparisons between pejibaye and other foods have been made. In Table 15.4 three studies are summarized. All essential amino acids are present, although at lower levels than in maize. The variability shown suggests that selection could change relative levels and probably improve these considerably.

Oil Quality

Oil quality has been studied more than protein quality, but only CIPRONA (1986) has studied a significant number of samples. Table 15.5

TABLE 15.4. AMINO ACID COMPOSITION DATA OF PEJIBAYE MESOCARP PROTEIN FROM SEVERAL SOURCES (PERCENT PER G NITROGEN)

	Reference		
Amino acid	Piedrahita and Velez (1982)	Zapata (1972)	Zumbado and Murillo (1984)
Lysine[1]	4.22	4.6	4.12
Histidine[1]	2.72	2.0	1.76
Arginine[1]	7.26	9.2	1.76
Aspartic acid	4.99	4.6	n.r.[2]
Threonine[1]	2.91	2.5	3.53
Serine	3.75	3.6	n.r.
Glutamic acid	4.68	6.3	n.r.
Proline	2.70	2.9	n.r.
Glycine[1]	3.21	4.5	5.29
Alanine	4.14	3.6	n.r.
Cystine	trace	—	n.r.
Valine[1]	2.76	2.7	3.73
Isoleucine[1]	1.96	1.7	3.14
Leucine[1]	2.63	2.6	5.49
Methionine[1]	1.46	1.3	1.57
Tyrosine[1]	1.65	1.4	2.75
Phenylalanine[1]	1.82	1.3	2.75
Tryptophan[1]	0.92	n.r.	n.r.
Protein % (dry wt)	9.00	5.7	5.10

[1]Essential amino acid.
[2]not reported.

TABLE 15.5. FATTY ACID COMPOSITION DATA OF PEJIBAYE MESOCARP OILS FROM SEVERAL SOURCES (% OIL)

Fatty acid	Zapata (1972)	Hammond *et al.* (1982)	Silva and Amelotti (1983)	CIPRONA (1986)	Oil palm[1]
Palmitic	40.2	29.6	44.8	32.2	42.2
Stearic	0.4	trace	1.5	1.5	4.9
Total SFA[2]	40.6	29.6	46.3	33.7	47.1
Palmitoleic	10.5	5.3	6.5	8.3	—
Oleic	47.5	50.3	41.0	45.5	40.6
Linoleic	1.4	12.5	4.8	11.6	11.2
Linolenic	—	1.8	1.0	2.0	—
Total UFA[2]	59.4	69.9	53.3	67.4	51.8

[1]Noiret and Wuidart (1976).
[2]SFA—saturated fatty acids; UFA—unsaturated fatty acids

outlines four of these studies and compares pejibaye with oil palm (*Elaeis guineensis* Jacq.). Although Zapata (1972), Hammond *et al.* (1982) and CIPRONA (1986) report high levels of unsaturated fatty acids, these are all from mesocarpa populations with relatively low oil levels. Silva and Amelotti (1983) report oil quality similar to that of oil palm from a sample that is probably from the Para microcarpa landrace, where Arkcoll and Aguiar (1984) found high oil levels.

Hammond *et al.* (1982) and Silva and Amelotti (1983) characterized the triglyceride structure in two samples. Arkcoll (personal communication) suggests that the triglyceride structures are extremely variable, since some samples from the same population are liquid at room temperature, whereas others are solid. Two samples remained liquid at 8°C. A detailed analysis of fatty acid composition and triglyceride structure should be included in the breeding program.

Vitamins

The mesocarp is frequently extremely rich in beta-carotene. Arkcoll and Aguiar (1984) reported enormous variation in this pro-vitamin. Other vitamins do not present such variation (Table 15.6), nor do they represent significant quantities in the diet.

POTENTIAL USES AND PRODUCTIVITY

Fruit for Direct Human Consumption

This is the traditional use for the pejibaye fruit, and in most of its geographic distribution, it is the only one known. The whole fruit is

TABLE 15.6. VITAMINS PRESENT IN PEJIBAYE MESOCARP, FROM SEVERAL SOURCES (PER 100 G FRESH MESOCARP)[1]

Reference	β-carotene	Niacin (mg)	Vitamin C (mg)	Riboflavin (mg)	Thiamine (mg)
Johannessen (1967)	315 mcg	0.12	1.19	—	—
I.N.N. (1959)	7300 I.U.	0.9	20.0	0.11	0.04
Leung (INCAP) (1961)	670 mcg	1.4	35.0	0.16	0.05
Arkcoll and Aguiar (1984)	0-70 mg	—	—	—	—

[1]Units in mcg = micrograms; I.U. = international units; mg = milligrams.

separated from the bunch and boiled in salted water for 30 to 60 min to eliminate the irritating oxalate crystals and trypsin inhibitor. Piedrahita and Velez (1982) report that cooking for 15 to 20 min deactivates the peroxidase enzyme. The fruit are then peeled, halved and pitted, and are ready for consumption. Pejibaye fruit are frequently consumed at breakfast, or as an appetizer at later meals. When the fruit are low in oil, they are tastier when acccompanied by mayonnaise or a sauce to help moisten them. The cooked fruit can also be used in stews or ground and used in a variety of preparations and pastries. About 40 pejibaye recipes have been collected by Calvo (1981).

While the flavor of the fruit are quite variable, the typical pejibaye has a distinct bland to strong flavor (depending upon carotenoid content) of its own. A liking for this is acquired fairly easily but it is not exciting enough to generate new markets without good quality control and clever marketing. Several authors have mentioned that pejibaye may taste like European chestnut (*Castanea sativa* M.Pl.) (Popenoe and Jimenez 1921; NAS 1975) but this is doubtful. Arkcoll and Aguiar (1984) found some similar to potato, others to maize, and some which were sweet. The latter were much more attractive than most fruit.

Mesocarp texture is also variable, ranging from that of a soggy potato to that of a good raw cashew (*Anacardium occidentale* L.). Texture appears to be determined by water and starch content; with watery, low starch types being soggy and starchy, and low water types being floury to crunchy even after cooking. Preferences vary from region to region: in Costa Rica the best fruit are dry, with high starch content and a nutty texture; in Manaus the best fruit are moderately dry and starchy, with a less nutty, though still firm texture. Moderate oil levels are also important to fruit quality in Manaus and may influence texture.

In Costa Rica whole or halved, pitted or unpitted, peeled or unpeeled fruit are prepared in salty water in 500 and 1000 g jars or cans. Quality is extremely variable but they sell well when fresh fruit are unavailable. In Colombia, dehydrated fruit have been prepared for the market.

Yields have been projected from small plots or population data. Corley (1983) points out the risks of this type of projection. Mora Urpi (1984) reported 25 MT/ha/yr of fresh bunches from nonselected, fertilized germplasm (mesocarpa group) growing in good edapho-climatic conditions in Costa Rica. Clement (1987) projected much lower yields from population data (microcarpa group) near Manaus, with 6 to 10 MT/ha/yr for unselected, unfertilized germplasm on poor soils with three months drought. Moreira Gomes *et al.* (1987) estimated potential yields of 24 MT/ha/yr from the Fonte Boa (mesocarpa group) population for unselected, unfertilized germplasm on poor soils with a favorable climate. Clement and Mora Urpi (1987) suggest that much higher yields should be easily attained within a cycle or two of any improvement program, although factors causing fruit drop must be eliminated, especially on poor soils.

Given these yields and strong market acceptance where it is well known, there are good prospects for producing good quality fruit for local and regional markets. Market saturation levels are, however, unknown. Excessive supply would lower prices from the current US$ 0.50-1.00 per kilogram range for best quality fruit. This might increase consumption somewhat, but could lower farm income, unless alternative uses are developed.

Fruit for Animal Ration

This is the major alternative use currently being studied. Because it is dried, fruit can partially or completely substitute for maize, sorghum or wheat. The pejibaye may be considered a tree cereal. This is especially true for animal rations, where pejibaye flour can substitute for part of the maize base that is generally used. Costa Rican researchers are currently leading this effort and Murillo and Zumbado (1986) have reviewed recent work with chicken feeds. Their team has studied pejibaye preparation (autoclaving, extrusion, sun drying, etc.) and maize substitution levels (using isoproteic preparations) for starter and primary rations for layers, pullets and broilers.

Soto (1983) studied pejibaye meal preparation for broiler starter ration and compared raw, sun-dried and autoclaved pejibaye to substitute for maize at different levels (0-100%) in the diet. Fifty percent substitution with autoclaved meal gave better weight gain and food consumption than the control diet. Raw and sun-dried meal gave results lower than the control at all levels.

Zumbado and Murillo (1984) found that whole ground fruit, including peel and seed, had somewhat more True Metabolizable Energy than did yellow maize. They suggested that this was due principally to the higher oil content of the pejibaye that was used (9% dry wt vs 4% dry wt for maize). They found that the total protein and essential amino acid composition of their fruit (Table 15.4) was inferior to that of maize and, therefore, could not be considered as a protein source.

Cooz (1984) studied extruded pejibaye meal to substitute for maize in starter ration for pullets, using second quality fruit, dried in a solar drier, and extruded at 150°C. From 25 to 75% substitution gave weight gains, food consumption and food conversion rates equivalent to, or slightly better than the control diet, while 100% substitution was only slightly inferior. The 25 and 50% substitution levels were cheaper per kilogram of weight than the control diet.

Loynaz (1985) studied extrusion temperatures (93°, 124° and 168°C) and substitution rates for broiler starter ration. Loynaz found that extrusion at 168°C and 34% substitution gave the best results for all three factors studied during the first 16 days. During the next 16 days the 124° and 168°C extrusion temperatures at 34% were better than the control, while 67% substitution was equivalent. These results show that very young birds do not tolerate pejibaye meal as well as older ones, as was also shown by Arkcoll and Aguiar (1984), but that after two weeks the pejibaye meal gave better results at moderate subsitution rates.

Facuseh (1986) studied temperature treatments, storage period and energy supplements with pejibaye meal at 50% substitution of maize for broiler starter ration. While he did not detect statistically significant differences among the treatments, there was a tendency for the heat-treated pejibaye meal to give better results, both with and without extra energy. Storage period did not affect acceptability of the meal.

Espinoza (1986) studied heat treatments and substitution levels for layers. He found that 30 and 60% substitution of heat-treated pejibaye gave statistically equivalent food consumption, egg production, egg weight and food conversion into eggs as the control diet, while giving better yolk color. Both these substitution levels were cheaper than the control diet.

These results show that pejibaye can be used as the principal energy source in primary rations, but should be used less intensively in starter rations. Heat treatment is essential, especially for starter rations, to deactivate the trypsin inhibitor. Heat treatment by extrusion is cheaper than other methods tested and is recommended. Moderate levels (30-60%

substitution of maize) also allows the production of a cheaper meal, both for starter and for primary rations, at Costa Rican costs for maize and second quality pejibaye.

Sangil (1985) studied chemical and biological treatments to deactivate the inhibitor in pejibaye meal. Saprophytic fungi (unidentified) were allowed to partially decompose the meal that was then used for pigs and rats. The other treatments were temperature and acetic acid (at 1 and 2%). Raw, partially decomposed meal, with 1 or 2% acetic acid, gave better results than the sorghum (*Sorghum vulgare* Pers.) based control diet, suggesting that fungal action deactivated the inhibitor.

In Brazil, Araujo (personal communication) reports that silaged pejibaye has given excellent results with pigs, although no experimental data are available. Silaging would be an excellent way to store pejibaye fruit for animal feeds and may give the additional benefit of deactivating the inhibitor. This method would be good for pigs and cattle, but less effective for chickens that must have dried meal.

Potential yields for animal ration should be similar to those for fresh fruit (previous section), especially if used in silage. For use in chicken feeds, the fruit would have to be dried, thus reducing the final product to 50% of the initial yield. A breeding program for fruit for animal ration would select starchy, low (10%) oil fruit from the Putumayo macrocarpa landrace for example, and should yield in excess of 25-30 MT/ha/yr of fresh bunches (Clement 1988).

In many areas of the humid tropics cereals do not yield well. For example, on the nutrient poor oxisols near Manaus, maize yields 800-1000 kg/ha/yr. This suggests that pejibaye could rapidly develop a market as a component of animal rations, if it can be produced cheaper than imported maize. Tracy (1985) was able to obtain a small profit when using sun-power to dry second quality pejibaye for animal ration in a rainy climate. A more efficient, low-cost drying method should improve profits. A 300 ha plantation in Costa Rica will soon provide information on the economics of pejibaye for fresh fruit and animal ration.

Fruit for Flour and Meal

Among Calvo's (1981) 40 recipes are several for making breads, cakes and other pastries. Tracy (1986) tested pejibaye flour, mixed with wheat flour, for bread in Costa Rica. Ninety percent wheat and 10% pejibaye gave bread dough with excellent baking characteristics, slightly less protein, more energy (from the oil) and more vitamin A (beta-carotene). Eighty-five percent wheat and 15% pejibaye gave a slightly heavier dough, similar to natural whole wheat bread. Both were acceptable to a small group of consumers. This product might develop well in the natural foods market.

Tracy (1986) also tested several cake recipes with good results, both in the kitchen and on the market. Pejibaye flour must be mixed with wheat flour to make a good cake in Costa Rica. In the Manaus region, however, some pejibayes can be used pure, with excellent results. This may be due to the higher oil content.

Clement and Mora Urpi (1987) point out that pejibaye meal is quite similar to yellow cassava or maize meal and could substitute for these in many areas, with nutritional advantages over the cassava flour. This was one of the alternative products developed by the Amerindians, who also domesticated the Putumayo and Vaupes macrocarpa landraces (Clement 1988). These landraces have extremely high starch levels and low oil levels in large fruit, excellent for making flour and meal that stores well for long periods. This meal can also be fermented to make 'chicha', a beer-gruel with good flavor and nutritional qualities. Flours can also be extruded to make a variety of attractive snacks that might find a large market (Arkcoll, personal communication).

As Tracy (1985) pointed out, for regions that do not produce bread cereals like the humid tropics, even 10% substitution of wheat can have a favorable effect on the local balance of payments, by reducing imports. Yields of 10-12 MT/ha/yr of dry flour are possible, if pests and diseases are controlled (Clement and Mora Urpi 1987) and would probably be an economic proposition. Piedrahita and Velez (1982) found 30% unacceptable fruit quality levels. Since this flour would go for human consumption and must be of high quality, phytosanitary quality control would be extremely important.

Fruit for Oil

Arkcoll and Aguiar (1984) were the first to point out pejibaye's oil potential. By searching for high oil fruit, these authors eventually found fruit with 62% oil in the dry mesocarp. Clement and Arkcoll (1985) later pointed out that the oilier fruited plants appear to be more frequent in the more primitive (i.e., less selected) populations, especially in the microcarpa group. This seems to be due to Amerindian selection for starch, which, being negatively correlated with oil, means that the more selected landraces have low oil levels in the mesocarp.

Table 15.5 shows that the unsaturated fatty acids are more common in pejibaye mesocarp oil than in oil palm, although the calaue (*Elaeis oleifera* (H.B.K.) Cortes) has similar oils and is now being introgressed with oil palm to improve oil quality in the species (Hartley 1977). The fact remains, however, that unsaturated fats have a good market value at present, and an alternative oil palm would be worth developing because of the increasing phytosanitary problems in oil palm, especially in South America (Hartley 1977).

As pointed out by Arkcoll and Aguiar (1984) and emphasized by Clement and Arkcoll (1985), most pejibayes have oil separation problems when pressure-extracted. The oil, starch and water form an emulsion that must be solvent extracted. There are plants that have good separation characteristics, but using only these would severely reduce the genetic base of any improvement program. The oiliest pejibayes are relatively dry (Arkcoll, personal communication) so that separation may not be a severe problem with improved materials. Research into the nature of the oil/starch complex is important for future industrialization of the pejibaye as an oil crop, however.

Clement and Arkcoll (1985) suggest that yields of 2-3 MT/ha/yr of oil are immediately feasible with in vitro cloning and could easily be raised to 5 or more tons in an improvement program. They suggest that another 2-3 MT/ha/yr of protein rich byproduct may be expected and that this may help defray extraction costs. The germplasm is available (Clement and Arkcoll 1985) and trial crosses have been made, but long-term financing is required for this breeding program. The problem of fruit drop must also be resolved before this product can become commercially viable.

Palmito

Palmito, or heart of palm, is the only product for which pejibaye is currently used by agrobusiness. There are more than 2000 ha planted in Costa Rica (Clement and Mora Urpi 1987) and the market appears to be firm. Pejibaye must compete for market share with the palmito from the agar palm (*Euterpe oleracea* Mart.), which is extracted from enormous natural populations in the estuary of the Amazon River. The agrobusinesses producing palmito of acai have little overhead compared to pejibaye, because they are essentially extractivist businesses. Quality control varies considerably from company to company and their intensive exploitation is devastating the natural populations. Both of these factors open room for pejibaye in the market, as quality control has been good and plantations are managed rather than devastated. Recent information (Mora Urpi, personal communication) suggests that this may be changing, with pejibaye gaining a competitive edge in the U.S. market because of quality control. The Costa Ricans are actively expanding their plantations at this time, after a period of stagnation.

Processing technology for the pejibaye palmito has been developed in Costa Rica and in Brazil. The most detailed studies are by Ferreira *et al.* (1982A,B) at the Instituto de Tecnologia de Alimentos (ITAL), in Campinas, Sao Paulo, Brazil. Ferreira (personal communication) has observed that the Costa Rican palmito may be stored without the salt solution changing appearance, while the Brazilian palmito becomes turbid within a short time. Research is underway to identify the causes of this turbidity, so as to be able to select against it in an improvement program. Clement *et al.* (1988) have recently reviewed the research results and necessities with respect to palmito of pejibaye and suggest a numerical ideotype more detailed than Clement's (1988).

Most pejibayes are spiny, both on the trunk and on the leaf petiole and rachis. Several spineless populations have been found in western Amazonia, especially around Yurimaguas, Peru (Clement *et al.* 1988). Many other western Amazonian populations show variable frequencies of spineless or reduced spinyness. There is also a spineless population in Costa Rica. This germplasm will form the genetic base for improvement programs currently planned in Brazil and Costa Rica.

Moreira Gomes (1983) reported yields of 1.2 MT/ha of market quality palmito on oxisols in Manaus at the first harvest, falling to 900 kg/ha at the second harvest. This decline is probably due to management of the plantation, rather than biological factors, as it has not been reported from Costa Rica. Zamora (1985) reported 3 MT/ha/yr of field harvested palmito in Costa Rica, of which 20 to 30% is market quality. Zamora (1985) also reported on a density trial whose highest yields were above 3.5 MT/ha/yr. With selected germplasm and good agronomic practices it should be possible to attain nearly 2 MT/ha/yr of market quality palmito.

Mora Urpi (personal communication) reports that Costa Rican palmito plantation managers are ignoring these research results in their rush to attend the newly expanded market and are managing three or four basal shoots on each plant, rather than only one or two as suggested by Zamora (1985), thus producing in excess of 10,000 stems/ha. How many of these are cut each year is still in doubt because of competition among plants, but yields are probably well in excess of 2 MT/ha/yr.

Both the apical and basal residue from palmito extraction have some potential use and require more intensive research to find new uses. The basal residue, just below the apical meristem, is a very tender trunk. This is fibrous, but not excessively so, and has a crispy texture. This could be thin-sliced (transversally) and deep fried to make chips, for example. The apical residue is slightly fibrous leaf and petiole material. In Peru the palmito is used as a fresh salad by separating the leaflets from the rachis, giving long, talharim-like strips that make a delightul presentation and have excellent flavor. In areas of production, this preparation could be introduced allowing a recover value from present waste. If the slightly tougher extreme apical leaflets could be softened without becoming flaccid, these could also serve as a vegetable. With the use of both of these residues, palmito costs could be lowered so as to compete more favorably with acai.

THE FUTURE OF PEJIBAYE

In the preceding sections pejibaye composition and potential uses have been examined. The pejibaye appears to have an enormous potential as a plantation crop in many areas or as an agroforestry component in others (Clement 1986B). Why is it not an important crop now?

Part of the answer is that the pejibaye was not spread around the world by the Spanish or Portuguese in the early years after European contact. If it had been, it would have been studied more widely and planted experimentally throughout the humid tropics. At present it is limited to its pre-Columbian distribution, where all of its co-evolved pests and diseases are also present. These may become limiting factors to its expansion as a plantation crop within tropical America, as has happened with rubber and cacao. Its expansion into Africa and Asia, using elite germplasm, should allow it to escape from its pests and to be introduced to new consumers. Since it is so well adapted to the humid tropics it should have no difficulty in adapting to new environments, both as a plantation crop and as an agroforestry component.

Another part of the answer is that the pejibaye is difficult to collect after the first few harvests because its spiny stem grows rapidly in height, attaining 15 to 20 m in 10 years. Work is underway to select smaller plants using both dwarf germplasm and growth parameters as selection criteria (Corley 1983), and spinelessness when desirable, as the germplasm exists.

A third part of the answer is that its fruits have a distinct flavor that is generally acquired rather than instantly desired, so that one must return several times to become a fan. This means that fresh fruit (after cooking) is not an immediately attractive export product, because it would require expert marketing to develop a demand for it. As it is introduced as a crop into other areas of the tropics, this will also be an initial problem but should be overcome without too much difficulty. Human populations in the tropics are more accustomed to exotic flavors than are the populations of the temperate zones but this situation is changing. Another way around this difficulty of acceptance is to develop new uses, which were extensively discussed above. The need for research into food technology and new products is clear and urgent to give this crop a needed push.

At the moment, only the palmito is an important agrobusiness, and only in Costa Rica. There appears to be an ample market for considerable expansion; market saturation levels are currently unknown.

This is not the case for the fruit. At the moment there is only one large (300 ha) plantation for fruit production in the world, in Costa Rica, and it is unknown how the market will react to such a concentration of fresh fruit. The owner of this plantation is also planning to go directly into animal production, so as to avoid market saturation, but new options for processing would probably allow him a better margin.

Although there are many unknowns to be resolved at present, the potential of the pejibaye appears to be enormous. Researchers and development planners around the humid tropics are enthusiastically considering the pejibaye for introduction into their regions. If a few tropical entrepreneurs pickup the pejibaye, the future is bright. Research is necessary to support these people and will require international financing, as Latin American institutions are currently in precarious financial straits.

REFERENCES

ARKCOLL, D. B. and AGUIAR, J. P. L. 1984. Peach palm (*Bactris gasipaes* H.B.K.), a new source of vegetable oil from the wet tropics. J. Sci. Food Agric. 35, 520-526.

CALVO, L. M. 1981. Culinary uses of the chontaduro. Cali (Colombia): Inst. Ciencias del alle Caucano. (Spanish)

CIPRONA 1986. Industrial uses of the pejibaye (*Bactris gasipaes*). Rep. Invest. Centro de Investigaciones de Productos Naturales, Univ. Costa Rica, San Jose. (Spanish)

CLEMENT, C. R. 1986A. Minimum descriptors for the pejibaye (*Bactris gasipaes* H.B.K.) and their phylogenetic implications. Thesis. Univ. Costa Rica, San Jose. (Spanish)

CLEMENT, C. R. 1986B. The pejibaye palm (*Bactris gasipaes* H.B.K.) as an agroforestry component. Agroforestry Sys. 4, 205-219.

CLEMENT, C. R. 1987. The pupunha, a domesticated tree. Cienc. Hoje 5(29) 42-49. (Spanish)

CLEMENT, C. R. 1988. Domestication of the pejibaye palm (*Bactris gasipaes*): past and present. Adv. Econ. Bot. 6, 155-174.

CLEMENT, C. R. and ARKCOLL, D. B. 1985. The *Bactris gasipaes* H.B.K. (Palmae) as an oil producing crop: potential and priority of investigation. In *Informe del Seminario-Taller Sobre Oleaginosas Promisorias*. L. E. Forero P. (Ed.) Bogota (Colombia): Programa Interciencias de Recursos Biologicos. (Spanish)

CLEMENT, C. R. and MORA URPI, J. 1985. Phenotypic variation of peach palm observed in the Amazon basin. In *Final Report: Peach Palm (Bactris gasipaes) H.B.K. germplasm bank*, 92-106. San Jose (Costa Rica): U.S. Agency of International Development.

CLEMENT, C. R. and MORA URPI, J. 1987. The pejibaye (*Bactris gasipaes* H.B.K., Arecaceae): multi-use potential for the lowland humid tropics. J. Econ. Bot. 41, 302-311.

CLEMENT, C. R., CHAVEZ, W. B. and MOREIRA GOMES, J. B. 1988. Consideration about the pupunha (*Bactris gasipaes* H.B.K.) as a source of palmito. In *Anais do Primeiro Encontro de Pesquisadores em Palmito*, 225-247. Curitiba (Brazil): Centro Nacional de Pesquisas Florestais/EMBRAPA. (Portuguese)

COOZ, S. A. 1984. Effect of the substitution of maize by pejibaye flour as a replacement diet for chickens during the initial stages of growth. San Jose (Costa Rica): Thesis, Univ. Costa Rica. (Spanish)

CORLEY, R. H. V. 1983. Potential productivity of tropical perennial crops. Exp. Agric. 19, 217-237.

ESPINOZA, J. A. 1986. Substitution of maize by thermally treated pejibaye flour in the diet for laying hens. San Jose (Costa Rica): Thesis, Univ. Costa Rica. (Spanish)

FACUSEH, J. E. 1986. Effect of storage time, thermal treatment and energy supplementation of pejibaye flour on the diet for broiling chickens. San Jose (Costa Rica): Thesis, Univ. Costa Rica. (Spanish)

FERREIRA, V. L. P. *et al.* 1982A. Comparison between some palmitos of *Guilielma speciosa* and *Euterpe edulis*. I. Physical, organoleptic and biochemical evaluation. Col. Inst. Tecnol. Aliment. 12, 255-272. (Portuguese)

FERREIRA, V. L. P. *et al.* 1982B. Comparison between some palmitos of *Guilielma speciosa* and *Euterpe edulis*. II. Physical and chemical evaluation. Col. Inst. Tecnol. Aliment. 12, 273-282. (Portuguese)

FONSECA, E. T. 1927. Brazilian vegetable oils. Campinas (Brazil): Inst. Tecnol. Aliment. (Portuguese)

HAMMOND, E. G., PAN, W. P. and MORA URPI, J. 1982. Fatty acid composition and glyceride structure of the mesocarp and kernel oils of the pejibaye palm (*Bactris gasipaes* H.B.K.). Rev. Biol. Trop. 30, 91-93.

HARTLEY, C. W. S. 1977. The Oil Palm. Longman, London.

I.N.N. 1959. Table of composition for Colombian foods. Bogota: Inst. Nacional de Nutricion. (Spanish)

JOHANNESSEN, C. L. 1967. Pejibaye palm: physical and chemical analysis of the fruit J. Econ. Bot. 21, 371-378.

LEUNG, W. T. 1961. Table of composition of foods for use in Latin America. Ciudad de Guatemala: INCAP. (Spanish)

LOYNAZ, B. A. 1985. Utilization of pejibaye flour extruded under different temperatures for the first feed to fatten chickens. San Jose (Costa Rica): Thesis, Univ. Costa Rica. (Spanish)

MORA URPI, J. 1984.. The pejibaye (*Bactris gasipaes* H.B.K.): Origin, flower biology and agronomic management. In *Palmeras Poco Utilizadas de America Tropical*, 118-160. Turrialba (Costa Rica): FAO/Centro Agronomico Tropical de Investigacon y Ensenanza. (Spanish)

MORA URPI, J. and CLEMENT, C. R. 1985. Races and populations of peach palm found in the Amazon basin. In *Final Report: Peach Palm (Bactris gasipaes H.B.K.) Germplasm Bank*, 107-141. San Jose (Costa Rica): U.S. Agency of International Development.

MOREIRA GOMES, J. B. 1983. Evaluation of the agronomic characteristics of pupunheira (*Bactris gasipaes* H.B.K.) for the production of palmito. Manaus (Brazil): Monograph, Univ. Amazonas. (Portuguese)

MOREIRA GOMES, J. B., CLEMENT, C. R., FERREIRA, S. A. N., and FONSECA, C. E. L. 1987. Phenotype variation of pupunha (*Bactris gasipaes* H.B.K.) selections of the population from Fonte Boa, Amazonas. Rep. Invest. Instituto Nacional de Pesquisas da Amazonia. Manaus (Brazil). (Portuguese)

MURILLO, R. M., and ZUMBADO A., M. 1986. Chemical composition and nutritive value of pejibaye flour in the nutrition of birds. Rep. Invest. Univ. Costa Rica/CONICIT, San Jose. (Spanish)

MURILLO, R. M. *et al.* 1983. Preliminary study on inhibitory factors for proteolytic enzymes in pejibaye (*Bactris gasipaes*) flour. Rev. Biol. Trop. 31, 227-231. (Spanish)

NAS. 1975. Underexploited Tropical Plants With Promising Economic Value. Nat. Acad. Sci., Washington, D.C.

NOIRET, J. M. and WUIDART, W. 1976. Possibilities to improve the fatty acid composition of palm oil. Results and perspectives. Oleagineux 31, 465-474.

PIEDRAHITA, G. and VELEZ, C. A. 1982. Methods for production and preservation of the flour obtained from the fruit of the chontaduro palm (*Bactris gasipaes* H.B.K.). Rep. Invest. Univ. del Valle, Cali, Colombia. (Spanish)

POPENOE, W. and JIMENEZ, O. 1921. The pejibaye, a neglected food plant of tropical Am. J. Hered. 12, 154-166.

SANGIL, J. 1985. Evaluation of the effects of different chemical and biological procedures on the inhibiting activity of proteolytic enzymes present in pejibaye of flour (*Bactris gasipaes*) used in animal nutrition. San Jose (Costa Rica): Thesis, Univ. Costa Rica. (Spanish)

SILVA, W. G. and AMELOTTI, G. 1983. Composition of the fatty substances from the fruit of *Guilielma speciosa* (Pupunha). Rev. Ital. Sostanza Grasse 60, 767-770.

SOTO, T. S. 1983. Use of pejibaye flour in diets for broiler chickens. San Jose (Costa Rica): Thesis, Univ. Costa Rica. (Spanish)

TRACY, M. D. 1985. The pejibaye fruit: problems and prospects for its development in Costa Rica. Thesis, Univ. Texas at Austin.

TRACY, M. D. 1986. Pejibaye flour, a hopeful option. Bul. Retadar 23, 3. (Spanish)

ZAMORA, F. C. 1985. Crop densities of pejibaye for palmito with single stalks. *In* Diversification Agricola, 75-78. San Jose (Costa Rica): Association Bananera Nacional. (Spanish)

ZAPATA, A. 1972. Pejibaye palm from the Pacific coast of Colombia (a detailed chemical analysis). J. Econ. Bot. 26, 156-159.

ZUMBADO, A., M. and MURILLO, R., M. 1984. Composition and nutritive value of pejibaye (*Bactris gasipaes*) in animal feeds. Rev. Biol. Trop. 32, 51-56.

16

Granadilla

Freddy Leal

The granadilla is a fruit native to the Central and South Americas. It belongs in the family Passifloraceae and in the genus *Passiflora*, which includes about 12 genera and 500 species of herbaceous and woody plants, usually climbing by tendrils. The giant granadilla (*P. guadrangularis* L.) is one of the more tropical species of the group and the one bearing the largest fruit. The sweet granadilla (*P. ligularis* Juss.) ranges from central Mexico through Central America and western South America. The giant and sweet granadillas are often merely called granadilla. Although belonging to the genus *Passiflora*, the granadillas are distinguishable from the commercially important passion fruits, namely, purple passion fruit (*P. edulis* Sims) and yellow passion fruit (*P. edulis* f. *flavicarpa* Deg.) (Morton 1987). Chemical and nutritional compositions of passion fruits have been extensively reviewed by Chan (1980) and will not be covered in this chapter.

Granadilla is also known by the names badea, parcha granadina, granadilla real, pasionaria, barbadine, giant granadilla, maracuja-acu, maracuja-vacu, maracuja-de-garapa, maracuja do para, maracuja-melao, maracuja-mamao, maracuja cascudo, rata pohol, timum belanda, sweet granadilla, parchita amarilla, and granadilla comun. The granadilla was brought into cultivation about the 18th century and can be found growing throughout the tropics, mostly in home gardens.

CHARACTERISTICS

Giant Granadilla

The giant granadilla is a strong vine with thick, succulent square shoots carrying large, entire, soft light green leaves about 10-20 cm long and 8-15 cm wide. The fleshy tuberous root is usually regarded as poisonous, although they are said to be eaten by Jamaicans as a substitute for yam (Purseglove 1968). The large, fragrant flowers are about 8-12 cm in diameter with three green basal bracts, with five fleshy sepals of green color on the outside and white, pink or purple inside. The petals are red internally and pink or white externally. The crown possesses 5 or 6 whorls of appendages.

The fruit is oblong-ovoid in shape, 10-30 cm in length and 12-15 cm wide, pale green in color, and becoming lighter colored when ripe. Its epicarp is thin and fragile. The mesocarp is 2-3 cm thick, fleshy, white or pink in color and insipid. The mesocarp encloses a cavity partially filled with juice, masses of whitish, yellowish, sweet-acid arils (referred to as pulp) and large, purplish-brown seeds (Leon 1968). Green, unripe fruits may be boiled and eaten as a vegetable (Kennard and Winters 1960; Purseglove 1968; Martin and Nakasone 1970).

There is great variability in size and weight when fruits are ripe; varying from 0.4 to 2.6 kg per fruit. The number of fruits per plant is from 16 to 50 (Haddad and Figueroa 1972). Usually the giant granadilla is propagated by seeds, but could be propagated by natural cuttings of 30-40 cm in length (Araque 1963). According to Araque (1963), plants propagated from seeds gave better yields and survived longer than from cuttings.

The giant granadilla bears fruit one year after planting and sometimes even after 9 months. It gives two crops per year; each plant produces from 16 to 50 fruits with a weight average of 1-1/2 to 2 kg per fruit. If the granadilla is planted at 3 x 6 m, 555 plants may be grown per hectare with a total yield of 20-25 tons. Fruits must be picked ripe as evidenced by changes from green to a pale yellow color. The giant granadilla must be handled with care because it is tender, soft and easily bruised.

Sweet Granadilla

The sweet granadilla is somewhat similar to the passion fruit (*P. edulis*). The vine is a strong grower; woody at the base and climbing by tendrils. Its leaves, sometimes lobed, are predominately heart-shaped and about 8-20 cm long. Flowers, usually 2 to a node, impart a sweet and musky odor. Petals and sepals are greenish-white, and the corona is white with zones of red-purple (Popenoe 1974). The fruit is larger than the passion fruit and is ellipsoid, 5-8 cm long and 4-5 cm in diameter. External color of the fruit is green, yellow or orange, and sometimes purplish. The pulp is white-yellow, mucilaginous, juicy and possesses a distinct aromatic flavor (Morton 1987). Seeds are black, flat and numerous.

FOOD USES

Granadilla pulp may be eaten from the shell or strained for making fruit drinks or nectars. When used to prepare fruit drinks, a portion of the pulp is taken and water is added to the seed mass. Arils are broken and the aromatic, acid juice is incorporated into the pulp. In this way, an agreeable refreshment is prepared that could be drunk or used in ice creams or milk-shakes. Another drink is prepared by mixing the fruit pulp with pieces of pineapple or peaches or papaya. To the granadilla juice is added crushed ice, sugar and dark wine to make a refreshing "sangaree".

COMPOSITION

The proximate composition of ripe, edible granadilla fruit is noted in Table 16.1, and the mineral and vitamin contents in Table 16.2.

Lipids

The lipid (fat) content of granadilla juice and flesh varies within a range of 0.1-2.6%, whereas the lipid value of arils and seeds ranges from about 1.3-3.2% (Munsell *et al.* 1950; Leung and Flores 1961; Morton 1987). The range in lipid values of granadilla is noted in Table 16.1. No report has been found related to the seed oil of granadilla.

TABLE 16.1. COMPOSITION (%), PH AND FOOD ENERGY (CAL) PER 100 G OF EDIBLE GRANADILLA

	Munsell *et al.* (1950)		Gongora-Lopez and Young-Lopez (1956)		Leung and Flores	Choucair	Purseglove
Constituent	A[1]	B[2]	A[1]	B[2]	(1961)[1]	(1962)[1]	(1968)[1,3]
Water	78.4	79.1	88	86	78.4	87.2	93.7
Proteins	0.3[4]	0.3[4]	0.9	1.1	1.9	0.9	0.7
Lipids	1.3	2.6	0.2	0.1	1.3	0.2	0.2
Carbohydrates	—	—	10.1	11.6	12.0[5]	10.1	4.3
Fiber	3.6	4.2	0.0	0.3	5.6	0.0	0.7
Ash	0.8	1.1	0.9	0.9	0.8	0.9	0.4
Food Energy	—	—	41	46	81	—	20

[1]Giant granadilla.
[2]Sweet granadilla.
[3]Flesh only; does not include seeds and arils.
[4]Nitrogen content; determine approximate protein by N x 6.25.
[5]Seeds and arils; value reported by Leung and Flores (1961) for total carbohydrates is 17.6%; that value represents both sugars (12.0%) and fiber (5.6%).

TABLE 16.2. MINERAL AND VITAMIN COMPOSITION OF RIPE GRANADILLA PULP (MG/100 G)

Constituent	Munsel *et al.* (1950)		Gongora-Lopez and Young-Lopez (1956)		Leung and Flores	Morton
	A[1]	B[2]	A[1]	B[2]	(1961)[2]	(1987)[2]
Minerals						
Ca	9.2	5.6	10	7	10	5.6-13.7
P	39.3	56.2	22	30	64	44-78
Fe	2.9	1.6	0.6	0.8	0.9	0.6-1.6
Vitamins						
Vitamin A	0.019	0.005	70[3]	—	0.005	0.0-.035
Thiamine	0.003	—	—	—	—	0.0-.002
Riboflavin	0.120	0.072	0.11	0.10	0.11	0.6-0.13
Niacin	1.897	1.237	2.7	2.0	1.6	1.42-1.81
Ascorbic acid	15.3	10.8	20	20	20	10.8-28.1

[1]Giant granadilla.
[2]Sweet granadilla.
[3]Vitamin A values in international units; 1 I.U. = 0.6 μg beta carotene.

Carbohydrates

The second most abundant components of granadilla fruit, next to water, are the carbohydrates which constitute between 4.3% (juice) to 12.0% (seeds and arils) (Gongora-Lopez and Young-Lopez 1956; Leung and Flores 1961; Purseglove 1968). Of nonsugar carbohydrates, Choucair (1962) reported 0% fiber, whereas Leung and Flores reported 5.6%. Pectin content was reported by Garces-Medina (1968) at about 0.40% (range 0.37-0.44) for the giant granadilla.

Enzymes

Garces-Medina (1968) determined pectinesterase activity in giant granadilla pulp and found a range of 0.36-0.56 (units as pectin-metoxilaso/ml of diluted pulp (1:1)). Heat inactivation of pectinesterase was studied. It was found that 91% inactivation of the enzyme occurred in 3 min at 80°C, whereas 100% inactivation occurred in 5 min at 85°C.

Nutrients

The data of Table 16.2 indicate that the granadilla is a good source of iron and niacin, a fair source of phosphorus and ascorbic acid, and a poor source of calcium, vitamin A, thiamine and riboflavin. Garces-Medina

(1968) reported the giant granadilla to contain an average 17 mg vitamin C per 100 g pulp (range 15-19), while Fonseca *et al.* (1969) reported a range of 12-17 mg/100 g pulp. These vitamin C values are within the range reported by other investigators (see Table 16.2).

FLAVOR AND COLOR

The most desirable characteristic of granadilla is its fragrant aroma and exquisite flavor. The fruit can be handled during processing without much damage to color, flavor and aroma (Rodriguez 1969). Apparently the cream-white pulp is quite stable.

PROCESSING

Harvesting and Peeling

Granadillas should be picked by hand just when mature to avoid the bruising that might occur if the overripe fruit fell from branches. At this moment, the external color is yellowish-green. Ripe fruits are washed to remove dirt, peeled by hand, and washed again (Rodriguez 1969).

Extraction of Pulp

The fruits are cut by hand and the pulp is disintegrated. Seeds enclosed within the juice arils are separated to avoid smashing the seeds. Both arils and seeds are blended together and passed through a 0.69 cm screen.

PREPARATION OF GRANADILLA NECTAR

The mixture is adjusted to 16-18% soluble solids with sugar, to 0.3% acidity with citric acid, and diluted 1:1. This mixture is pasteurized at 90°C for 5 minutes, packed in tin cans and stored for 20 days. Four different samples were prepared varying the acidity, the soluble solids and the dilution ratio. After 20 days, they were evaluated. The panelists selected as "best" the granadilla nectar possessing 0.3% acidity, 17% solids and diluted 1:1. No changes in flavor, aroma and color were noted during the period of storage.

NECTAR QUALITY

After 15-20 days of storage, the nectar quality as reported by Rodrigues (1969) was as follows:

Taste:	Good
pH:	3.45
Acidity as citric acid:	0.3%
Ascorbic acid:	7.4 mg/100 g of nectar
Soluble solids:	17%
% of nectar was 45% structure phase	
Viscosity:	40 cP
Organoleptic properties:	good color, smell, taste, texture and appearance

REFERENCES

ARAQUE, R. 1963. The Granadilla. MAC-IAN-BAP-CBR, No. 4 Serie de Cultivos, Venezuela. (Spanish)

CHAN, H. T. 1980. Passion fruit. *In* Tropical and Subtropical Fruits. S. Nagy and P. E. Shaw (Editors). AVI Publishing Co., Westport, CT.

CHOUCAIR, K. 1962. Columbian Fruit Culture, 2 vols. Bedout Editorial, Medellin, Columbia. (Spanish)

FONSECA, H., NOGUEIRA, J. N. and MARCONDES, A. M. S. 1969. Ascorbic acid and beta carotene contents of Brazilian fruits and vegetables. Arch. Latinoam. Nutr. 19(1) 9-16. (Portuguese)

GARCES-MEDINA, M. 1968. Pectin, pectinesterase and ascorbic acid in pulp of tropical fruits. Arch. Latinoam. Nutr. 18(4) 401-412.

GONGORA-LOPEZ, J. and YOUNG-LOPEZ, N. 1956. Composition Tables of Colombian Foods, Bogota, Colombia. (Spanish)

HADDAD, O. and FIGUEROA, M. 1972. Study of flowering and fruiting in parcha granadina (*Passiflora guadrangularis* L.). Agron. Trop. 22(5) 483-496. (Spanish)

KENNARD, W. and WINTERS, H. F. 1960. Some Fruits and Nuts of the Tropics. USDA, Agric. Res. Serv. Publ. 801.

LEON, J. 1968. Botanical Fundamentals of Tropical Culture. Interramerican Institute of Agricultural Sciences of OEA. San Jose, Costa Rica. (Spanish)

LEUNG, W. T. W. and FLORES, M. 1961. Food Composition Table for Use in Latin America. National Institutes of Health, U.S. Govt. Printing Office, Washington, D.C.

MARTIN, F. and NAKASONE, H. Y. 1970. The edible species of Passiflora. Econ. Bot. 24(3) 333-343.

MORTON, J. F. 1987. Fruits of Warm Climates. Julia F. Morton Publisher, Miami, FL.

MUNSELL, H. E., WILLIAMS, L. O., GUILD, L P., TROESCHER, C. B. and HARRIS, R. S. 1950. Composition of food plants of Central America. IV. El Salvador. Food Res. 15, 263-296.

POPENOE, W. 1974. Manual of Tropical and Subtropical Fruits. Hafner Press, New York.

PURSEGLOVE, J. W. 1968. Tropical Crops: Dicotyledons, Vol. II. John Wiley & Sons, New York.

RODRIGUEZ, A. 1969. Preparation of nectar from parcha granadina (*Passiflora guadrangularis* L.) and its canning. Ph.D. Thesis. Caraca Universidad Central de Venezuela, Caracas. (Spanish)

17

Pomegranate

G. N. Mohan Kumar

The pomegranate, native to Persia (De Candolle 1967), had its first domestication in Iran about 2000 B.C. (Evreinoff 1949). The name, pomegranate, owes its origin to Latin; "pome" meaning apple, and "granate" meaning many seeded and it was, initially called *Malum granatum*, meaning seeded apple.

From the place of its origin, the pomegranate spread to nearby areas, such as the Mediterranean and subsequently reached eastern countries like India and China (Purseglove 1981). Presently, it is cultivated in Spain, Morocco, Egypt, Israel, Iran, Afghanistan, Arabia, Baluchistan, Pakistan, India, Saudi Arabia, U.S.A. (California), Argentina, northern parts of Chile, and to a lesser extent, in Burma and China.

BOTANY

The pomegranate belongs to the family Punicaceae which is represented by only one genus and two species; *Punica granatum* L. and *P. protopunica* Balf. Only *P. granatum* is cultivated for its edible fruits. It is a small tree or a densely branched woody shrub growing to a height of 6 to 7 m when left to itself. Under cultivation, it is maintained as a low headed bush of 2 to 4 m. The leaves are opposite or in whorls of 5 or 6, entire, oblong-lanceolate, short petioled and dark green. Flowers are hermaphrodite, male or intermediate. Sex ratio has an important bearing on productivity. Cultivars producing hermaphrodite flowers are capable of good fruit set and, therefore, are well suited for commercial cultivation while those that produce male or intermediate flowers are only of ornamental value.

The fruit is spherical and thick skinned (Fig. 17.1). Botanically, it is called balusta. This is a special type of fruit developing from an inferior ovary characterized by two whorls of basal carpels lying within the receptacle. With the development of the ovary, the outer carpels become tilted up and superposed. This results in the formation of two layers of chambers with outer or upper ones occupying a parietal position. The pericarp of the fruit is rough and leathery and the inside chambers are made up of thin walled carpels (Fig. 17.2). The testa of the seed is filled with a whitish pulp (aril) acid juice (Dutta 1982).

ADAPTATION AND CULTURAL CONSIDERATIONS

The cultivation of the pomegranate is mainly confined to tropics and subtropics and it grows well in arid and semi-arid climates. The best climatic conditions are found in Iran, Afghanistan and Baluchistan (Singh 1976). Favorable growth takes place where winters are cool and summers are hot. It is evergreen in the tropics and deciduous in the subtropics. It has the abilty to withstand frosty conditions, but below –10°C the hardiness is poor. A temperature of 38°C and dry climate during fruit

FIG. 17.1. POMEGRANATE FRUITS

FIG. 17.2. POMEGRANATE FRUITS IN CROSS SECTION

development produces best quality fruits. Areas with high relative humidity or rain are totally unsuitable for its cultivation as fruits produced under such conditions tend to taste less sweet.

The pomegranate has an unique ability to adapt itself to diverse soil conditions. Soils generally regarded as unsuitable for other fruit crops can be considered for pomegranate. However, it can perform better in well drained, deep loamy or alluvial soils of sufficient depth. It can tolerate alkaline conditions better than any other fruit crop.

The pomegranate is a cross-pollinated crop and seldom breeds true to type. In many parts of India it is propagated through seeds. Low yields, poor quality and hard-seeded fruits are usually associated with plants propagated through seeds. Vegetative propagation by means of hardwood stem cuttings and air-layering are preferred to seed propagation.

The pomegranate is relatively free from the attack of diseases. The fruit borer (*Virachola isocrates*) is a serious pest. Fruit cracking is another limitation in its successful cultivation.

PRESENT IMPORTANCE AND FUTURE PROSPECTS

Data on area and production are not well documented in many producing countries. Available data on area and production indicate that Iran, India and United States are important producers. In the United States, it is grown on 600 ha in California. Iran produces about 40,000 MT annualy, half of which is processed into syrup (Firuzeh and Ludders 1979). The total area under pomegranate cultivation in India is around 1200 ha (Patil and Wavhal 1980). However, the Indian demand is overwhelmingly high and this is met by importing pomegranate from Afghanistan. Because of steady and high yields, excellent table and therapeutic values, good keeping quality, low maintenance cost and drought tolerant ability, India is steadily increasing its acreage for this fruit crop under dry land farming systems.

Potentials for development of cultivars resistant to fruit cracking and fruit borers exist, and such development would enhance the value of the crop. Presently, cultivars selected for soft seeds, higher total soluble solids and richer color, are being made available to producers.

USES

The pomegranate is cultivated for its fruit which can be consumed as dessert or can be processed into juice, grenadine or wine (Hodgson 1917; Chace *et al.* 1930; La Rue 1969). In India, pomegranate is mainly used as a table fruit, while in Iran and United States a part of the produce is processed. The juice is an excellent appetizer. The rind of the fruit contains about 30 percent tannins which can be successfully utilized in tanning leather, either alone or mixed with synthetic tannins (Nasacheva *et al.* 1974). In fact, it was once used for tanning Morocco leather. Pomegranate fruit is also valued for its therapeutic properties. Nadakarni (1976) in 'Indian Materia Medica' has defined its medicinal properties.

COMPOSITION

The juice content of pomegranate fruits accounts for about 45-65% of the whole fruit or 76-85.5% of the arils (Veres 1976). Table 17.1 lists general compositional data of pomegranate juice. Water constitutes the major part of juice and values of 82.9% (Tressler and Josyln 1961) to 85.4% (Paul and Southgate 1978) have been reported. Sugars are the major organic components and values ranging from 9-20 g/100 ml juice have been reported (Table 17.1; also Herrmann 1981). Major sugars are the reducing sugars, namely, glucose and fructose (Lee *et al.* 1974A).

TABLE 17.1. COMPOSTION OF POMEGRANATE JUICE

Component		Reference 1	2	3	4
Total Solids	(%)	—	13.3 -17.2	18.0	18.9
Total sugars	(%)	—	—	14.6	14.9
Reducing	(%)	13.9	9.6 -15.0	—	—
Nonreducing	(%)	—	0.6 - 2.2	—	—
Protein	(%)	0.2	0.2 - 0.5	0.8	0.4
Pectin	(%)	—	0.02- 0.04	—	—
Ash	(%)	0.4	0.2 - 0.3	0.3	0.3
Acids	(%)	1.5	—	0.9	0.7
pH		—	—	3.1	3.6
Vitamin C	(mg %)	—	4 -10	11	12

1 Source: Tressler and Joslyn (1954, 1961).
2 Source: Benk (1985).
3 Source: Herrmann (1981).
4 Source: Seppi and Franciosi (1980).

Benk (1985) reported total reducing sugar values of 9.6-15.0 g/100 ml, and Seppi and Franciosi (1980) reported 7.2 g glucose and 7.9 g fructose per 100 ml of Italian pomegranate juice. Sucrose (nonreducing) was not detected by Tressler and Joslyn (1961), whereas Benk (1985) reported a range of 0.6-2.2 g per 100 ml, and Seppi and Franciosi (1980) reported less than 1 g per 100 ml.

Organic acids are the second most abundant components of juice and amounts ranging from 0.7-1.5% have been reported (Table 17.1). Seppi and Franciosi (1980) reported (per 100 ml) citric acid (0.5 g), isocitric acid (0.003 g) and malic acid (0.11 g). Small amounts of oxalic and succinic acids have also been reported.

Minor amounts of protein and pectin are present in juice. Minerals reported by Paul and Southgate (1978) and Seppi and Franciosi (1980) in mg per 100 ml juice include potassium (150-200), phosphorus (8-11), magnesium (3-4), calcium (3) and iron (0.2). Vitamin C is the major vitamin and generally ranges from about 4-12 mg/100 ml (Table 17.1); however, values as high as 44 mg/100 ml for Yugoslavian pomegranate juice have been cited. Values for thiamine, riboflavin and niacin are similar to those reported for the total edible portion of the fruit (Table 17.2).

Table 17.2 lists the composition of pomegranate fruits as per 100 g edible portion. Water is the major component and shows a range of about 76-82%. Total soluble solids (includes sugars, acids, minerals, vitamins but not fiber) have been reported in a range of 9-18.7% (Kader *et al.* 1984; Sayed *et al.* 1985). Total carbohydrates (includes soluble sugars

TABLE 17.2. COMPOSITION OF POMEGRANATE FRUITS (PER 100 G EDIBLE PORTION)

Component	Reference 1	2	3	4	5	6
Calories (per 100 g)	66.0	67.0	63.0	—	78.0	72.0
			%			
Water	76.2	81.8	82.3	78.0	78.1	80.0
Protein	1.9	0.8	0.5	1.6	1.6	1.0
Fat	0.2	0.7	0.3	0.1	0.3	0.6
Red. sugars	13.5	—	—	—	—	—
Total carbohydrates	—	16.2	16.4	19.6	19.6	17.7
Fiber	6.4	2.0	0.2	5.1	4.5	1.1
Ash	0.5	0.5	0.5	0.7	0.4	0.7
			mg/100 g			
Ca	14.0	10.0	3.0	10.0	12.0	13.0
P	—	34.0	8.0	70.0	37.0	23.0
Fe	0.5	0.6	0.3	0.3	1.0	0.7
Na	4.0	—	3.0	—	—	7.0
K	210.0	—	259.0	133.0	—	379.0
Carotene	0.09	0.07	0.03	0.06	—	0.07
Thiamine	0.01	0.07	0.03	0.06	—	0.07
Riboflavin	0.02	0.03	0.03	0.01	—	0.01
Niacin	0.4	0.9	0.3	0.3	—	0.3
Ascorbic acid	14.0	8.0	4.0	14.0	—	7.0

1 Source: Wills *et al.* (1986).
2 Source: Wu Leung and Flores (1961).
3 Source: Watt and Merrill (1963).
4 Source: Anon. (1948-76).
5 Source: Wu Leung *et al.* (1968).
6 Source: Wu Leung *et al.* (1972).

and insoluble fiber) range from about 16-20% (Table 17.2; also Morton 1987). Reducing sugars are the primary soluble components and generally vary from about 9-15% (Table 17.2; also Siddappa and Bhatia 1954; Sayed *et al.* 1985). Sucrose has been reported in pomegranate fruit at nil to negligible levels, namely 0% (Wills *et al.* 1986) and 0.18% (Siddappa and Bhatia 1954). Benk (1985), however, reported sucrose values of 0.6-2.2 g/100 ml for juice. These sucrose differences might be due to geographical growing area, horticultural practices and/or different varieties. Fiber (insoluble carbohydrate) is high and shows a range of about 1-6%.

Titratable acidity is found in a range of 0.3-1.6% (Herrmann 1981; Kader *et al.* 1984; Sayed *et al.* 1985). Proteins, fats and most vitamins are found at relatively low levels. Vitamin C is the major vitamin and is found in a range of about 4-14 mg/100 g edible fruit.

In addition to the constituents presented in Tables 17.1, 17.2 and 17.3, the fruits also contain pigments responsible for color, as well as tannins (polyphenols) that impart a bitter taste. Presence of tannins in higher quantities is, therefore, undesirable. The arils and the fruit peel are rich in anthocyanins. These include pelargonidin-3-glucoside and pelargonidin-3,5-diglucoside which are found in small amounts in arils and in larger concentrations in the peel. In addition, the peel also contains cyanidin-3-glucoside and cyanidin-3,5-diglucoside in appreciable amounts (Du *et al.* 1975). Kriventsov and Arendt (1985) also examined anthocyanins in the arils and found them to be rich in cyanidin-3,5-diglucoside, delphinidin-3-glucoside and delphinidin-3,5-diglucoside. Among the 16 cultivars they studied, cv. Wonderful had the highest anthocyanin content (600 to 765 mg per 100 g).

Lee *et al.* (1974B) found both depside and flavonol type polyphenols in the fruit. Coumaryl quinic acid, chlorogenic acid isomer, neochlorogenic acid isomer and neochlorogenic acid are depside type and D-gallocatechin, L-epicatechin and L-catechin are flavonol type polyphenols.

The seeds contain about 15% oil having high refractive index as well as high iodine value and low melting point. The saturated fatty acids account for 8%, monounsaturated 10%, diunsaturated 10% and conjugated fatty acids account for 7% (probably punicic acid). The oil is reported to have potential for industrial use (El-Shaarawy and Nahapetian 1983).

TABLE 17.3. COMPOSITION OF SOME VARIETIES OF POMEGRANATE FRUITS

Cultivar	TSS[1] (°Brix)	Acidity (as % citric acid)	Total sugars (%)	Ascorbic acid (mg/100 g)
Anar Shirin-e-Mohamad Ali	12.00	2.734	9.70	10.23
Bedana	13.00	1.796	8.33	13.40
Dholka	13.00	0.512	9.62	10.20
Kazil Anar	10.50	2.820	8.27	9.78
Kandhari	13.00	1.120	8.62	8.08
Nabha	11.50	0.496	9.26	12.24
Russian seedling	12.50	1.892	7.81	12.10
Shirin-Anar	12.00	3.079	6.75	12.60
Srinagar-Special	14.00	0.416	10.40	21.50
Surakh Anar	11.50	2.063	6.90	9.80

Source: Misra *et al.* (1983).
[1]Total soluble solids.

CULTIVARS

By and large, pomegranate cultivars are named after the place where they are extensively cultivated. Most of the pomegranate cultivars have been selected for high yields and fruit qualities such as soft seededness, high total soluble solids (TSS), better color of the fruit and arils. Fruits with hard seeds possess poor edible quality and, therefore, cultivars producing soft seeded fruits are preferred. Hard seeded fruits, having higher juice content and intense red color can be utilized for processing.

List of Important Cultivars

Alandi.—It is named after a village in Maharastra, India, where it was first selected. Fruits are medium in size with blood red or deep pink arils, sweet, but slightly acidic pulp. Seeds are hard.

Bassein Seedless.—Fruits are medium sized with greenish yellow skin. Arils pinkish white in color and have soft seeds.

Dholka.—It is named after the place in Gujarath, India, where it is cultivated on a large scale. Fruits are large, rind greenish white, arils fleshy, white or pinkish-white, sweet. Seeds soft; medium copper.

Early Wonderful.—A California bud mutant of the cv. Wonderful, an important cultivar of California. Fruits large, 7-8.2 cm diam, round-oblate in shape, skin deep red, attractive, smooth and glossy. Flesh firm. Ripening in mid-September, two weeks before Wonderful. Tree medium sized, spreading, drooping. Tolerant to most diseases and insects (Brooks and Olmo 1982).

Ganesh (GBG-1).—This is an important commercial variety of Maharashtra in India. It is an excellent, high yielding selection from the cv. Dholka. Fruits are medium sized with pink arils, soft seeded and high in total soluble solids (TSS).

GKVK-1.—It is a seedling progeny selected at the Division of Horticulture, University of Agricultural Sciences, Bangalore, India, and released in 1985. The fruits are medium to large in size, have attractively colored arils. High in TSS and pulp content. Seeds are soft.

Kabul.—This variety is grown in Afghanistan. Fruits large, rind thick, deep in color and has a tinge of pale yellow. Arils blood red in color and slightly bitter in taste.

Kandhari.—Fruits large, rind deep red, arils fleshy and blood red. Juice slightly acidic, but sweet. Hard seeded.

Muskat Red.—Fruits small to medium in size, thick rind, arils fleshy and sweet. Seeds are not very hard.

Paper Shell.—Fruits medium in size, rind thin. Arils fleshy and deep pink. Seeds soft. A prolific bearer.

Rasel Baghi.—An early variety from Palestine, bearing sweet fruits.

Roman Chokab (black pomegranate).—It is grown in Baghdad. Fruits dark in color with large seeds; sweet-sour in taste.

Selimi.—A variety of Baghdad. It bears very big fruits weighing up to 1 kg. Fruits are juicy and sweet.

Spanish Ruby.—Fruits small to medium in size, soft seeded, thick skinned. A prolific bearer.

Wellissi.—Soft seeded fruits with a tendency for early bearing. Fruits are of very good quality.

Wonderful.—(Fig. 17.1 and 17.2). It originated as a cutting in Florida and was propagated in California in 1896 (Morton 1987). Wonderful is an important cultivar of California. It produces large fruits sometimes up to 12.7 cm diam and bright crimson red in color. Pulp highly colored, very juicy with fine flavor. It is early in bearing and a good shipper (Bailey 1925).

Yercaud-1 (YCD-1).—A clonal selection from the Horticultural Research Station, Yercaud, Tamil Nadu, India. Fruits are medium sized, aril deep purple in color, soft seeded. Fruits weigh 200-450 g. Mean yield per plant is 120 fruits. TSS 9.1 to 15.2, acidity 0.7% and reducing sugars 12.2% (Sayed *et al.* 1985).

HARVESTING AND HANDLING

The pomegranate fruits are generally ready for harvest 5 to 7 months after flowering. The fruits are clipped from tree using clippers. It is important to harvest properly matured fruits since pomegranates do not ripen off the tree. The practice of picking early in the season to avoid fruit cracking is known to result in poor quality fruits (Hays 1953). In California, harvesting begins in the middle of September and continues until mid October (Kader *et al.* 1984).

Harvest Indices

The development of an intense red color on the fruit skin generally serves as harvest index in many cultivars. At this stage, the arils inside would have developed a bright red color. Some of the cultivars such as Bassein Seedless and GKVK-1, however, do not develop an intense red color on the fruit skin. In such cultivars, a change in color from green to yellow is suggestive of proper maturity. Experienced growers decide maturity of the fruits based on the cracking sound of the grains inside the fruit when the fruit is pressed slightly. Harvest indices of minimum maturity for California grown Wonderful pomegranates include titratable acidity less than 1.85% acid and red color of the juice which should be equal to or darker than Munsell color chart 5 R-5/12 (Anon 1975).

Handling

Nailed wooden lug boxes similar to the ones used for persimmons, table grapes and peaches are used for handling pomegranates in United States (Ryall and Pentzer 1974). In India, bamboo baskets or gunny bags are commonly used for handling and transport of pomegranates.

STORAGE

Response Of Pomegranate Fruit To Storage Conditions

Mukherjee (1958) reported that among all tropical and subtropical fruits, pomegranate has a long shelf life. However, the presence of numerous pores in its thick peel permits free movement of moisture from the fruit leading to loss in fresh weight. Storage conditions such as temperature and relative humidity not only influence such losses but also affect the postharvest behavior of pomegranate fruits.

Work done on postharvest physiology and handling of pomegranates is scarce and limited to a single cultivar, Wonderful. The following is based on the studies of Elyatem and Kader (1984) and Kader *et al.* (1984) on postharvest behavior of pomegranate fruits.

Response To Temperature And Relative Humidity

Loss Of Fruit Weight.—Temperature and relative humidity of the storage atmosphere generally dictates the rate of moisture loss from the fruit. A storage temperature of 5° C and relative humidity of 95% reduced the weight loss to 1.4% as against 6.1% loss in fruits stored at the same temperature but at 75% humidity. Fruits that lose more than 5% weight appear shriveled and such fruits are difficult to market. Though pomegranate fruits can safely be stored at 5° C for 2 weeks, longer storage results in chilling injury. If the fruits are intended to be stored longer, the minimum safe storage temperature is 10° C (Tables 17.4 and 17.5).

Quality.—Kader *et al.* (1984) found a reduction in soluble solids content and titratable acidity and an increase in pH of fruit in storage compared to their content at harvest (Table 17.4). Storage temperature had no effect on fruit color.

Respiration.—Pomegranate is a nonclimacteric fruit which means that there is no sudden rise in the respiration rate leading to a climacteric. An increase in temperature does enhance respiration rate (Fig. 17.3) but does not culminate in a climacteric peak. Because of this

TABLE 17.4. EFFECT OF STORAGE TEMPERATURE AND DURATION ON POMEGRANATE CV. WONDERFUL

Storage (weeks)	°C	Weight loss (%)	SSC[1] (%)	TA[2] (%)	pH
Initial	—	0.00	18.70	1.58	3.09
4	0.0	0.80	18.70	1.33	3.24
	2.2	0.80	18.40	1.29	3.20
	5.0	0.80	18.50	1.43	3.18
	10.0	0.90	18.70	1.35	3.08
8	0.0	0.90	18.30	1.44	3.25
	2.2	1.00	18.20	1.39	3.26
	5.0	1.10	18.40	1.20	3.36
	10.0	1.20	18.10	1.18	3.36
12	0.0	0.90	18.20	1.27	3.29
	2.2	1.10	18.00	1.09	3.38
	5.0	1.20	18.10	1.06	3.42
	10.0	1.40	18.00	1.08	3.32
16	0.0	1.00	NA[3]	NA	NA
	2.2	1.30	17.80	1.16	3.42
	5.0	1.50	17.60	1.24	3.39
	10.0	1.90	17.40	1.11	3.36

Source: Kader *et al.* (1984).
[1]SSC = Soluble solids content.
[2]TA = Titratable acidity.
[3]NA = Not analyzed.

nonclimacteric nature, physiological changes leading to improvement in the quality of the fruit does not occur subsequent to harvest. Increased respiration at 0° C is due to chilling injury response.

Response To Exogenous Ethylene Application

Respiration and Ethylene Production.—Exogenous ethylene application stimulates respiration, the magnitude of increase depending on the concentration of applied ethylene (Fig. 17.4 and 17.5). Ethylene application is also known to temporarily stimulate production of ethylene by the fruit (Fig. 17.6). Response of pomegranate fruits to applied ethylene provides further proof for its nonclimacteric nature.

Quality.—Exogenous ethylene application at 10, 100 or 1000 ppm concentration did not affect external color, juice color, soluble solids content, pH or titratable acidity.

TABLE 17.5. EFFECTS OF STORAGE TEMPERATURE AND DURATION ON POMEGRANATE FRUITS

		Chilling injury[1]				
		External		Internal		
Weeks	°C	Fruit score	Affected (%)	Fruit score	Affected (%)	Decay %
Initial	—	1.0	0.0	1.0	0.0	0.0
4	0.0	1.0	0.0	1.0	0.0	0.0
	2.2	1.0	0.0	1.0	0.0	0.0
	5.0	1.0	0.0	1.0	0.0	0.0
	10.0	1.0	0.0	1.0	0.0	0.0
8	0.0	2.3	73.3	2.9	60.0	0.0
	2.2	1.4	26.7	2.5	60.0	0.0
	5.0	1.0	0.0	1.1	6.7	0.0
	10.0	1.0	0.0	1.0	0.0	0.0
12	0.0	2.9	80.0	2.9	80.0	26.7
	2.2	2.1	66.7	3.9	100.0	13.3
	5.0	1.3	33.3	1.8	40.0	13.3
	10.0	1.2	20.0	1.0	0.0	6.7
16	0.0	3.3	93.3	3.3	100.0	60.0
	2.2	2.4	60.0	4.2	93.3	60.0
	5.0	1.5	33.3	2.1	40.0	26.7
	10.0	1.3	40.0	1.1	6.7	20.0

Source: Kader *et al.* (1984).

[1]Evaluations were made after 4 days at 20° C following the storage treatment using the following scoring system: 1 = none, 2 = slight, 3 = moderate, 4 = severe and 5 = extreme injury.

POSTHARVEST DISEASES AND DISORDERS AFFECTING FRUIT QUALITY

Diseases

Soft Rot.—Soft rot is caused by *Rhizopus arrhizus* and *R. stolonifer*. Small water soaked lesions appear on the affected fruits which increase in size with advancement of the disease. The fruit is converted into a pulpy mass inside, though the disease itself is confined to the rind. Affected fruits crack open from the point of infection. A temperature of 30° C and 80% relative humidity is congenial for maximum incidence. This disease can be prevented by treating the packing straw with SO_2 producing chemicals. Coating the fruits with linseed, mustard or castor oil is reported to reduce the incidence of soft rot caused by *R. arrhizus*. Dipping the fruits in 2,4-D at 500 ppm concentration is known to give complete protection to the fruit from this disease (Kanwar and Thakur 1972A,B; 1973A,B).

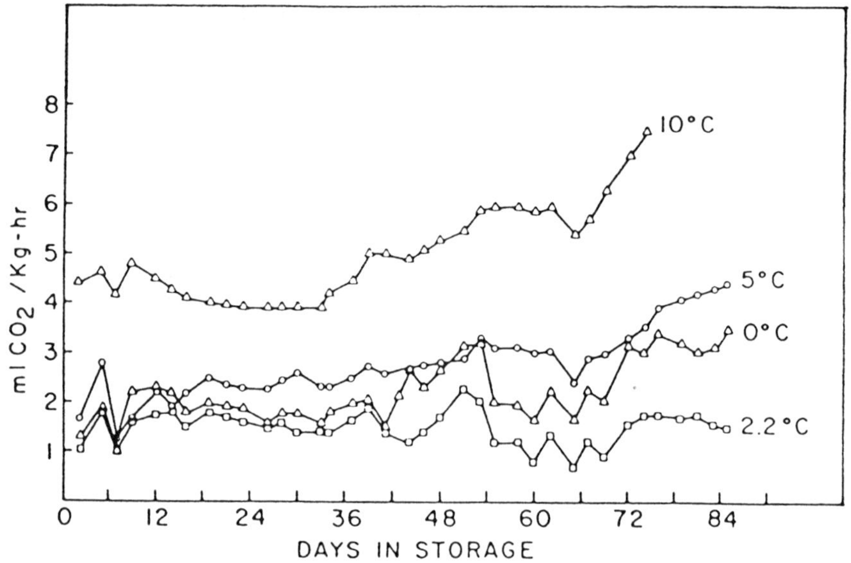

From Elyatem and Kader (1984)

FIG. 17.3. EFFECT OF STORAGE TEMPERATURE ON RESPIRATION RATE OF POMEGRANATE FRUITS. THE INCREASED RATE OF RESPIRATION AT °0 C IS ATTRIBUTED TO THE PHYSIOLOGICAL DAMAGE DUE TO CHILLING INJURY

Gray Mold Rot.—This disease is caused by the fungi *Botrytis cinerea.* The decay begins at the calyx end of the fruit. With the advance of the disease, the fruit skin turns light brown, tough and leathery. Disintegration of the grains into a dark mass follows. Gray-colored mycelial growth will be evident on the affected surface under moist conditions.

Heart Rot.—It is caused by *Alternaria alternata* and several species of *Aspergillus.* Heart rot is otherwise known as Core rot. The disease is characterized by an abnormal skin color and blackened arils inside. *A. alternata* is serious on cracked fruits. Light gray spots that appear on the affected fruits in the beginning, enlarge later to form irregular patches. The fruits rot and a fermented odor emanates from them (Kanwar and Thakur 1973A; Vyas and Panwar 1976). Salunkhe and Desai (1984) feel that the disease develops while the fruit is on the tree. Such fruits can be detected while sorting and it is important to eliminate them from commercial packing.

Penicillium Rot.—This disease can be detected by blue or green spore masses on the fruits. Cracked fruits and fruits infected by the borers are highly vulnerable to this disease.

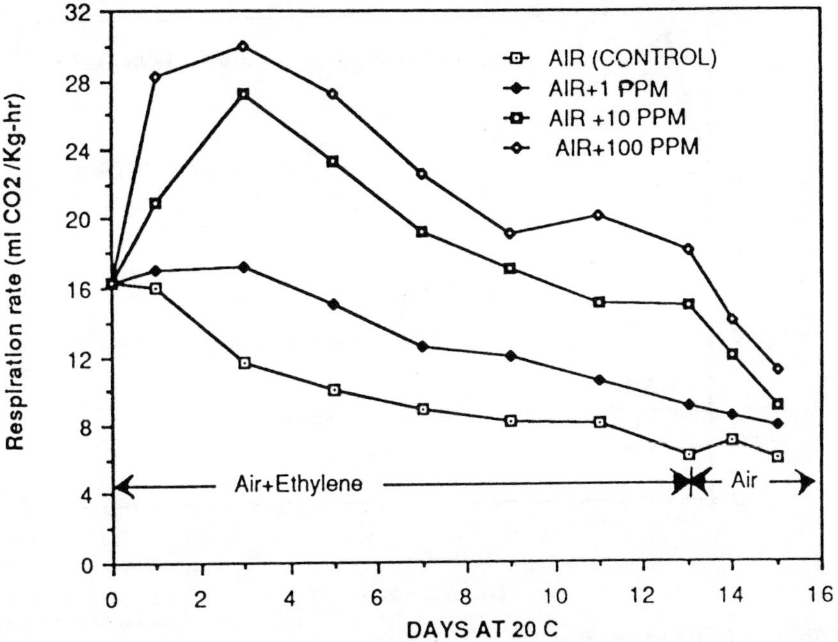

From Kader *et al.* (1984)

FIG. 17.4. EFFECT OF ETHYLENE CONCENTRATION ON THE RATE OF RESPIRATION OF POMEGRANATE FRUITS CV. WONDERFUL

Disorders

Shriveling.—This can occur when the fruits are stored under conditions of low relative humidity and high temperature. When fruits lose more than 5% of their fresh weight, shriveled symptoms become evident. This disorder can be overcome by reducing the vapor pressure difference of the storage atmosphere through increasing relative humidity and maintaining the recommended storage temperature (Kader *et al.* 1984).

Internal Breakdown.—The affected fruits appear normal externally, but the arils inside are poorly developed and depict a flattened appearance in contrast to well-developed plumpy arils of healthy fruits. Affected fruits also lack desirable flavor and color. Grains inside the fruit present a streaked appearance due to the radiating white lines from the seed (Ryall and Pentzer 1974).

Fruit Cracking.—This is a pre-harvest disorder that affects the quality of the fruit in storage. Though the exact cause for this disorder is unknown, it appears that fluctuation of soil moisture and relative humidity, dry wind, rain or heavy irrigation following a dry spell may lead to

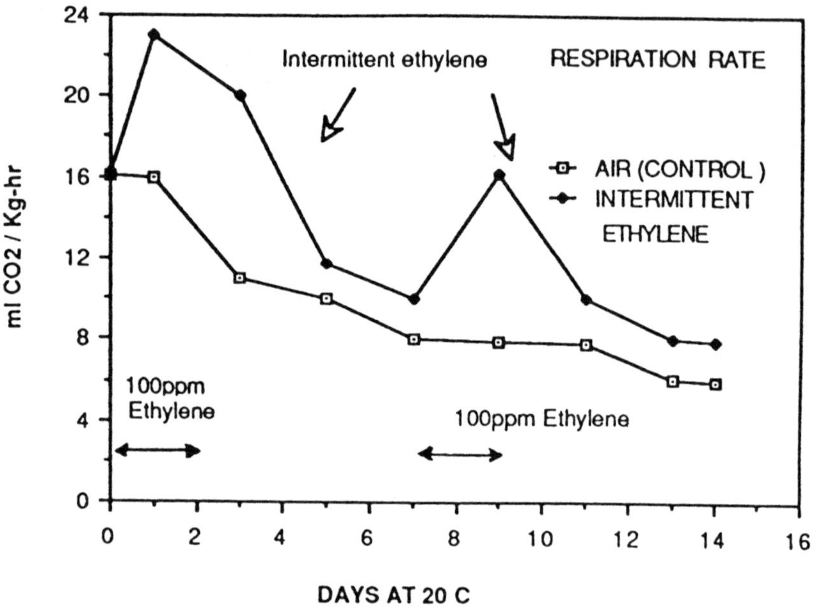

From Kader *et al.* (1984)

FIG. 17.5. EFFECT OF ETHYLENE APPLICATION ON THE RESPIRATION RATE OF POMEGRANATE FRUITS

fruit cracking. Sharma and Chauhan (1973) suggested boron deficiency as a probable cause. Harvesting fruits when ripe, avoiding greater fluctuations in soil moisture and supplying boron, when deficient, might help in mitigating this disorder.

Chilling Injury.—Since pomegranate is a crop of warmer climates, it is sensitive to low temperature in storage. Chilling injury symptoms include brown discoloration of the skin and the locular septums separating the arils. The arils themselves turn pale in color. Kader *et al.* (1984) found that the fruits at 0°, 2.2° and 5° C were affected by chilling injury after 12 weeks of storage while those at 10° C did not develop the symptoms. Chilled fruits become more susceptible to surface molds.

Enzymatic Browning.—Browning is otherwise known as husk scald. This disorder which affects fruits in storage resembles superficial scald of apples in appearance. Scald affected areas may pick up secondary infection by molds. Generally, the arils remain unaffected. Browning of the skin is mediated by the enzyme polyphenol oxidase (PPO) which catalyzes the oxidation of O-dihydroxy phenols leading to a brown color. PPO can be inactivated by dipping the fruits in boiling water for 2 minutes. The incidence of browning can also be reduced by dipping the fruits

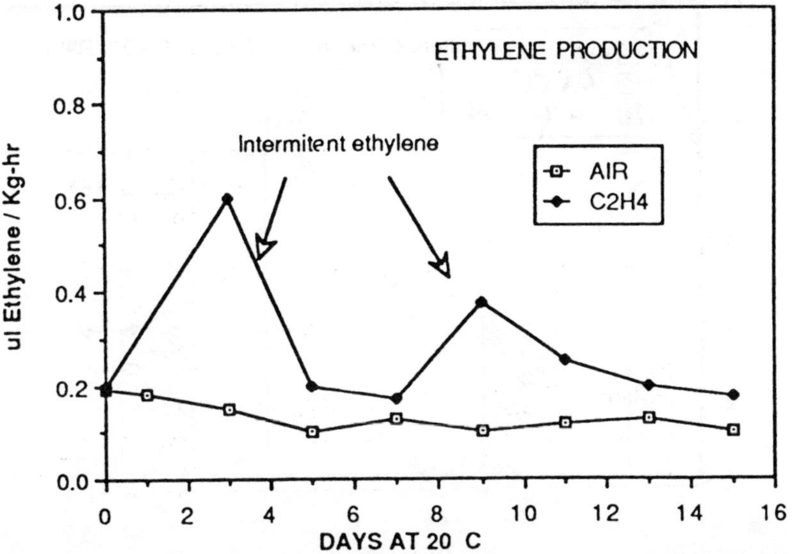

From Kader *et al.* (1984)

FIG. 17.6. EFFECT OF ETHYLENE APPLICATION ON ETHYLENE PRODUCTION BY POMEGRANATE FRUITS

in Maneb at 5 mM concentration for 30 seconds (Fig. 17.7) or by storing the fruits in modified atmospheres having low O_2 concentration. Oxygen concentration of 2% and a temperature of 2° C are effective in controlling this disorder. Low oxygen concentration leads to the accumulation of products of glycolysis such as ethanol, causing off flavors. The off flavor, however, is dissipated when the fruit is transferred to air at 20° C (Ben-Arie and Or 1986).

PROCESSING

Pomegranate Juice

Selection Of The Fruit.—High yields of juice sweet in taste and rich in color and flavor can be obtained only from well ripened fruits. Cracked and sunburnt fruits can also be used if they are well ripe (Tressler and Joslyn 1961). Cultivars such as Khandharm, Wonderful, Mangulati, Ahmar, Aswad, Halwa, Paper Shell and Spanish Ruby are used for processing (Salunkhe and Desai 1984). The cultivar Wonderful is very rich in anthocyanins (Kriventsov and Arendt 1985) and, therefore, may be more desirable for producing highly colored juice.

Extraction Of The Juice.—Pomegranate juice can be extracted using a cloth and rag hydraulic press. Either whole fruits or halves or quarters

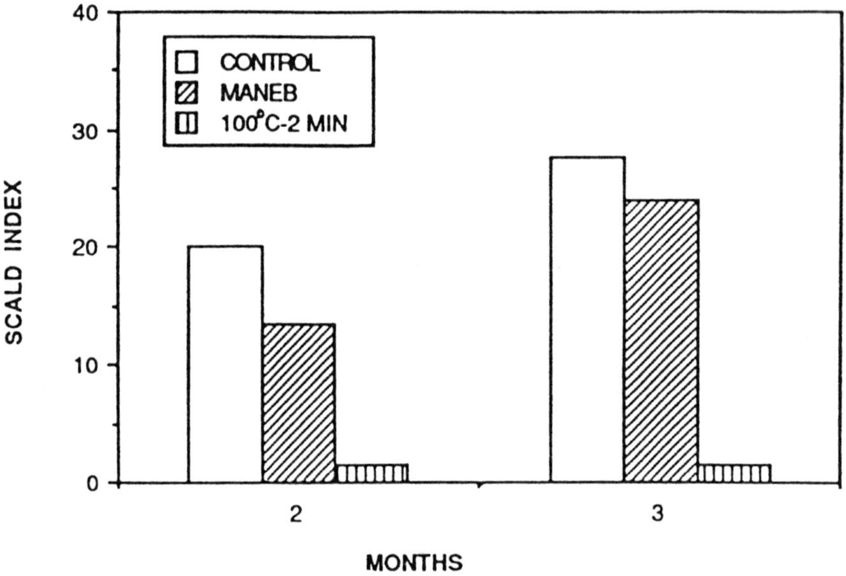

From Ben-Arie and Or (1986)

FIG. 17.7. EFFECT OF HOT WATER AND MANEB (BIS-DITHIOCARBAMATE 5 mM) DIPS ON THE DEVELOPMENT OF HUSK SCALD ON WONDERFUL POMEGRANATE STORED AT 10° C

are fed to the press to express the juice. One ton of fruit can yield about 300 to 340 liters of juice. Tressler and Joslyn (1961) reported the development of a special press by Sutton of the Laboratory of Fruit and Vegetable Chemistry, U. S. Bureau of Agricultural Chemistry for extraction of pomegranate juice.

Clarification Of Juice.—Chace *et al.* (1930) reported the following procedure for clarification of juice.

"The juice is placed in 50 gallon barrels, and one fourth pound of gelatin, previously dissolved in a small quantity of hot water, is added to each barrel. The barrels are allowed to stand about three hours, after which the clear juice is siphoned from them. The juice is then mixed with one percent of its weight of kieselguhr and heated to 175° F. After it has stood overnight, it is filtered cold and sterilized in bottles or in a continuous pasteurizer at 175° to 180° F.

The residue from which the juice has been siphoned can also be filtered with kieselguhr, after which it is treated according to the process just described for juice. When this scheme of treatment is used, the juice will keep brilliant for six months and sediment in objectionable quantities will not deposit within a year. If desired, the juice can be sweetened, but this should be done before it is filtered, as unfiltered commercial sugar solutions are seldom brilliant."

The juice can be preserved by using sodium benzoate (Salunkhe and Desai 1984). Potassium meta bisulphite (KMS) may not be suitable as it can bleach the color of the juice.

Methods To Reduce Tannin Content In The Juice.—Tannins impart bitter taste to the juice and, therefore, are undesirable in the juice. Addition of gelatin as described above is effective to reduce the bitterness. Allowing the fruit to shrivel for a few days or separating the grains from the fruit prior to expressing the juice can reduce tannin content in the juice appreciably. A reduction of nearly 30% in tannin content is possible if the juice is extracted from the separated grains (Salunkhe and Desai 1984). The peel and the rag (locular membranes) are rich in tannins and contaminate the juice when the whole fruit is used. It is also possible to keep the tannins at low level by maintaining the pressure of the hydraulic press around 100 psi (Rodriguez *et al.* 1975).

Reconstitution Of The Juice.—Pomegranate juice is rich in acid and, therefore, it is preferable to dilute and sweeten to improve its palatability. Because of its good flavor, pomegranate juice blends well with other juices.

Grenadine

A product known as grenadine can be prepared using pomegranate juice. Hot sugar syrup is dissolved into the juice and filled into sterile bottles at 82° to 91° C. This product has a good keeping quality. However, it is advisable to refrigerate the product in order to retain better color (Tressler and Joslyn 1961). The true grenadine syrup is essentally a pomegranate product. But most of the so called grenadine syrups are synthetic preparations (Cruess 1938).

Jelly

Hays (1953) reported that a jelly can be made from pomegranate. A detailed procedure, however, is not described.

REFERENCES

ANON. 1948-1976. The Wealth of India. Council of Scientific and Industrial Research, New Delhi.

ANON. 1975. Pomegranates. Calif. Adm. Reg. 75(4) 150.253-150-255.

BAILEY, L. H. 1925. Standard Cyclopedia of Horticulture. The Macmillan Co., New York.

BEN-ARIE, R. and OR, E. 1986. The development and control of husk scald on 'Wonderful' pomegranate fruit during storage. J. Am. Soc. Hortic. Sci. 111(3) 395-399.

BENK, E. 1985. Preparation of fruit nectars and fruit juice drinks using tropical and subtropical fruits and juices. Fluess. Obst. 52, 358-364. (German)

BROOKS, R. M. and OLMO, H. P. 1982. Register of new fruit and nut varieties. List 32. HortScience 17(1) 17-23.

CHACE, E. M., CHURCH, C. G. and POORE, H. D. 1930. The 'Wonderful' variety of pomegranate. U. S. Dep. Agric. Circ. 98.

CRUESS, W. V. 1938. Commercial Fruit and Vegetable Products. McGraw Hill Book Co., London.

DE CANDOLLE, A. 1967. Origin of Cultivated Plants. Hafner Publishing Co., New York.

DU, C. T., WANG, P. L. and FRANCIS, F. J. 1975. Anthocyanins of pomegranate, *Punica granatum*. J. Food Sci. 40, 417-418.

DUTTA, A. C. 1982. Botany for Degree Students. Oxford University Press, New Delhi, India.

EL-SHAARAWY, M. I. and NAHAPETIAN, A. 1983. Studies on pomegranate seed oil. Fette Seifen Anstrichm. 85(3) 123-126. (German)

ELYATEM, S. M. and KADER, A. A. 1984. Postharvest physiology and storage behavior of pomegranate fruits. Sci. Hortic. 24, 287-298.

EVREINOFF, V. A. 1949. The pomegranate. Fruits Outre. Mer. 4, 161-170. (French)

FIRUZEH, P. and LUDDERS, P. 1979. Pomegranate growing in Iran. Erwerbsobstabu. 21(12) 260-262. (German)

HAYS, W. B. 1953. Fruit Growing in India. 2nd Rev. Edition. Kitabistan, Allahbad, U.P., India.

HERRMANN, K. 1981. Review on chemical composition and constituents of some important exotic fruits. Z. Lebensm. Unters. Forsch. 173, 47-60. (German)

HODGSON, R. W. 1917. The pomegranate. Calif. Agric. Exp. Stn. Bull. 276.

KADER, A. A., CHORDAS, A. and ELYATEM, S. 1984. Response of pomegranates to ethylene treatments and storage temperature. Calif. Agric. 38(7,8) 14-15.

KANWAR, Z. S. and THAKUR, D. P. 1972A. Postharvest fungal diseases of pomegranate in Haryana. Indian J. Myco. Plant. Path. 2(2) 189-191.

KANWAR, Z. S. and THAKUR, D. P. 1972B. Controlling soft rot of pomegranate fruits by treatment with packing straw. Sci. Cult. 38(10) 450-452.

KANWAR, Z. S. and THAKUR, D. P. 1973A. Some new fruit rots of pomegranate in Haryana. Sci. Cult. 39(6) 274-276.

KANWAR, Z. S. and THAKUR, D. P. 1973B. Controlling soft rot of pomegranate (*Punica granatum* L.) fruits caused by *Rhizopus arrhizus* Fischer, by using growth regulators and antibiotics. Haryana J. Hortic. Sci. 2(1/2) 50-55.

KRIVENTSOV, V. I. and ARENDT, N. K. 1985. Anthocyanins of pomegranate juice. Trudy Gosudarstvennogo Mikitskogo Botanicheskogo Sada. 83, 110-116. (Ukranian)

LA RUE, J. H. 1969. Growing pomegranates in California. Univ. Calif. Agric. Exp. Stn. Leaflet 305.

LEE, S. W., KIM, K. S. and KIM, S. D. 1974A. Studies on the composition of pomegranate fruit during maturation. 1. Changes in sugars, organic acids, amino acids and respiration rate. J. Korean Soc. Hortic. Sci. 15(1) 57-63. (Korean)

LEE, S. W., KIM, K. S. and KIM, S. D. 1974B. Studies on the changes in the composition of pomegranate fruits during ripening. 2. Changes in phenol compounds and anthocyanin pigments. J. Korean Soc. Hortic. Sci. 15(1) 64-71. (Korean)

MISRA, R. S., SRIVASTHAVA, R. P. and KUKSAL, R. P. 1983. Evaluation of some pomegranate cultivars for valley areas of Garhwal hills. Prog. Hortic. 15(1-2) 24-26.

MORTON, J. F. 1987. Fruits of Warm Climates. Julia F. Morton Publisher, Miami, FL.

MUKHERJEE, P. K. 1958. Storage of pomegranates (*Punica granatum* L.). Sci. Cult. 24, 94.

NADAKARNI, A. K. 1976. Indian Materia Medica. Popular Prakashan, Bombay, India.

NASACHEVA, E. P., ANISIMOVA, K. I. and SHEINBOK, S. D. 1974. Tannins in the rind of pomegranate and prospects of utilization by tanning industry. Rastitel'nye Resury. 9(2) 267-270. (Russian)

PATIL, A. V. and WAVHAL, K. N. 1980. Improvement of fruit crops in Maharastra. Punjab. Hortic. J. 20(1,2) 2-7.

PAUL, A. A. and SOUTHGATE, D. A. T. 1978. The Composition of Foods, Part 4. Elsevier Press, Amsterdam.

PURSEGLOVE, J. W. 1981. Tropical Crops-Dicotyledons. The English Language Book Society and Longman, London.

RODRIGUEZ, R., RAINA, B. L., PANTASTICO, E. B. and BHATTI, M. B. 1975. Quality of raw materials for processing. *In* Postharvest Physiology, Handling and Utilization of Tropical and Subtropical Fruits and Vegetables. E. G. Pantastico (Editor), AVI Publishing Co., Westport, CT.

RYALL, A. L. and PENTZER, W. T. 1974. Handling, Transportation and Storage of Fruits and Vegetables. Vol. 2, Fruits and Nuts. AVI Publishing Co., Westport, CT.

SALUNKHE, D. K. and DESAI, B. B. 1984. Postharvest Biotechnology of Fruits. Vol. II. CRC Press, Boca Raton, FL.

SAYED, S., RAMADOSS, S., NANJAN, K. and MUTHUSWAMY. 1985. YCD-1 Pomegranate. South Indian Hortic. 33(1) 452.

SEPPI, A. and FRANCIOSI, A. 1980. Chemical composition of pomegranate juice (*Punica granatum*). Riv. Soc. Ital. Sci. Aliment. 9(3) 211-212. (Italian)

SHARMA, V. and CHAUHAN, K. S. 1973. Pomegranate cultivation. Farmer and Parliament 8, 11-12 and 24-25.

SIDDAPPA, G. S. and BHATIA, B. S. 1954. The identification of sugars in fruits by paper chromatography. Indian J. Hortic. 11, 19.

SINGH, R. D. 1976. Pomegranate. Agric. Agro-Ind. J. 9(9) 23-26.

TRESSLER, D. K. and JOSLYN, M. A. 1954. The Chemistry and Technology of Fruit and Vegetable Juice Production. AVI Publishing Co., Westport, CT.

TRESSLER, D. K. and JOSLYN, M. A. 1961. Fruit and Vegetable Processing Technology. AVI Publishing Co., Westport, CT.

VERES, M. 1976. A contribution to the study of composition of cultivated pomegranate. Hranai Ishrana 17, 426-432. (Yugoslavian)

VYAS, N. L. and PANWAR, K. S. 1976. A new postharvest disease of pomegranates in India. Curr. Sci. 45(2) 76.

WATT, B. K. and MERRILL, A. L. 1963. Composition of Foods. Agriculture Handb. No. 8., Agr. Res. Serv., U.S.D.A., Washington, DC.

WILLS, R. B. H., LIM, J. S. K. and GREENFIELD, H. 1986. Composition of Australian foods. 31. Tropical and subtropical fruit. Food Technol. Aust. 38(3) 118-123.

WU LEUNG, W., BUSSON, F. and JARDIN, C. 1968. Food Composition Table for Use in Africa. FAO and U. S. Dept. of Health, Education and Welfare, Washington, D. C.

WU LEUNG, W., BUTRUM, R. R. and CHANG, F. H. 1972. Proximate composition of mineral and vitamin contents of East Asian food. Part 1. Food Composition Table For Use In Africa. FAO and U. S. Dept. of Health, Education and Welfare, Washington, D.C.

WU LEUNG, W. and FLORES, M. 1961. Food Composition Table for Use in Latin America. National Institute of Health. Bethesda, MD. (Spanish)

18

Calamansi or Calamondin

Linda B. Mabesa

Calamansi (also known as kalamansi or calamondin) is indigenous to the Philippines and was formerly identified as *Citrus mitis* Blanco (or *C. microcarpa* Bunge) and as *C. madurensis* Lour. It is believed to be a hybrid of *Citrus reticulata* Blanco X *Fortunella* spp. It is normally considered as a substitute for lemon. The calamansi plant is somewhat spiny, with up-right branches and grows from three to five meters high (PCARRD 1980; Gutierrez 1982). The leaves are broadly oval, dark green above and pale green below. The stalks are very narrow and scarsely winged and measure almost a centimeter. The white, rather short-stalked flowers are axillarly, solitary and rarely paired (Gutierrez 1982).

The fruits are subglobose ranging from 3-3.5 cm long and 2-4.5 cm in diameter (PCARRD 1980). Each fruit has about 7 to 10 segments with few seeds enclosed in each segment. The peel adheres loosely to the skin. Figure 18.1 shows a portion of the branch showing the fruit and flower, and Figure 18.2 shows the different sizes of the fruit and its cross section.

HARVESTING

A study conducted by Pantastico and Mendoza (1976) revealed that methods and conditions during harvesting have great influence on the quality of the calamansi. It was noted that skin blemish on the fruit surface due to finger pressing (known as oleocellosis which is indirectly measured as Rind Oil Release Pressure or RORP) was related to the prevailing relative humidity (RH) and light intensity at the time of harvest. Oleocellosis was minimal at 2 p.m. during which time the recorded RH was only 36%. Table 18.1 shows the effect of climate conditions on

FIG. 18.1. PORTION OF THE BRANCH OF CALAMANSI SHOWING THE FRUIT AND THE FLOWER

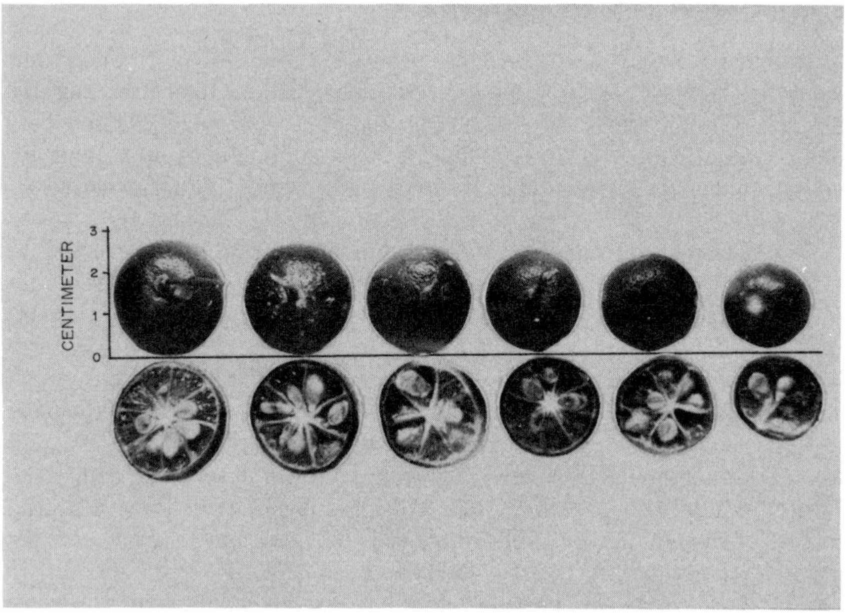

FIG. 18.2. CALAMANSI FRUIT AND ITS CROSS SECTION

RORP and degree of oleocellosis in calamansi. It was further suggested by these researchers that harvesting of calamansi should be carried out using clippers and polyurethane padded handgloves to minimize damage to the fruits.

TABLE 18.1. CLIMATIC CONDITIONS AFFECTING RIND OIL RELEASE PRESSURE (RORP) AND DEGREE OF OLEOCELLOSIS DURING HARVESTING OF CALAMANSI

Time	Temp. (°F)	RH	Light intensity (foot candle)	RORP	Degree of oleocellosis
8 (a.m.)	78.0	66.5	500	104	2.90
9	78.8	63.9	1351	147	2.47
10	80.0	59.0	3025	183	2.02
11	85.1	59.0	6150	274	1.35
12	86.9	44.2	10700	280	1.30
1 (p.m.)	87.5	41.7	9525	292	1.17
2	90.0	36.2	5825	294	1.05
3	82.4	39.5	4675	197	1.65
4	80.6	48.5	2475	197	2.35

Source: Pantastico and Mendoza (1976).

POST HARVEST HANDLING

Calamansi fruits should be green to light green in color. The fruits should be heavy and juicy. Surface blemishes can be tolerated. Lightly discolored, puffy and badly creased fruits are of poor quality (Food Terminal, Inc. 1979). Fruits should contain a minimum of 40% juice by weight to meet the minimum maturity requirement (Pantastico *et al.* 1975).

Pantastico and Mendoza (1976) recommended that calamansi should be stored at 8°-10°C at 90% relative humidity. The expected storage life of the calamansi fruits will be about 2 weeks with a weight loss of approximately 6.5%.

The prevailing market cost of calamansi is computed not only from the farm-gate price, transportation, handling and packaging expenses. Middle men also add the amount that will compensate for the damaged fruits which can be as high as 73% during transport from the farm to the market (Table 18.2). Most of the fruits damaged were located at the bottom of the container (Table 18.3). This observation was based on the survey conducted by Pantastico and Mendoza (1976).

Commercial trials on postharvest physiology, handling and storage of calamansi were conducted by Mendoza and Pantastico (1979). The

TABLE 18.2. COST AND RETURN ANALYSIS OF CALAMANSI FRUITS FROM A FARM IN VICTORIA, OR. MINDORO TO DIVISORIA MARKET (PRICE IN PHILIPPINE CURRENCY)

Item		Cost/crate
A. Wholesale price Divisoria		P110.00
B. Farm price		P50.00
C. Transportation and handling costs		13.50
1. Packaging materials	P0.50	
2. Labor in packing	1.00	
3. Crate (reusable 4x) at P2.00/crate	0.50	
4. Truck fare	3.00	
5. Loading and unloading ship	2.00	
6. Ship fare	2.00	
7. Truck fare to Divisoria	2.50	
8. Unloading	1.00	
9. Stall fee	1.00	

Source: Pantastico and Mendoza (1976).

Farm price + transportation and handling costs	– P63.50
Handlers supposed profit if fruits were undamaged (P110.00 – 63.50)	– 46.50
Usual profit of middlemen/crate	– 20.00
Increase in price due to damaged fruits	– 26.50

TABLE 18.3. DEGREE (%) AND NATURE OF DAMAGE ON CALAMANSI TRANSPORTED AT DIFFERENT HANDLING POINTS USING THE KAING[1] AND WOODEN CRATES

	Major handling points					
	Calapan Pier[2]		Batangas Pier[3]		Divisoria Pier[4]	
Type of injury	Kaing	Wooden crates	Kaing	Wooden crates	Kaing	Wooden crates
Oleocellosis						
a. upper	11.50	5.5	13.25	6.0	25.25	6.0
b. bottom	13.60	9.5	22.50	9.5	42.00	9.5
Pressure						
a. upper	3.05	2.0	4.50	4.0	5.25	3.5
b. bottom	3.75	2.0	13.75	3.5	16.00	2.5
Abrasion						
a. upper	3.33	2.50	6.50	3.00	9.25	4.0
b. bottom	8.75	4.50	11.25	5.50	15.75	8.5
Total						
a. upper	17.80	10.00	24.50	13.00	39.75	13.50
b. bottom	26.10	16.00	47.50	17.50	73.00	20.50

Source: Pantastico and Mendoza (1976).
[1]Kaing is a type of woven bamboo basket.
[2]Calapan Pier, 38 km from the production source, transport jeep.
[3]Batangas Pier, 18 hr in ship hold, include loading and unloading.
[4]Divisoria Pier, 111 km from Batangas pier, transport trucks.

calamansi fruits three days after harvesting were packed in several containers called "kaing" (woven bamboo baskets). Each "kaing" was lined with one of the following: newsprint; sealed polyethylene; sealed polyethylene provided with tunnel and perforated polyethylene.

The fruits were transported from the farm to the laboratory (16 hr total transport time). Temperatures at the top, middle and bottom portions of the containers were monitored during transport. Fruit damage was determined upon arrival of the fruits.

Results of this study indicated that the temperature inside the "kaing" increased from bottom to top (Table 18.4) and the "kaing" lined with polyethylene provided with a tunnel in the middle exhibited the lowest average temperature. This implied that the tunnel provided in the middle of the "kaing" dissipated the heat. Temperature in the unlined "kaing" showed the same trend.

Table 18.5 summarizes the nature of calamansi discards from the different locations inside the "kaing" with different liners after delayed transport. Discards due to punctures resulted because of the stalks at-

TABLE 18.4. TEMPERATURE (C) AT THE TOP, MIDDLE AND BOTTOM PORTIONS OF "KAING" WITH DIFFERENT LINERS DURING TRANSPORT OF CALAMANSI

Type of liner	Top	Middle	Bottom	Average
No liner	28.0	28.0	26.4	26.8
Newsprint	29.7	26.6	26.5	27.6
Polyethylene, sealed	30.9	28.3	28.2	29.1
Polyethylene, sealed plus tunnel	28.3	23.3	25.8	25.8
Polyethylene, perforated	30.6	27.9	27.3	28.6

Source: Mendoza and Pantastico (1979).

TABLE 18.5. DISCARDS IN CALAMANSI FRUITS AFTER DELAYED TRANSPORT USING DIFFERENT LINERS

	Type of liner				
Nature of discard location	No liner	News-print	Poly-ethylene, perforated	Poly-ethylene, sealed	Poly-ethylene, sealed + tunnel
Top					
Punctures	0.7	0.6	0.5	0.4	0.4
Compression	1.1	1.7	1.7	1.0	0.7
Abrasion	0.8	0.9	1.0	0.9	0.8
Overripening	22.0	12.6	16.2	31.9	36.9
Middle					
Punctures	1.6	1.6	0.3	1.6	1.6
Compression	3.6	0.5	0.6	0.4	1.6
Abrasion	1.4	1.6	1.1	2.5	2.2
Overripening	9.6	12.6	15.5	29.3	24.7
Bottom					
Punctures	0.9	1.2	0.8	1.0	1.0
Compression	2.9	2.5	1.8	2.6	3.7
Abrasion	1.2	2.9	1.9	3.6	2.7
Overripening	20.9	11.7	10.9	35.6	31.3
Total mechanical damage	14.2	13.5	10.7	14.0	14.7
Total overripening	52.5	36.9	42.6	96.5	92.9

Source: Mendoza and Pantastico (1979).

tached to the fruits, whereas abrasion and compression were due to the location of the fruits in the container. It was speculated further that overripening or yellowing of calamansi inside the "kaing" was brought about by the build up or accumulation of ethylene. This was supported by the observation that there was a higher amount of overripe fruits in "kaing" lined with sealed polyethylene bags compared with the number of overripe fruits in "kaing" lined with newsprint and perforated polyethylene.

The effect of storing calamansi packed in "kaing", with liners described previously, at 10°C was also evaluated by Mendoza and Pantastico (1979). In their study, calamansi fruits were stored and then treated with Benlate. The sorted and the Benlate-treated calamansi were packed in different "kaing" lined with different liners. Results are shown in Table 18.6. It was deduced from the data that sealed and perforated liners induced the development of yellow fruits regardless of whether or not the fruits were sorted and treated with Benlate. However, Benlate seemed to minimize yellowing of the fruits. The researchers suggested that calamansi should be stored green and treated with Benlate prior to long term storage.

In the same study, Pantastico and Mendoza (1976) determined the percentage discard caused by injuries in the calamansi fruits with different liners stored for 5 days at 10°C. Results are presented in Table 18.7.

Data *et al.* (1975) evaluated the effect of temperature, wax treatment, maturity and growth regulator on the soluble solids, titratable acidity and ascorbic acid content of calamansi fruits stored under refrigerated conditions. The total soluble solids of green-mature, small and green mature, medium calamansi fruits increased as storage time progressed. However, the reverse trend was evident in ripe calamansi fruits. Although titratable acidity of the calamansi decreased as storage progressed, this trend was more prominent for green fruits. Ascorbic acid also decreased regardless of initial maturity prior to storage.

When green-mature and medium-sized fruits were stored at different temperatures, total soluble solids increased during storage and this increase was pronounced at higher storage temperature. Total titratable acidity tended to decrease at storage temperatures lower than 10°C. The same trend was observed for ascorbic acid content.

Soluble solids of calamansi fruits during storage increased markedly when treated with Ethrel but gibberellic acid had no effect at all. A similar trend was observed for the titratable acidity and ascorbic acid. Ethrel further enhanced the yellowing of stored calamansi fruits. Gibberellic acid on the other hand maintained the green color of the calamansi fruit throughout storage.

TABLE 18.6. COLOR DEVELOPMENT OF CALAMANSI FRUITS PACKED IN "KAING" WITH DIFFERENT LINERS AFTER 5 WEEKS AT 10°C

Treatment	Green				50% Yellow				Yellow			
	Top	Middle	Bottom	Ave.	Top	Middle	Bottom	Ave.	Top	Middle	Bottom	Ave.
I Unsorted without Benlate												
Traditional (no liner)	25.2	31.4	40.4	32.3	50.2	33.0	43.2	41.9	24.6	35.6	16.4	25.5
Newsprint	52.2	39.8	39.4	43.8	23.3	48.1	42.9	38.1	24.6	12.0	17.7	18.1
Polyethylene, sealed	36.5	43.7	39.6	23.9	23.9	21.6	49.6	31.7	39.6	34.8	10.8	28.5
Polyethylene, sealed + tunnel	21.7	16.4	22.0	20.0	52.1	67.7	66.3	62.0	26.2	15.8	11.6	17.9
Polyethylene, perforated	23.3	36.8	42.7	35.3	47.1	48.0	44.6	46.6	29.6	15.1	12.8	19.2
II Sorted with Benlate												
Traditional (no liner)	46.5	43.7	52.8	47.6	22.2	34.1	27.9	28.0	31.3	22.2	19.3	4.3
Newsprint	48.0	56.2	56.2	53.6	41.0	28.5	38.2	39.2	11.0	5.3	5.2	7.2
Polyethylene, sealed	37.6	17.3	19.0	24.6	30.4	47.5	58.7	45.5	32.0	35.2	22.3	29.8
Polyethylene, sealed + tunnel	30.1	32.2	21.5	27.9	35.8	47.0	45.1	42.5	31.2	17.5	33.4	27.4
Polyethylene, perforated	28.1	38.0	42.7	36.3	35.0	35.2	38.8	36.2	38.9	25.7	19.0	27.5
III Sorted without Benlate												
Traditional (no liner)	52.2	52.6	47.6	50.8	35.9	28.0	41.7	35.2	11.8	19.4	10.7	14.0
Newsprint	54.1	62.5	57.7	58.1	28.4	32.5	29.6	30.2	17.5	5.0	12.7	11.7
Polyethylene, sealed	12.5	23.1	23.7	19.8	69.4	59.1	39.9	56.1	18.2	17.9	36.4	24.2
Polyethylene, sealed + tunnel	20.2	19.3	35.4	25.0	62.6	36.4	59.8	52.9	17.1	44.3	4.8	22.1
Polyethylene, perforated	31.5	50.2	41.7	41.1	32.2	41.1	36.2	36.8	35.4	8.8	20.7	21.6

Source: Mendoza and Pantastico (1979).

TABLE 18.7. PERCENTAGE DISCARD CAUSED BY INJURIES IN CALAMANSI FRUITS ON CRATES WITH DIFFERENT LINERS AND STORED AT 10°C

Treatment	Punctures				Abrasions				Compressions			
	Top	Middle	Bottom	Ave.	Top	Middle	Bottom	Ave.	Top	Middle	Bottom	Ave.
I Unsorted without Benlate												
Traditional (no liner)	5.7	7.4	7.1	6.7	22.7	15.0	16.6	18.1	15.1	5.6	10.2	10.3
Newsprint	2.6	2.2	2.9	2.6	2.2	2.1	2.9	2.4	6.6	5.2	5.7	5.8
Polyethylene, sealed	1.0	1.2	1.3	1.2	0.9	2.8	8.4	4.0	1.2	2.0	2.5	1.9
Polyethylene, sealed + tunnel	4.3	3.6	5.7	4.5	16.7	9.1	15.2	13.7	8.0	7.0	9.8	8.3
Polyethylene, perforated	1.6	1.3	1.0	1.3	2.0	2.4	3.8	2.7	1.3	1.7	1.3	1.4
II Sorted with Benlate												
Traditional (no liner)	0.6	0.4	0.9	0.6	7.4	6.7	4.7	6.3	7.9	2.0	2.6	4.2
Newsprint	1.6	1.0	0.8	1.1	5.9	3.7	3.0	4.2	3.8	2.1	1.3	2.4
Polyethylene, sealed	2.7	0.9	1.6	1.7	7.0	3.7	4.1	4.9	2.5	1.3	6.1	3.3
Polyethylene, sealed + tunnel	1.3	1.5	0.7	1.2	2.0	1.8	2.3	2.0	3.8	3.3	2.9	3.3
Polyethylene, perforated	2.1	1.7	1.4	1.7	4.2	2.3	4.8	3.8	3.0	3.1	1.5	2.5
III Sorted without Benlate												
Traditional (no liner)	3.3	0.5	0.7	1.5	9.3	2.7	2.7	4.9	3.4	1.1	1.2	1.9
Newsprint	0.8	1.2	1.5	1.2	2.1	2.1	2.0	2.1	1.3	1.9	1.2	1.5
Polyethylene, sealed	1.9	0.4	0.9	1.1	3.4	2.0	1.5	2.3	2.5	2.5	2.4	2.5
Polyethylene, sealed + tunnel	3.0	0.7	0.8	1.5	6.0	3.0	1.7	3.6	3.9	3.3	2.8	3.3
Polyethylene, perforated	1.1	0.6	2.4	1.4	1.0	2.0	1.6	1.5	2.4	2.1	2.2	2.2

Source: Mendoza and Pantastico (1979).

Wax treatment of the calamansi fruits with Sta-Fresh 215 + TBZ prevented any changes in the total soluble solids which was evident in unwaxed fruits and those treated with Johnson's Prima Fresh. Titratable acidity was not affected by wax treatment. Loss of ascorbic acid during the first two weeks was prevented by wax treatment. Thereafter, wax treatment was not effective in preventing losses of ascorbic acid.

Del Rosario *et al.* (1976) measured the respiration of stored calamansi as affected by maturity, temperature, growth regulators and wax treatments. The findings of these researchers indicated that calamansi fruits exhibited a typical climacteric pattern which was more pronounced in small fruits. It was observed that the respiration rate was higher in small calamansi fruits than in big ones regardless of storage temperatures. Results implied that maturity of calamansi fruits differed with size.

Respiration rates increased directly with storage temperature regardless of size or maturity. This could be attributed to the rapid deterioration in quality of the calamansi fruits when stored at temperatures higher than 10°C.

Ethephon-treated calamansi fruits showed an increasing respiration rate throughout the entire storage at 10°C. Those fruits that were

TABLE 18.8. COMPOSITION OF CALAMANSI OF DIFFERENT SIZES

	Fruit diameter, mm				
	35-above	32-34	28-31	25-27	20-24
Size distribution (%)					
Based on number	5.10	19.99	57.49	15.42	1.83
Based on weight	7.40	24.48	53.71	11.39	0.96
Weight per fruit (g)	18.8	15.8	12.5	9.5	6.8
Number of fruits per kg	23	63	80	105	148
Physical composition (% by weight)					
Seeds	9.60	11.32	12.09	10.60	8.49
Peel	24.63	23.05	23.79	26.63	28.49
Juice	19.21	17.70	19.82	21.19	15.89
Juice characteristics					
Total solids, °Brix	8.65	8.50	8.70	8.75	8.60
Acidity, % citric acid (anhydrous)	6.17	6.22	6.71	6.61	6.79
Essential oil content (ml)					
Per kg of fruit	6.3	6.1	6.3	6.8	*
Per kg of peel	25.6	26.5	26.5	25.4	

Source: Mauricio *et al.* (1967).
*Insufficient essential oils for analysis.

treated with gibberellic acid maintained a lower respiration rate and the same trend was noted for the control.

Among the waxes used, only Tag and Banana waxes were effective in maintaining a lower respiration rate when stored at 10°C. Prima Fresh was not effective and Sta-Fresh even induced respiratory activity in the stored fruits.

Chilling injury of calamansi during cold storage was evaluated by de Castro (1973). It was noted that chilling injury was severe at low temperature storage (1°C) and this effect was highly apparent in immature fruits. It was suggested that chilling injury can be prevented by coating the calamansi fruits with a wax emulsion like Sta-Fresh 215 + Thiabendazole and Prima Fresh. Intermittent high and low temperature during storage can also minimize chilling injury of the calamansi fruits.

Mayuga (1973) studied the possible methods to control decay of calamansi fruits during storage. Storing the fruits at 10°C and treating them with vitamin K and gibberellic acid extended the marketability for three and two weeks, respectively. Also, among the fungicides tested, Benlate was the most effective in controlling decay, thus improving the saleable life of calamansi for 2 weeks. Mechanical injury during transport enhanced the incidence of decayed fruits during storage.

COMPOSITION

Co *et al.* (1979) studied the chemical constituents of the different parts of the calamansi fruit. Results revealed that the peel contained the highest ascorbic acid (61 mg %), followed by pulp (39 mg %) and juice (34 mg %). The total carotenoid extracted from peel ranged from 50-75 mg %.

The composition of the different parts of calamansi was also analyzed by Mauricio *et al.* (1967). Results of the study are shown on Tables 18.8 to 18.10. It was noted that the yield of seeds, pulp and juice did not vary much among lots with different sizes. However, as calamansi fruits got smaller, the weight of the peel was higher than the peel from the bigger fruits. Limonene was also found to be the predominant fraction of the steam-distilled oil and this amount was not affected by the maturity of the fruit. The essential oil obtained after steam distillation of one kilogram of leaves yielded about 2.5 ml. The hexane-extracted oil from the seeds exhibited a specific gravity of 0.9161, a refractive index at 25°C of 1.4709, an acid number of 4.81, a saponification number of 189, an iodine number of 147 and an unsaponifiable matter of 18.88%. Sutaria and San Diego (1982) disclosed that based on their studies, calamansi fruit contained 89.93% moisture, 0.204% nitrogen, 1.275% protein and 1.653% α-amino nitrogen (dry weight basis).

TABLE 18.9. ESSENTIAL OILS OF CALAMANSI[1]

		Immature	Ripe
Weight of fruits analyzed	(g)	6748	4470
Number of fruits		795	260
Weight per fruit	(g)	8.2	17.2
Essential oils:			
Actual yield	(ml)	34.5	24.1
Volume per kg of fruit	(ml)	5.3	5.4
Terpeneless fraction (% by volume of raw oil)		10.5	8.6
Limonene (% by volume of raw oil)		88.5	87.7

Source: Mauricio *et al.* (1967).
[1]Steam-distilled essential oils.

FNRI (1980) reported that 100 g edible portion of calamansi fruit contains 89.8% moisture, 0.4 g protein, 1.0 g fat, 8.3 g total carbohydrates, traces of fiber, 0.5 g ash, 18 mg calcium, 12 mg phosphorus, 0.8 mg iron, 0.02 mg thiamine, 0.01 mg riboflavin, 0.2 mg niacin and 45 mg ascorbic acid.

The amino acid composition of the edible portion of calamansi was also investigated by Sutaria and San Diego (1982). These researchers revealed that the amino acid composition of the edible portion of calamansi on a percent dry weight basis were as follows: 2.148 aspartic acid, 0.128 theonine, 0.341 serine, 0.662 glutamic acid, 0.226 proline, 0.04 tyrosine, 0.06 phenyl alanine, 0.063 lysine, 0.016 histidine and 0.026 arginine.

The pectin content, as pectic acid, of fresh calamansi peel and juice was 4.64% and 0.06%, respectively (Cruel 1979). On the other hand, Villasana (1985) reported that the yield of pectin from dried calamansi peels was 26.78% which is comparable to that of lemon (28.70%) and

TABLE 18.10. PROXIMATE COMPOSITION[1] OF VARIOUS PARTS OF CALAMANSI FRUIT

	Proximate composition (%)				
	Crude protein	Crude fat	Crude fiber	Crude ash	N-free extract
Peel	6.38	2.95	22.05	4.41	64.21
Pulp	5.30	1.82	15.68	6.99	70.21
Whole seed	3.87	37.36	19.55	14.35	24.87
Seed hull	2.18	19.22	42.13	8.04	28.43
Seed meal[2]	7.16	3.75	16.71	27.16	45.22

Source: Mauricio *et al.* (1967).
[1]Dry basis.
[2]Obtained after extracting the seeds with hexane.

grapefruit (22.90%) with a methoxyl content of 12.19-12.49%. The setting time at 30°C of the pectin extracted from calamansi peel was 17-18 sec which is considered a rapid-set pectin. The moisture and ash contents of the pectin extracted by Villasana (1985) were 20.44-20.92% and 3.49-3.69%, respectively. The viscosity of the pectin sols (0.1-0.3%) from calamansi peels ranged from 1.40-1.49 cp.

Darwiz (1981) isolated, fractionated and characterized the alcohol-insoluble solids (AIS) of calamansi peels. The alcohol-insoluble solids of fresh calamansi peels were extracted using 80% ethanol and the values ranged from 9 to 11% depending on the month when sampling was conducted. Table 18.11 shows the AIS of calamansi peels.

Pectin, hemicellulose and cellulose were obtained by extracting the AIS with water, hydrochloric acid and potassium hydroxide, respectively. Table 18.12 summarizes the different fractions of the AIS of the calamansi peel. Composition of the different fractions were further analyzed and results are shown in Table 18.13. Based on these findings, it can be concluded that calamansi peel is a good source of pectin.

The volatile flavor constituents of fresh and processed calamansi juice were determined by Mina (1980). Limonene was found to comprise 88.08% of the total volatile components of fresh calamansi juice. In the bottled juice, the volatiles consisted of acetaldehyde, acetone, ethanol, propanol, acetic acid and limonene. The presence of linalool, terpinene-4-ol, terpineol, citronellol and geraniol were detected in pasteurized and stored juices. The increasing concentration of terpineol and terpinene-4-ol contributed to the more pronounced cooked and chemical flavor in processed juices.

TABLE 18.11. COMPOSITION[1] OF ALCOHOL-INSOLUBLE SOLIDS OF CALAMANSI PEELS (DRY BASIS)

Sampling time	Sample no.	Protein (%)	Crude fiber (%)	Ash (%)
October	1	8.40	60.76	4.24
	2	8.44	60.68	4.56
	3	7.91	61.10	4.37
November	1	8.35	60.67	4.86
	2	7.96	60.39	4.41
	3	7.67	60.90	4.19
December	1	7.92	60.75	4.51
	2	7.62	60.59	4.37
	3	7.94	60.77	4.10

Source: Darwiz (1981).
[1]N = 2.

TABLE 18.12. COMPOSITION[1] OF THE VARIOUS FRACTIONS OF THE ALCOHOL-INSOLUBLE SOLIDS OF CALAMANSI PEEL (DRY BASIS)

Data	Sample no.	Fractions				
		Pectin			Hemi-cellulose	Cell-ulose
		WS[2]	AS[3]	Total		
October	1	6.26	17.22	23.48	11.02	49.74
	2	6.23	17.28	23.51	11.22	49.46
	3	6.34	16.89	23.23	11.47	49.63
November	1	6.40	17.45	23.85	11.23	49.45
	2	5.99	17.56	23.55	10.99	49.40
	3	6.45	17.05	23.59	11.53	49.37
December	1	6.42	17.47	23.89	11.44	49.31
	2	6.38	17.20	23.58	11.32	49.27
	3	6.47	17.12	23.61	11.43	49.34

Source: Darwiz (1981).
[1]N = 2.
[2]Water-soluble pectin fraction.
[3]Acid-soluble pectin fraction.

Using thin layer chromatography, Anzaldo (1980) was able to detect the presence of β-pinene, α-pinene, nerol and terpinene from calamansi oil. Hydrosteam distillation was employed to extract the calamansi oil. The yield was only 0.77%.

PROCESSING

Product Development

Due to the popularity of calamansi juice, various processed products have been developed to produce a processed juice with high acceptability. BPI (1983) developed bottled calamansi juice, calamansi syrup, calamansi nip and preserves. The process for bottled juice involved addition of one cup sugar per gallon of juice, hot filled and processed at 80°- 82°C for 20-30 min.

Calamansi syrup is prepared by adding two parts sugar to one part juice and kept in the refigerator without prior heat processing. Another product similar to calamansi syrup called nip was also formulated. Calamansi juice is mixed with sugar until the final concentration is 60°Brix. The mixture is allowed to stand undisturbed for three days. The lower layer is called nip which is subsequently siphoned and bottled.

The process for calamansi preserves calls for the cutting of slits on the stem end of the green calamansi fruits. The juice and seeds are squeezed out. The remaining pulp and peel are soaked in water for three

TABLE 18.13. COMPOSITION[1] OF DIFFERENT FRACTIONS OF ALCOHOL-INSOLUBLE SOLIDS OF CALAMANSI PEEL

Fractions	Neutral sugar							Not accounted for (%)
	Rhamnose (%)	Arabinose (%)	Xylose (%)	Mannose (%)	Galactose (%)	Glucose (%)	AGA[2] (%)	
Fraction A (pectins)	0.52	15.52	—	2.89	1.89	2.75	70.60	4.87
Fraction B (hemicellulose)	—	—	77.97	—	6.31	13.20	—	2.60
Fraction C (cellulose)	—	—	13.90	1.11	—	82.62	—	2.37

Source: Darwiz (1981).

[1]N = 2.

[2]AGA = anhydrogalacturonic acid.

days. After soaking, the calamansi is boiled in water 3-4 times, changing the water after each boiling. The prepared calamansi are then boiled for 15 min in syrup (2 parts sugar and 1 part water) and, subsequently, soaked overnight in the same syrup. The process is repeated until the syrup is thick after which the calamansi preserve is ready for bottling.

Foam mat drying was used by Espiritu (1968) in the preparation of ready-to-drink calamansi juice from dried powder. Calamansi puree was acidified with 7% citric acid and total soluble solids adjusted to 50°Brix. Egg albumin was used as the foaming agent. The resulting foam was dried at 60°-71°C for 80 min. Based on the quality of the reconstituted juice, it was noted that the best dried material can be produced from a foam with 2% egg albumin and a foam density of 0.33 g/ml. The final moisture content should be 1.8%.

Lim (1977) prepared a vitamin C enriched-calamansi syrup. Different ratios of pasteurized calamansi juice and 60°Brix syrup were tried to attain the best flavor blend. It was reported that 40:100 (juice:syrup) ratio was the best.

Standardization for the preparation of candied calamansi peels was conducted by Batulan (1977). Results indicated that the peel should be boiled three times in water for 10 min to remove the bitter principle. The average equilibrium plumping time was about 3 hr and the resulting product was dried for 7 hr in a cabinet drier which is equivalent to 24 hr sun drying.

Magivili (1971) prepared powdered pectin from calamansi seeds. The resulting pectin contained 0.32-0.47% calcium pectate with the methoxyl content of 0.035-0.52%. Percent sag of the test jellies prepared from calamansi peels was 21-25.9%. The values obtained were comparable to percent sag obtained from the test jellies using the commercial pectin which was 27.2%.

Freezing was used by Soliven (1958) to concentrate calamansi juice. The method involved squeezing of the juice through a cheese cloth and placing the filtered juice into glass containers. The juice was then transferred into a deep-freezer until frozen. The slush frozen juice was broken up using a spatula and the ice-concentrate mixture centrifuged to separate the ice from the juice. The process of freezing and centrifuging was repeated until the desired density for the concentrated juice was obtained. Warm sugar cane syrup or corn syrup was added to the concentrated juice.

Factors Affecting Quality of Processed Calamansi

Pineda (1980) evaluated the effect of crushed seeds on the sensory qualities of processed calamansi juice. Crushed seeds and emulsifier were added to the juice at the rate of 1-10% and 2%, respectively. The resulting mixture was homogenized, heated to 70°C for one minute then filled into sterilized bottles. The bottled juice was pasteurized for 15 min. Results of this study indicated that bottled juice without added seeds was judged not bitter. Intensity of bitterness perceived in juices increased proportionally with the amount of seeds added to the juice. It was suggested therefore, that incorporation of the seeds during extraction of calamansi juice should be minimized if not totally prevented.

Addition of sugar to calamansi juice without heat treatment was found to be very effective in preserving the flavor of calamansi (Tinsay 1978). It was reported that processed calamansi juice developed bitterness on storage. On the other hand, calamansi juice was found to develop a shoe polish or painty character note on standing even at refrigeration temperatures. These undesirable changes were prevented with the addition of sugar to calamansi juice and then stored at freezing temperatures. The reconstituted juice exhibited a mellow flavor better than the freshly extracted juice which had been reconstituted and then sweetened.

Fortification of calamansi juice mixed at the rate of 40:100 (juice:60°Brix syrup) with ascorbic acid improved its flavor. Sulfiting of the juice prior to bottling was suggested to prevent browning of the juice during storage (Lim 1977). Addition of sulfite did not affect the pH, total soluble solids and titratable acidity of the processed juice.

Alcid (1981) studied cloud stabilization of calamansi juice during storage. It was noted that addition of 0.1% citrus pectin resulted in a juice with the most stable cloud during storage. It was also noted that pH had no significant effect on the cloud stability of the juice.

The effect of extraction method and addition of sulfites on the quality of calamansi juice stored at 55°C was evaluated by Abarquez (1972). It was found that the pulped samples were more viscous than hand-reamed juices. This was due to the higher amount of pulp per unit volume of juice. Furthermore, the pulped samples exhibited greater degree of separation than the hand reamed samples. Also, the pulped samples showed pronounced aftertaste, thus it was less acceptable than the calamansi juice extracted by hand reaming.

Addition of sulfites did not affect the chemical properties of the calamansi juice on storage at 55°C. However, better color retention in the calamansi juice was observed when sulfite was added. A yellowish brown color developed in hand-reamed juice in contrast to the initial bright yellow after 3 weeks storage at 55°C. Viscosity of the juices also decreased as storage time progressed.

Gatchalian and Santos (1970) evaluated the effect of blanching and extraction methods on the yield and quality of fresh calamansi juice. Two

TABLE 18.14. EXTRACTION METHODS EMPLOYED IN THE STUDY OF GATCHALIAN AND SANTOS (1970)

Symbol	Description
I A	Fruits were peeled and juice squeezed through a muslin bag
I B	Fruits cut 1/3 from the stem end, hand squeezed and filtered
I C	Fruits cut half-way, hand squeezed and filtered
II A	Fruits cut in both ends and machine-pulped
II B	Fruits cut 1/3 from stem end and machine-pulped
II C	Fruits cut half-way and machine-pulped

methods of extraction were employed: hand squeezing or machine pulping using a Reeves pulper provided with a 0.90mm sieve. Table 18.14 summarizes the different extraction procedures used in this study. The chemical properties of calamansi juice obtained from the different extraction methods are presented in Table 18.15. Data revealed that hand squeezing of calamansi resulted in juices with higher soluble solids than machine-pulped fruits regardless of whether or not the fruits were blanched.

Viscosity of the juice was affected by the method of extraction. Peeling of the calamansi fruits before hand squeezing resulted in a very viscous juice. The juice did not show distinct pulp particles. On the other hand, juices from machine-pulped fruits were also viscous but appeared to contain a considerable amount of very fibrous particles which were likely to be fragments of albedo. Blanching did not significantly affect the viscosity of the calamansi juice after hand squeezing or machine pulping.

Sensory evaluation data revealed that hand-squeezed samples were bitter, orange in color and judged as the least acceptable. Sliced,

TABLE 18.15. CHEMICAL PROPERTIES OF CALAMANSI JUICE USING DIFFERENT EXTRACTION METHODS

Extraction methods[1]	Soluble solids, °Brix		Titratable acidity, %		pH	
	Unblanched	Blanched	Unblanched	Blanched	Unblanched	Blanched
I A	8.3	8.0	5.71	4.76	2.50	2.51
I B	7.6	7.4	6.13	5.03	2.35	2.50
I C	7.5	7.5	6.03	5.22	2.35	2.40
II A	5.5	5.5	4.63	4.78	2.35	2.50
II B	6.0	5.8	5.42	4.37	2.40	2.45
II C	5.7	5.7	4.70	4.19	2.35	2.45

Source: Gatchalian and Santos (1970).
[1]See Table 18.14.

blanched and hand-squeezed calamansi juice was yellow orange in color, no bitterness and considered most acceptable.

Payumo *et al.* (1965) investigated the effect of the method of extraction on the quality of bottled calamansi juice stored at ambient and at refrigerated temperature (4°-5°C). The extraction procedures employed were:

I. Cutting small portions at one end of the fruit and squeezing the juice by hand into a muslin bag to remove the seeds;
II. Same as (I) except the fruits were peeled before cutting;
III. Same as (I) except the fruits were blanched in boiling water for 1 min and immediately dipped in cold water before cutting;
IV. Same as (III) except that the fruits were blanched for 2 min.

Results of this research are presented in Tables 18.16 and 18.17. It was noted that blanching the fruits for one to two minutes in boiling water resulted in a juice that was more acceptable than the juice prepared from unblanched or peeled fruits after 6 months storage at room temperature. However, freshly prepared juice from unblanched fruits was found to possess the best sensory quality attributes.

Juices that were stored at refrigeration temperatures retained ascorbic acid better than those stored at room temperature. Method of extraction had no effect on the retention of ascorbic acid during storage.

Lim (1977) attempted to establish the mechanism involved in the ascorbic acid oxidation in the nonenzymatic browing of heat-treated calamansi juice. Freshly extracted juice was filtered, heated to 96°C for 3 min and then refiltered. The prepared juice was either treated with ion exchange with subsequent deaeration or deaerated only. The treated juices were pasteurized at 82°C and filled hot into sterilized 60 ml narrow-mouth bottles. Black plastic caps were used to seal the bottles. The processed juices were stored at 37°C for 40 days.

Results of this study indicated that ion-exchange treated juice retained the most ascorbic acid during storage, whereas the control and deaerated samples lost 100% ascorbic acid after 30 days storage.

Furfural was detected as early as the fifth day of storage which indicated that catalytic oxidation was already occurring. The absence of furfural at the end of storage of the untreated calamansi implied that it had already reacted to form the dark pigment.

The hydrogen peroxide content of the juice declined continously during storage with the subsequent increase of the dissolved oxygen. The dissolved oxygen was used during formation of the metal ascorbate com-

TABLE 18.16. RETENTION (%) OF ASCORBIC ACID IN CALAMANSI JUICE PREPARED BY DIFFERENT METHODS OF EXTRACTION AND STORED AT AMBIENT TEMPERATURE AND AT 5°C

Extraction methods[1]	Storage period months	Ascorbic acid (mg/100 ml)		Retention	
		ambient	5°C	ambient	5°C
I	0	37.25	37.25	—	—
	1	37.14	37.39	99.70	100.38
	2	32.74	34.84	87.89	93.63
	4	29.52	30.07	79.25	80.56
	6	29.14	29.40	78.23	78.93
II	0	36.74	36.74	—	—
	1	34.81	36.65	94.75	99.76
	2	29.52	31.61	80.35	86.04
	4	27.53	31.37	74.94	85.38
	6	25.45	27.85	69.27	75.80
III	0	37.04	37.04	—	—
	1	33.73	35.59	91.06	96.08
	2	31.34	31.58	84.61	85.26
	4	29.65	31.05	80.05	83.83
	6	28.82	30.40	77.81	82.07
IV	0	37.02	37.02	—	—
	1	31.75	33.28	85.76	89.90
	2	31.09	33.27	83.98	89.87
	4	27.42	29.44	74.07	79.52
	6	21.25	27.39	57.40	73.99

Source: Payumo *et al.* (1965).
[1]See text for detailed descriptions.

TABLE 18.17. PHYSICOCHEMICAL COMPOSITION OF CALAMANSI JUICE EXTRACTED BY DIFFERENT METHODS OF EXTRACTION

Physicochemical characteristics	Extraction method[1]			
	I	II	III	IV
Yield of juice, percent (w/v)	39.0	29.0	40.0	39.0
Total soluble solids, °Brix	7.0	7.0	6.5	6.5
Acidity, as percent anhydrous citric acid	7.65	7.61	7.58	7.57
pH	2.4	2.4	2.5	2.5
Physical appearance of juice	Bright orange color	Dull orange color slimy	Bright orange color	Bright orange color

Source: Payumo *et al.* (1965).
[1]See text for detailed descriptions.

plex as evidenced by the total absence of the dissolved oxygen at the end of the storage period.

Absorbance and loss of ascorbic acid were significantly correlated in samples treated with ion exchange with or without deaeration. It appeared that partial removal of trace metals and dissolved oxygen from the juice promoted another reaction which lead to the increase in absorbance and ascorbic acid.

Based on these findings, there are several mechanisms of ascorbic acid oxidation going on simultaneously in the stored juice. In the presence of trace metals and dissolved oxygen, the metal-catalyzed ascorbic acid oxidation by molecular oxygen predominated. However, in the absence of trace metals and dissolved oxygen, the loss of ascorbic acid with subsequent increase in browning may have proceeded through an undetermined mechanism.

Ascorbic acid, total carotenoid, provitamin A and sensory quality attributes were determined by Nisperos *et al.* (1982) after various process-

TABLE 18.18. MEAN[1] SENSORY SCORES FOR PURE CALAMANSI JUICE (PASTEURIZED AND STORED FOR ONE WEEK AND 1 1/2 YEARS) DILUTED WITH DIFFERENT AMOUNTS OF WATER

	Dilutions				
Sensory Qualities	1:6	1:7	1:8	1:9	1:10
Color					
Stored 1 week	3.35	3.31	3.12	2.81	2.77
Stored 1 1/2 yr	3.57	3.47	3.47	3.23	3.17
Flavor blend					
Stored 1 week	3.12	2.88	2.88	2.73	2.38
Stored 1 1/2 yr	2.87	3.10	3.00	2.83	3.00
Presence of stale flavor					
Stored 1 week	3.42	3.62	3.54	3.54	3.38
Stored 1 1/2 yr	3.73	3.63	3.87	3.67	3.87
Presence of cooked flavor					
Stored 1 week	3.19	3.62	3.46	3.54	3.27
Stored 1 1/2 yr	3.77	3.70	3.70	3.50	3.73
General Acceptability					
Stored 1 week	2.77	3.04	2.81	3.12	2.77
Stored 1 1/2 yr	2.83	2.93	3.10	3.00	3.07

Source: Nisperos *et al.* (1982).

[1]N = 26 for pasteurized and stored for 1 week. N = 30 for pasteurized and stored for 1 1/2 years. Range of scores for color: 5—bright yellow; 1—brownish yellow. Flavor blend: 5—very good blending; 1—poorly blended. Presence of stale flavor: 5—none; 1—stale flavor. Presence of cooked flavor: 5—none; 1—cooked flavor. General acceptability: 5—acceptable; 1—unacceptable.

ing operations and storage of calamansi juice. Results of this study are summarized in Tables 18.18 to 18.20. It was observed that after homogenization and pasteurization, ascorbic acid content of the juice was not affected. However, after 12 days storage at room temperature, ascorbic acid content was reduced from 52.84 mg/100 ml to 38.64 mg/100 ml.

Homogenization reduced total carotenoids from 0.37 mg/100 ml to 0.18 mg/100 ml and the provitamin A from 0.26 mg/100 ml to 0.17 mg/100 ml. Titratable acidity also changed from 7.02 mg/100 ml to 5.44 mg/100 ml, pH from 2.22 to 3.33, soluble solids from 8 to 8.73°Brix and dry matter from 8.58% to 9.58 during homogenization. There were no observed changes in these parameters during subsequent processing steps.

Sensory qualities of the processed juice stored for 1 week up to 1 1/2 years were not significantly different from each other. Cooked flavor was not detected in pasteurized juice nor was stale flavor apparent in stored samples.

TABLE 18.19. TITRATABLE ACIDITY[1], PH, SOLUBLE SOLIDS AND DRY MATTER OF CALAMANSI JUICE DURING PROCESSING AND STORAGE

Processing step and length of storage	Titratable acidity (mg citric acid/100 ml)	pH	Soluble solids (°Brix)	Dry matter (%)
After squeezing	7.02a	2.22c	8.00f	8.58e
After homogenization	5.44c	2.33d	8.73bc	9.58bc
After heating to 70°C and de-aeration	5.59bc	2.43c	8.67bc	9.79b
After pasteurization	5.50bc	2.42c	9.00abc	10.03a
After 24 hours	5.54bc	2.63a	9.20a	9.84b
After 5 days	5.60b	2.62a	8.80abc	9.21d
After 10 days	4.86cd	2.63a	8.60cd	9.21d
After 12 days	4.84cd	2.52b	8.93bc	9.11d
After 17 days	4.86d	2.56ab	9.13ab	9.35cd
After 27 days	4.85d	2.42c	8.93ab	9.00d
After 37 days	4.85d	2.40cd	8.93ab	9.18d
After 47 days	4.81cd	2.48bc	9.00ab	9.26cd
After 7 months	5.47bc	2.62a	8.80abc	9.00ab
After 8 months	4.77de	2.60ab	8.60cde	9.11d
After 9 months	4.73de	2.55b	8.20ef	9.10d
After 12 months	4.71e	2.55b	8.50de	9.10d

Source: Nisperos *et al.* (1982).

[1]Means folowed by a common letter are not significantly different (P = 0.05) from each other.

TABLE 18.20. ASCORBIC ACID[1], TOTAL CAROTENOIDS AND PROVITAMIN A OF CALAMANSI JUICE AFTER EVERY PROCESSING STEP AND LENGTH OF STORAGE

Processing step and length of storage	Ascorbic acid (mg/100 ml)	Total carotenoids (mg/100 ml)	Provitamin A (mg β-carotene 100 ml)
After squeezing	52.84a	0.37a	0.26a
After homogenization	42.20abc	0.18bc	0.17b
After heating to 70°C and de-aeration	45.10abc	0.17bc	0.11c
After pasteurization	42.35abc	0.21b	0.11c
After 24 hours	45.68ab	0.18bc	0.10c
After 5 days	43.92abc	0.18bc	0.096cd
After 10 days	40.30abc	0.17bc	0.08cd
After 12 days	38.64bc	0.20bc	0.08cd
After 17 days	40.32abc	0.20bc	0.05d
After 27 days	24.72cd	0.21b	0.09cd
After 37 days	32.00cd	0.21b	0.07cd
After 47 days	24.72cd	0.21b	0.07cd
After 7 months	23.60cd	0.13bc	0.04d
After 8 months	16.80c	0.11c	0.04d
After 9 months	18.40e	0.10c	0.05d
After 12 months	14.40e	0.10c	0.05d
HSD	13.20	0.096	0.06
CV (%)	12.64	17.78	23.70

Source: Nisperos *et al.* (1982).

[1]Means followed by a common letter are not significantly different (P ≤ 0.05) from each other.

Uses

Calamansi has many uses. Its juice finds its way in many food preparations. It is used to add flavor to noodle dishes as well as to broiled fish. It can also be used to marinate meat as substitute for lemon juice and as flavoring for iced tea. Calamansi juice is also utilized as an acidulant in the preparation of jams and jellies.

Tenmatay (1965) evaluated the use of calamansi juice as flavoring for "pinipig-bibingka" (a type of rice cake) and coconut candy. Results of this study revealed that calamansi juice can be used as flavoring for these two products.

Aside from its juice, calamansi peels can be glazed or made into candies (TRC 1986). The rind and young leaves can be used as flavoring for native cakes and pastries giving the product a lemony aroma (Reyes 1982).

The use of calamansi extract as a coagulant in the preparation of white cheese was reported by Gomez *et al.* (1983). Extract was prepared from the combination of pure juice and washings of the pulp. Results indicated that 20 ml of the extract can coagulate 100 ml of the carabao's milk, whereas only 15 ml of the extract was utilized to coagulate cow's milk after heating the mixture at 50°C. The curd obtained from cow's milk was soft and spreadable.

Calamansi possesses also some medicinal and pharmaceutical properties. The juice can be used as a hair-wash, deodorant and can also be used to remove stain from clothes. Other applications include bleaching agent for skin blemishes and freckles, and cure for acne and cough. The oil from the leaf has a carminative action even in quantities less than 1 ml (Quisumbing 1962; Gutierrez 1982).

LITERATURE CITED

ABARQUEZ, P. C. 1972. Effect of extraction methods and sulfiting on qualities of kalamansi juice stored at accelerated temperature. Thesis. University of the Philippines, Diliman, Q.C.

ALCID, G. 1981. Cloud stabilization of kalamansi (*Citrus microcarpa* Bunge) juice. Thesis. University of the Philippines, Diliman, Q.C.

ANZALDO, F. E. 1980. Essential oil production in the Philippines: Problems and Prospects. NSDB Tech. J. Apr.-Jun., 67-70.

BATULAN, G. 1977. Technical feasibility for pilot scale production of candied calamansi (*Citrus microcarpa* Bunge) peel. Thesis. University of the Philippines, Diliman, Q.C.

BPI. 1983. Preservation of Fruits and Vegetables. Bureau of Plant Industry, Manila.

CO, J. L., LUIS, E. D. and AMIGLEO, S. Q. 1979. Calamansi peel is more nutritious. NIST News Release. Metro Manila.

CRUEL, C. 1979. pH and pectin determination of ten Philippine fruits. Thesis. University of the Philippines, Diliman, Q.C.

DARWIZ, A. A. 1981. Isolation, fractionation and characterization of alcohol insoluble solids of calamansi peels. Thesis. University of the Philippines at Los Baños, College, Laguna.

DATA, E. S., MENDOZA, D. B. and PANTASTICO, ER. B. 1975. Storage of calamansi fruits (*Citrus mitis* Blanco). I. Chemical changes. Phil. Agric. 59 (3 & 4) 119-125.

DE CASTRO, F. A. 1973. Storage studies on calamansi fruits. II. Physiological disorder with particular reference to chilling injury and transportation. Thesis. University of the Philippines at Los Baños, College, Laguna.

DEL ROSARIO, R. P., MENDOZA, D. B. and PANTASTICO, ER. B. 1976. Storage studies of calamansi fruits. IV. Respiration. Phil. Agric. 60, 249-255.

ESPIRITU, J. 1968. Foam mat drying of calamansi puree. Thesis. University of the Philippines, Diliman, Q.C.

FNRI 1980. Food Composition Table. Food and Nutrition Res. Inst., Manila.

FOOD TERMINAL, INC. 1979. Qualities of good tropical fruits. Marketday (Phil.), April, p. 25.

GATCHALIAN, M. M. and SANTOS, L. M. 1970. The effect of pretreatment and extraction method on yield and quality of fresh kalamansi (*Citrus microcarpa* Bunge) juice. Phil. J. Sci. 99, 241-253.

GOMEZ, I. V, MOMONGAN, L. D., LAWAS, F. R. and PARDUCHO, R. R. 1983. Use of calamansi extract in dairy products manufacture. An. Hus. Agric. J. 17(10) 14-17.

GUTIERREZ, H. G. 1982. Manual of Philippine Materia Medica. 5(1) 78-79.

LIM, J. S. 1974. Ascorbic acid oxidation in the non-enzymatic browning of heat-treated kalamansi (*Citrus microcarpa* Bunge) juice. Thesis. University of the Philippines, Diliman, Q.C.

LIM, Y. B. L. 1977. Vitamin C-enriched kalamansi (*Citrus microcarpa* Bunge) syrup. Thesis. University of the Philippines, Diliman, Q.C.

MAGIVILI, T. H. 1971. Pectin preparation from calamansi peels and watermelon rind. Thesis. University of the Philippines, Diliman, Q.C.

MAURICIO, T. C., DE LEON, J. T., SERRANO, L. B. and CARANGAL, A. R. 1967. Physico-chemical properties of some commercial varieties of citrus in the Philippines. Phil. Agric. 57(6) 524-535.

MAYUGA, D. A. 1973. Storage studies on calamansi fruits. III. Decay control. Thesis: University of the Philippines at Los Baños, College, Laguna.

MENDOZA, D. B. and PANTASTICO, ER. B. 1979. Post-harvest physiology, handling and storage of fruits and vegetables: Semicommercial scale. UP-NSDB Integrated Research Program Report. Phase II. College, Laguna.

MINA, C. 1980. Volatile flavor constituents of fresh and processed calamansi (*Citrus microcarpa* Bunge) juice. Thesis. University of the Philippines, College of Home Econ., Diliman, Q.C.

NISPEROS, M. O., RAYMUNDO, L. C. and MABESA, L. B. 1982. Ascorbic acid color, provitamin A and sensory qualities of calamansi (*Citrus mitis* Linn) juice after various processing operations and lengths of storage. Phil. Agric. 65, 353-361.

PANTASTICO, ER. B. and MENDOZA, D. B. 1976. Post harvest physiology, handling and storage of fruits and vegetables. UP-NSDB Integrated Research Program Report. Phase I. College, Laguna.

PANTASTICO, ER. B., SUBRAMANYAM, H., BHATTI, M. B., ALI, N. and AKAMINE, E. K. 1975. Harvest indices. *In* Postharvest Physiology, Handling and Utilization of Tropical and Subtropical Fruits and Vegetables. Er. B. Pantastico (Editor). Avi Publishing Co., Westport, CT.

PAYUMO, E. M., ABDON, A. A., PILAC, L. M., MANIQUIS, P. L. and ALBASTRO, V. Q. 1965. Effect of methods extraction of the quality of processed kalamansi (*Citrus microcarpa* Bunge) juice. Phil. J. Sci. 94, 449-457.

PCARRD. 1980. The Philippine Recommends for Citrus. Philippine Council for Agriculture Resources Research and Development. Los Baños, Laguna.

PINEDA, N. J. 1980. Effect of crushed seeds on the sensory qualities of processed calamansi. Thesis. University of the Philippines at Los Baños, College, Laguna.

QUISUMBING, E., 1962. Medicinal Plants of the Philippines. Bureau of Printing. Metro Manila.

REYES, A. 1982. Miracle giant calamansi. Bulletin Today. Metro Manila.

SOLIVEN, F. A. 1958. The concentration of calamansi juice by freezing. J. Phil. Pharm. Assoc. 45(3) 67-69.

SUTARIA, P. B. and SAN DIEGO, L. 1982. Essential amino acid analyses of selected Philippine vegetable and fruits. Phil. J. Sci. 3(1-2) 45-55.

TENMATAY, L. 1965. The preparation and utilization of calamansi (*Citrus mitis* Blanco) fruit pulp. Thesis. University of the Philippines, College of Home Economics, Diliman, Q.C.

TINSAY, E. 1978. Flavor difference between freshly extracted sweetened kalamansi juice and frozen kalamansi syrup. Thesis. University of the Philippines, Diliman, Q.C.

TRC. 1986. Kalamansi. Technology Resource Center, Makati. 19.19.

VILLASANA, G. E. 1985. Comparative study of pectin extracted from calamansi (*Citrus madurensis* Linn.) and pomelo (*Citrus grandis* (L.) Osbeck) peels. Thesis. University of the Philippines at Los Baños, College, Laguna.

19

Lucuma

L. Antonio Lizana

Lucuma fruit is known for its delicate flavor and is used in pastries, ice creams, milk shakes, pies, tarts and other dessert specialties. The fruit has very high acceptance in local markets of Chile and Peru, but less so in Ecuador. In addition to uses as a dessert fruit, some technology has been developed to extract, preserve and utilize the fruit pulp. A dehydrated, powdered product is being produced by a tomato cannery in Peru (Morton 1987).

The generic name "lucuma" is generally used to describe this fruit, but it is also referred to by the following common names: lucumo, lucmo, lucma, rucma and mamon (Morton 1987). The lucumo (lucuma) is a subtropical tree that grows in frost-free valleys along with avocados, cherimoyas and mountain papayas. The tree apparently originated on the slopes of the Ecuadorian and Peruvian mountains (Popenoe 1934; Baehni 1942). Distribution from that area to Chile and Bolivia to the south and Colombia and Costa Rica to the north is apparent.

Very few studies have been conducted on lucuma tree growth, and only recently has information become available on grafting, budding and selection of fruit types. There have been several attempts to grow the tree in southern Florida, but success has been marginal. The tree flowers between June and August in Florida, and the fruit matures in about 180 days (Martin *et al.* 1987).

BOTANY

The botanical classification is: order—Ebenales; family—Sapotaceae; genus—*Lucuma* or *Pouteria* (after Baehni). The genus, *Pouteria*, encompasses about 318 species distributed in the Americas, Africa, Asia and the tropical Pacific. In Chile there are two species; the first is *Lucuma valparadisea* Mol. or *Pouteria splendens* (A.DC.) O. Ktze. which is commonly known as 'palo colorado' due to the color of its wood and its nonedible fruits. The second species is *Pouteria lucuma* Ruiz & Pavon, and this species produces edible fruit generally known as the lucuma. Other synonyms used for the lucuma taxonomic classification include: *Achras lucuma* Ruiz & Pavon, *Lucuma bifera* Mol., *Lucuma turbinata* Mol., *Lucuma biflora* Gmel., *Lucuma obovata* H.B.K. (*Lucuma obovata* Baehni), *Lucuma obovata* var. Ruiz A.DC. and *Richardella lucuma* (Ruiz & Pavon) Aubr. The current most valid name for lucuma is *Pouteria lucuma* Ruiz & Pavon (Munoz 1987).

A related species is *L. mammosa* Gaertn. (*Calocarpum sapota* Merr.) or *Pouteria mammosa* (L.) Cronquist, sapote or mamey sapote. The lucuma is also related to canistel or egg fruit (*Pouteria campechiana* (H.B.K.) Baehni). The egg fruit is native to the tropics of Central America and has a small fruit with pulp consistency and color similar to a hard-boiled egg yolk (Popenoe 1934, 1963). Lucuma fruits have the same characteristics as the egg fruit but are bigger in size. Bailey and Bailey (1977) no longer recognize *Lucuma* as a valid genus name and suggest the genus designation *Pouteria*.

HORTICULTURE

In Chile, the lucuma fruit grows well in two areas, namely, Quillota (Aconcagua River valley) and La Serena (Elqui River valley). Both locations are frost-free and near sea level or no higher than 100 m in altitude. The valleys are somewhat influenced by the ocean, but are not directly exposed to the ocean. Field reports from these main growing areas of Chile indicate that the behavior of the lucuma tree is quite similar to the Hass avocado tree (Lopez 1984) with respect to lower temperature tolerance. The tree can also grow inland at up to 1800 m in some of the northern valleys of Chile. In countries with a more tropical climate, the fruit tends to grow better under milder subtropical conditions, as for example, in dry areas of Bolivia (2000 m altitude) and up to 3000 m in some locations in Peru. Besides fruit consumption, lucuma wood is used for fine furniture manufacturing in Peru.

The lucuma is a tall tree (8-15 m) similar in size to the avocado, but in general has a darker green color. The leaves are smaller than avocado leaves and more coriaceous. Most trees in full production come directly from seeds. This indicates a very broad variation in fruit types, especially in size, pulp color and number of seeds (Popenoe 1934). Also important is the variation in pulp texture, time of harvest and fruit yield.

The juvenile stage might take from 15-20 years under natural growing conditions. Together with low and slow seed germination, and slow growth of the tree, many fruit growers have been discouraged from starting commercial plantings.

Some research on seed germination has been encouraging. A temperature period of 26°C and application of 2000 ppm GA_3 for 10 hr increased seed germination and seedling growth (Saavedra 1975A). Also, successful grafting or budding has been shown to be affected by the vegetative condition of the rootstock, the physiological condition of the scion or budwood used, and the lack of adequate techniques to propagate the tree. Saavedra (1975B) has shown that the most appropriate time to obtain scion wood in Quillota, Chile was from August to October. Also, it was not necessary to use NAA (100 ppm) to speed up cambium development. In Peru, the month of August is suggested for this practice (Anon. 1968). Callus formation is very slow and it is recommended that the wrapping be kept on the graft or bud for 3-4 months (Saavedra 1975B). Temperature fluctuations appear critical in the failure of the wood to take after grafting. Therefore, a controlled greenhouse operation might be the answer to this dilemma (Saavedra 1975B).

In Chile during the last few years, interest in lucuma commercial plantings has increased and some research has been conducted with some progressive fruit growers in La Serena and the Azapa valley in the north (CORFO 1984, 1987). Newer orchards have used selections of the best fruit types, and trees have been grafted or budded on select seedlings which resulted in greater fruit production uniformity and, also, the precocity of coming into bearing. Controlled trials with these tree types have shown up to 7 kg fruit per plant and some fruit types have shown about 80% pulp yield (Alache *et al.* 1988). Lucuma trees are self-fertile and insect pollinated; they take about 8-9 months from fruit set to commercial harvest.

HARVEST MATURITY

Fruit can be harvested in Chile from April to December depending on the production area. One maturity index commonly used for many fruits is the color change of the peel. This index is based on the relationship between color and physiological activities of the fruit, as occurs in lemons, bananas, papayas, mangos and other tropical and subtropical fruits. Here, the change from a green to yellow color is very noticeable (Popenoe 1934). However, this index cannot be effectively used for lucuma because changes in color are not always apparent.

Fruits produced during the first months of the season, April and May, usually show peel color changes from leaf green to yellow green at harvest time. Fruits of mid to late season, harvested from June to December, show very slight color changes. Peel color changes of dark green to light green or no color changes at all make the proper fruit harvest date very difficult to ascertain (Lopez 1984). Moreover, very intense respiratory activity and sugar accumulation are noticeable during the fruit's ripening process without any detectable changes in peel color (Lizana *et al.* 1986). A similar situation occurs with pulp color, but with green fruit there is also latex exudation that can be used as a complementary index besides the light yellow color.

The lack of identifiable fruit selections results in considerable variability in fruit types on the market. Variability exists in peel and pulp colors, ranging from green to yellowish-green in peel color, and light yellow to orange yellow in pulp color. On the other hand, there is a very definitive development pattern associated with soluble solids and pulp firmness that can be utilized as a more precise index of a maturity stage. Since the lucuma fruit has a very dense and dry pulp (drier than mature banana pulp), a simple aqueous method to directly determine soluble solids is not feasible. It is necessary to dilute lucuma pulp with water and disrupt the pulp cells by mechanical means. If pre-mechanical disruption of cells is not undertaken, the soluble solids value will be lower than the true value (Lizana *et al.* 1986).

A simpler index uses pulp softening, and some data are presented employing a presionometer with a 1.1 cm diam head (similar to the one used on apples) applied to pulp without peel (Fig. 19.1). If the fruit ripens on the tree, it becomes soft and very delicate. Careful handling of ripened fruit is advised during harvesting to avoid mechanical injury.

FRUIT

The fruit is oblate, ovate or elliptic; sometimes with a small pointed end at the calyx end and a glabrous-resistant skin of an attractive bright dark to light green color that changes little with ripening. The pulp is hard and has a sticky latex when green, but during the maturation process, the latex disappears, the pulp turns dry and is very sweet, and the typical and attractive flavor appear.

Lucuma has a very distinct climacteric maturation. It produces the typical increase in CO_2 evolution which reaches a maximum of 3-4 fold in five days while other changes are occurring. Pulp resistance decreases from about 13.6 to 0.9 kg; sugars (soluble solids) increase from about 9 to 16%; also, there are distinct changes in pulp color from light yellow to orange yellow. The change in peel color is not well defined in all fruits; some show a very noticeable change from dark green (leaf color) to a yellowish-green color; some change little from dark green to light green,

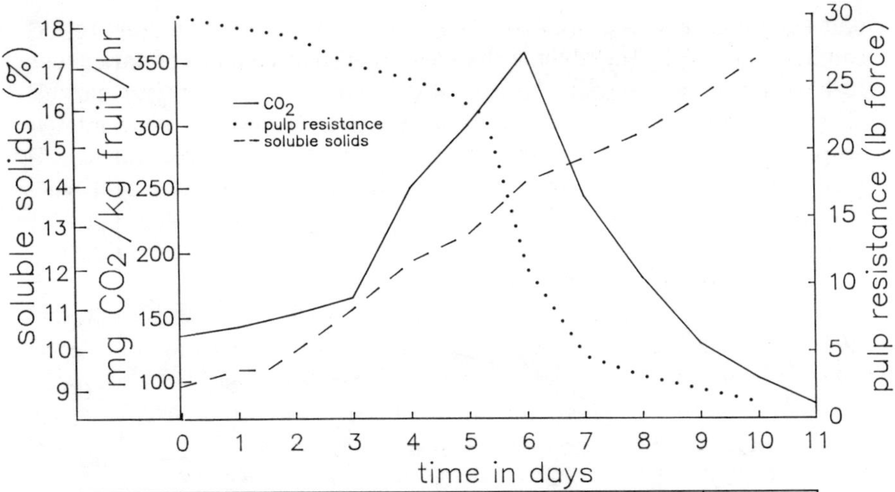

peel color	dark green	light green	yellow green	green yellow	green yellow
pulp color	light yellow	creamy yellow	yellow	dark yellow	orange yellow
maturity	stage 1	stage 2	stage 3	stage 4	stage 5

FIG. 19.1. RESPIRATORY PRODUCTION OF CO_2; PEEL COLOR CHANGES; SOLUBLE SOLIDS, COLOR CHANGES AND SOFTENING OF PULP DURING DISTINCT MATURITY STAGES OF LUCUMA FRUITS (*LUCUMA BIFERA* MOL.)

while others do not change at all. Nevertheless, the maturation period is defined by five stages with distinct differences in the evolution of ripening parameters (Fig. 19.1).

Sugars of lucuma are glucose, fructose, sucrose and inositol (Lizana *et al.* 1986). At the green stages only sucrose was detectable, but as maturation progresses the two monosaccharides—glucose and fructose—increased (Fig. 19.2). Inositol was detected in small amounts, about 0.06%. At full maturity the sugars were present in the following order of importance (g/100 g dry wt), namely, glucose (8.4%), fructose (4.7%), sucrose (1.7%) and inositol (0.06%). The major organic acids were citric and succinic. Apparently, these two acids were present in much greater amounts than other acids.

NUTRITIONAL COMPOSITION

The characteristic dry, mealy pulp consistency of lucuma (similar to a hard-boiled egg yolk) is reflected in its chemical composition (Table 19.1). The medium water content (64-72%) and high caloric value (99-122 cal) of lucuma are comparable to values reported for other tropical fruits,

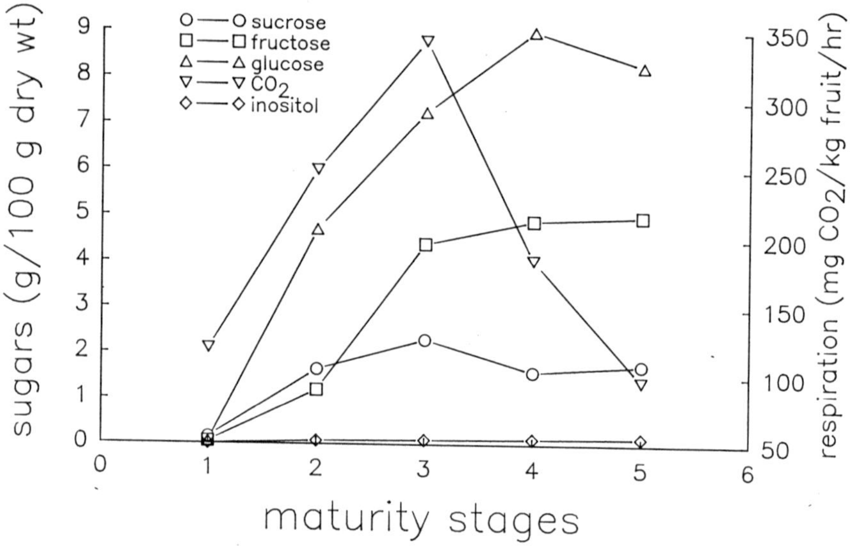

FIG. 19.2. PRODUCTION OF CONSTITUENT SUGARS AND RESPIRATORY CO_2 DURING MATURATION OF LUCUMA FRUITS (*LUCUMA BIFERA* MOL.)

TABLE 19.1. CHEMICAL COMPOSITON OF LUCUMA PER 100 G OF EDIBLE PULP

		Chile[1]	Peru[2]
Calories		122.00	99.00
Water	(g)	64.80	72.30
Proteins	(g)	1.40	1.50
Lipids	(g)	0.50	0.50
Fiber	(g)	1.00	1.30
Ash	(g)	0.90	0.70
Na	(mg)	6.00	—
K	(mg)	470.0	—
Ca	(mg)	—	16.00
P	(mg)	—	26.00
Fe	(mg)	—	0.40
Thiamine	(mg)	0.01	0.01
Niacin	(mg)	2.20	1.96
Ascorbic acid	(mg)	25.00	5.40
Riboflavin	(mg)	—	0.14

[1]Source: Schmidt-Hebbel and Pennacchioti (1985).
[2]Source: Calzada (1972).

namely, avocado, banana, cherimoya and breadfruit (Wenkam and Miller 1965; Nagy and Shaw 1980). Lucuma fruits contain higher amounts (per 100 g serving) of riboflavin, niacin and ascorbic acid than apples and bananas (Wenkam and Miller 1965; Lopez 1984).

PROCESSING

The pulp of the fruit is very dry when ripe and it is usually not eaten directly. The pulp is mashed and mixed with cream or added to food, cakes, pastries, ice creams, tarts and other desserts.

In Chile, the pulp (without peel) is processed and preserved frozen (–8 to –10°C) in plastic containers until used. In Peru, some trials have been conducted to produce lucuma flour. Pulp is sliced into small pieces, dried in a warm air system at 42°C for 24 hr and, finally, ground when it reaches a 10% moisture level. The resulting flour is golden yellow and very sweet with a strong aroma, but with a somewhat rough texture (Lopez 1984).

REFERENCES

ALACHE, J., OSOARIO, A. and OLAVE, J. 1988. Preliminary results concerning the behavior of the lucumo (*Pouteria lucuma*) in the Azapa valley. Simiente 58 (1-2) 7. (Spanish)

ANON. 1968. Cultivation of the lucumo (*Lucuma obovata* H.B.K.). Informative No. 5. Program of Native Fruits. Univ. Agraria, La Molina, Peru. (Spanish)

BAEHNI, C. 1942. Notes on the Sapotaceaes II. The genus *Pouteria*. Candollea 9, 147-476. (Spanish)

BAILEY, L. H. and BAILEY, E. Z. 1977. Hortus Third. Macmillan Publishing Co., New York.

CALZADA, J. 1972. Cultivation of the lucumo. Tec. Bol. No. 2, National Agriculture University, Lima, Peru. (Spanish)

CORFO 1984. Cultivation of the cherimoyo, papaya and lucumo in Chile. Manager of Development. Corporation of Development of Production. Santiago, Chile. (Spanish)

CORFO 1987. Introduction of fruit species and varieties. Soc. Agr. SACOR Ltd., Santiago, Chile. (Spanish)

EBEL, G. 1935. The lucumos in Chile. Rev. Chil. Hist. Nat. 34, 183-203. (Spanish)

LIZANA, L. A., REGINATO, G. and LAVANDEROS, J. C. 1986. Characterization of the maturation of the lucuma (*Lucuma bifera* Mol., *L. obovata* H.B.K.). Proc. Interam. Soc. Trop. Hortic. 30, 105-120. (Spanish)

LOPEZ, C. 1984. Cultivation of the lucumo and its industrialization. Rev. Campesino 115(9) 38-53. (Spanish)

MARTIN, F. W., CAMPBELL, C. W. and RUBERTE, R. M. 1987. Perennial Edible Fruits of the Tropics. U.S. Dept. Agric. Handb. 642.

MORTON, J. F. 1987. Fruits of Warm Climates. Julia F. Morton Publisher, Miami, FL.

MUNOZ, M. 1987. Nomenclature of the lucumo cultivated in Chile. Agric. Tec. (Chile) 47(4) 416-418. (Spanish)

MUNOZ-PIZARRO, C. 1959. Synopsis of Chilean flora. University of Chile, Santiago, Chile. (Spanish)

NAGY, S. and SHAW, P. E. 1980. Tropical and Subtropical Fruits. AVI Publishing Co., Westport, CT.

POPENOE, W. 1934. Manual of Tropical and Subtropical Fruits. Macmillan and Co., New York.

POPENOE, W. 1963. Lucuma. *In* L. H. Bailey's Standard Cyclopedia of Horticulture 2, 1918-1920, Macmillan and Co., New York.

SAAVEDRA, E. 1975A. The effect of temperature and growth regulators on the germination and plant growth of the lucumo (*Lucuma bifera* Mol.). Inv. Agr. (Chile) 1(3) 147-152. (Spanish)

SAAVEDRA, E. 1975B. Grafting of the lucumo (*Lucuma bifera* Mol.). Inv. Agr. (Chile) 1(3) 191-193. (Spanish)

SCHMIDT-HEBBEL, H. and PENNACCHIOTI, M. I. 1985. Table of chemical composition of Chilean foods. Universitaria (Editor) Santiago, Chile. (Spanish)

WENKAM, N. S. and MILLER, C. D. 1965. Composition of Hawaii fruits. Hawaii Agric. Exp. Stn. Bull. 135.

Index